Revision of the
Nostocaceae
With Cylindrical Trichomes

Revision of the Nostocaceae With Cylindrical Trichomes

(Formerly Scytonemataceae and Rivulariaceae)

Francis Drouet

Research Fellow
Academy of Natural Sciences of Philadelphia

Hafner Press
A Division of Macmillan Publishing Co., Inc.

New York

Collier Macmillan Publishers

London

Copyright © 1973 by Francis Drouet

Published by Hafner Press
A Division of Macmillan Publishing Co., Inc.
866 Third Avenue
New York, N. Y. 10022

Library of Congress Catalog Card Number: 72-97442
ISBN: 02-844060-9

Printed in the United States of America, by
Noble Offset Printers, Inc. New York, N.Y. 10003

Acknowledgments

To the following people I am indebted for solicitously providing access to literature and specimens during the progress of this work and for making suggestions concerning this manuscript: Mr. Richard W. Hildebrand, Dr. Charles W. Reimer, Dr. Joséphine Th. Koster, Dr. J. B. Hansen, Dr. H. Radclyffe Roberts, Mr. Harold B. Louderback, Mr. and Mrs. William A. Daily, Dr. Roy E. Cameron, Dr. A. Orville Dahl, Mrs. Marjorie P. Kraus, Dr. Robert L. Adams Jr., Mr. Kutbul R. Khan, Prof. Dr. Per Størmer, Dr. Herman S. Forest, Prof. Dr. P. Bourrelly, Dr. Herbert Habeeb, Mr. Gary Breen, Miss Margaret Greenwald, Mr. James Price, Dr. Luis R. Almodóvar, Prof. Dr. Heinrichs Skuja, Dr. David J. Hibberd, Mrs. Pamela Cambridge, Miss Patricia Sims, Dr. Dennis C. Jackson, Mr. Ian Tittley, Dr. Emil Müller, Mr. P. W. James, Dr. Cornelis den Hartog, Dr. W. F. Prud'homme van Reine, Dr. Heinz Körner, Dr. C. G. G. J. van Steenis, Dr. M. A. Donk, Prof. Dr. K. H. Rechinger, Mr. Ronald C. Phillips, Mrs. Linda Newton Irvine, Dr. H. Riedl, Dr. Forrest M. Begres, Dr. Harold J. Humm, Miss Ruth Brown, Mrs. Mary Margeson, Mrs. Lillian Jones, Mrs. Naomi Smith, Dr. A. R. Perry, Dr. Sten Ahlner, Mme. Dr. S. Jovet-Ast, Mrs. Susan Brooke, Prof. Dr. Carlo Cappelletti, Mrs. Grace Selser, Dr. Yves Le Gal, Dr. Carlo Steinberg, Dr. I. Mackenzie Lamb, Dr. Clark T. Rogerson, Mrs. Florence Lovejoy, Dr. P. Compère, Prof. Dr. A. Pitot, Dr. Ove Almborn, Dr. William Randolph Taylor, Dr. Elizabeth McClintock, Mrs. Dorothy Hubbard, Dr. M. B. E. Godward, Dr. J. W. G. Lund, Dr. Heinrich Kuhbier, Dr. S. L.

Everist, Dr. Torgny von Wachenfeldt, Dr. S. K. Goyal, Prof. D. A. Webb, Mr. M. Zigliara, Dr. Elzada U. Clover, and Miss M. F. Servin. Mrs. Helen Wu Loos finished and inked my sketches for Figures 16, 52, and 58. Mr. Hildebrand prepared a large majority of the illustrations and inked certain sketches made by myself. Mrs. Loos assembled the figures into plates.

The National Science Foundation provided funds during the period 1966-1973 to make much of this project possible, and the American Philosophical Society gave money for almost half the expenses of the trip to study Type specimens in European herbaria in 1970.

Abstract

Abstract.—This research comprised much field observation and laboratory culture in conjunction with microscopic study of some 20,000 living and preserved specimens (including a great majority of the almost 1,700 Types upon which more than 3,000 specific and infraspecific names depend) of these blue-green algae, collected over a large part of the earth. Previous classifications ascribed prime importance to the structure of sheath materials. It became obvious, as this study progressed, that these sheath materials, excreted as colloids by protoplasts (trichomes), often disperse immediately in the water; or they congeal or solidify, frequently adjacent to trichomes, and are then further modified, under influences of the environment. Also, it became obvious that the only criteria of systematic value in this group are trichomatal characteristics. It is concluded that four wide-ranging species—*Scytonema Hofmannii* Ag., *Calothrix parietina* (Näg.) Thur., *C. crustacea* Schousb. & Thur., and *Raphidiopsis curvata* Fritsch & Rich—constitute the Nostocaceae with cylindrical trichomes.

Contents

Revision of the
Nostocaceae
With Cylindrical Trichomes

Introduction

The purpose of this study was to investigate the range of morphological variation in each of the species of Nostocaceae formerly segregated as Scytonemataceae and Rivulariaceae, in which cylindrical trichomes are commonly encountered, and to place these species in their appropriate systematic order. Myxophyceae of which the uniseriate trichomes may develop heterocysts, but do not branch, constitute the Nostocaceae. This paper is another in a series of revisions of the classification of blue-green algae, the previous ones of which are by Drouet & Daily (1956) and by Drouet (1968).

Some 20,000 living and preserved specimens of plants concerned in this revision were studied microscopically by me during more than 40 years. Stored in the herbarium, these same specimens were studied over and over during the past six years. Large numbers of them, dried from the fresh state, are still viable. Many were collected by myself, in fresh-water and marine habitats, at localities through much of North America, Brazil, and Europe. It was of special interest to document in the herbarium plants which occupied diversified habitats, particularly where the availability of water varied within distances of a few centimeters. Collectors in many parts of the world sent both living and preserved algae to me. Material in herbaria of numerous countries was examined. As a result, the species of this group seem to be adequately represented in this paper by specimens from major parts of their geographical ranges.

Many specimens, including those of my own collecting, were put

1

into laboratory culture of appropriate kinds here. These cultures were studied microscopically at frequent intervals over several years, and perceptible changes in them were documented in the herbarium. Of special interest were cultures in which spores were observed in the process of germination and where the subsequent growth of the germlings was noted. Hundreds of cultures received from researchers on algae in various parts of the world have been grown here over periods of several years, and specimens of them displaying morphological changes have been filed in the herbarium. Laboratory culture, impure (xenic) and allegedly "pure" (axenic), produces changes in appearance of these algae similar to those effected in analogous natural environments.

The literature was sedulously reviewed to discover the almost 1,700 specific and subspecific taxa at one time or another ascribed to this group. Attempts were made to find and study microscopically, for each taxon, the original (Type) material from which authors made their descriptions. These attempts involved visits at some 45 North American and European herbaria, and loans of specimens from many other herbaria and private collections. A large majority of the Type specimens was found and examined. Other Type specimens were unavailable to me because they had been lost or destroyed in wars, in fires, through neglect, or because of lack of appropriate storage facilities. Many of my requests for loans, especially from Russia, India, and central Europe, remain unanswered.

It is taken for granted in this paper that a species of blue-green algae consists of all individuals in the world which are clonally related to each other, no matter how disparate their morphology or physiology may be. It were irrational to view such organisms as objects devoid of past and future histories or to forget that their trichomes are in a continuous progression of structural and functional change which achieves differentiation of very few hard parts. The observer-collector sees, collects, and preserves part of a population (clone) as representative of the resultant product of growth during what may have been many years of existence. The descriptions and discussions of species in the systematic part of this revision include not only those morphological characteristics heretofore arbitrarily considered "typical" or "normal," but also all other variations produced during growth in every kind of environment.

The morphology and physiology of trichomes of these blue-green algae alter continuously until the processes of growth or degradation have been almost or completely stopped by catastrophic physical or chemical changes of the environment or by thickening of cell walls. They are halted abruptly and conclusively by many phycologists who kill and preserve algae with formalin and other reagents before microscopic examination. Growth (division, enlargement, and differentiation of cells) of vegetative segments of trichomes is manifestly and profoundly influenced by changes in availability of water and in quality of impinging light, which appear to be intimately associated with elaboration of structural and functional diversity. No two populations and no two trichomes can at any moment be expected to be either morphologically or physiologically exactly congruous.

The unsophisticated researcher may initiate morphological or physiological investigation of presumedly axenic cultures of Nostocaceae on the truly false premise that he is dealing with organisms of minimal diversity of structure and function. As noted of the Oscillatoriaceae (Drouet, 1968, p. 10), these plants in laboratory culture are genuine ecophenes produced, usually for short periods of time, in relatively unchanging environments. It can be difficult to apply statistical methods perspicaciously to studies of individual species, where size (Fig. 22) and modes of growth and degradation of cells vary widely in different, or even in identical, microenvironments. Similar difficulties may be encountered by those who employ methods of numerical taxonomy and molecular biology. Workers in all these disciplines have hitherto used trichomes and populations in small numbers from yet smaller numbers of select or fortuitously chosen localities. The use of vast numbers of trichomes or many thousands of clones from all kinds of habitats throughout the geographic range of a species may be found profitable.

The last revision of the Nostocaceae involving large numbers of collections from all over the world and reviewing comprehensively the systematic literature and the original material of species and subspecific taxa was Bornet & Flahault's "Revision des Nostocacées hétérocystées" (1886–88). This revision elaborated and refined the system of Thuret (1875), wherein this family embraced also those species which we now segregate in the Oscillatoriaceae and the Stigonemataceae. Kirchner (1900) distributed the genera of Nostocaceae as circumscribed by Bornet & Flahault into five families: *Oscillatoriaceae* Harv., *Scytonemataceae* Kütz., *Rivulariaceae* Kütz., *Nos-*

tocaceae Kütz., and *Stigonemataceae* Kirchn. The Oscillatoriaceae, in the species of which none of the uniseriate cells of the unbranched trichomes differentiates as a heterocyst (see the discussion of the trichome below), were revised recently by Drouet (1968). The Stigonemataceae include those species wherein the trichomes branch, the cells become multiseriate, and some cells mature as heterocysts.

The Nostocaceae, as treated here, include the remaining three of the above families, where the cells of the several species are uniseriate and some of them are capable of developing into heterocysts; but the trichomes never branch. This revision treats those species and genera which since 1900 have been placed in the families Scytonemataceae and Rivulariaceae, wherein the trichomes are, in general, cylindrical. It is planned in the near future to revise the classification of the remaining genera and species of the Nostocaceae, in which the trichomes are commonly torulose or moniliform. A detailed description and a review of the synonymy of the family *Nostocaceae* Kütz., along with keys to the genera of the entire family, will have to await the publication of the next revision.

The specific and subspecific taxa of Bornet & Flahault (1886–88) prove in large part to be growth forms (ecophenes), confined to restricted habitats, of species of global distribution, each capable of flourishing or surviving in a wide variety of habitats.

Species of Nostocaceae with cylindrical trichomes range all over the surface of the earth, or at least in broad zones about it. They occupy habitats varying from permanent bodies of water many meters deep to aerial sites where water is received in the form of occasional rains or dews. They thrive in the blazing sunlight of the hottest parts of deserts, in the dimmest of indirect lighting of caves, and in hot springs. One species is confined to marine and almost marine habitats, as far as present knowledge goes. Two grow almost solely in fresh-water habitats. One is found in fresh-water, but occasionally in marine, habitats.

Macroscopic and submacroscopic habits of plants have, in the past, furnished the criteria upon which generic taxa in the group were based. However, since congealed or solidified colloidal material, produced by trichomes and comprising much of a plant, functions, as in the Oscillatoriaceae (Drouet, 1968, p. 7–9), as a deposit of excretory matter whose structure depends, during its accretion and degradation, upon influences of the environment rather than those of a trichome, we must discount the value of a plant's macroscopic and submacroscopic structure in the systematics of this group. As in the Oscillatoriaceae, the morphology of trichomes affords the only dependable criteria upon which to base a classification.

Plants

A plant of a species discussed in this paper may be a cushion, crust, slime, gelatinous mass of any shape, or an aggregation of naked or almost naked trichomes, attached to a substrate or free in the water. Congealed or solidified colloidal material in the vicinity of the trichomes usually becomes to some extent impregnated with or surrounded by silt, calcium carbonate ($CaCO_3$, forming marl and tufa), calcium sulfate ($CaSO_4$), or (in some hot springs) silica; and the plants thereby often develop mealy or stony consistencies. In certain kinds of investigations, the observer may wish to designate a single trichome, with or without adjacent excreted material, as a plant; or he may wish to refer to a plant as a population of trichomes. In the systematic part of this paper, the term "plant" is used to specify a mass (clone) of conspecific trichomes, with any associated sheath material, which occupy a single habitat.

Trichomes

Trichomes are the protoplasts of plants of this group. Each is more or less cylindrical and can grow to an indeterminate length. It is divided at intervals into cells by cross walls. In cell division, as in the Oscillatoriaceae (Drouet, 1968, p. 3), a new cross wall grows centripetally from the middle of the outer wall toward the center. A plasmodesma, demonstrable by mounting the trichome in chlor-zinc-iodide, extends through a hole in the center of each cross wall between adjacent cells. All cells of an undifferentiated trichome divide as a rule in unison, unless restricted by factors of the environment. Enlargement of a cell may involve solely increase in diameter, or elongation and decrease in diameter, or elongation or shortening and increase in diameter. The terminal cell at either end of each trichome is of a characteristic shape (hemispherical to almost spherical or conical) until it begins to develop thickened walls or is damaged by an outside agent. In species of Calothrix, the several or many cells at one end or at both ends of a trichome may elongate, shrink in diameter, become vacuolated, and gradually lose the pigments in the protoplasts and often the capability of further division. Strictly cylindrical trichomes are broken with the disintegration of intercalary cells. Trichomes constricted at cross walls break in this manner also but more commonly by further constriction or by splitting of cross walls.

Except for the obvious plasmodesmata, undifferentiated cells of a trichome are similar in structure to those of the Oscillatoriaceae (Drouet, 1968, p. 3–6). A firm outer membrane (termed "wall" here, although it is in no way homologous with walls of other green plants) incloses the more viscous protoplasm. Granules of various kinds and several pigments (Blinks, 1951; Lewin, 1962; Lang & Fisher, 1969) are distributed in the protoplasm. Among the granules may be pseudovacuoles, seemingly small reservoirs of gas (Klebahn, 1895; Bowen & Jensen, 1965; Walsby, 1969) which appear black in transmitted light and red in reflected light; these are rather generally seen in protoplasts of blue-green algae in the plankton and in habitats productive of sulfur bacteria. They appear and disappear as factors of the environment change.

In species of Nostocaceae, vegetative parts of a trichome are sometimes encountered outside the sheath material. When one finds them thus they are often short (Figs. 6, 54), and their movement is of a rotary sliding kind. This same type of movement is recorded (Dodd, 1954; Schwabe, 1964) for those planktonic ecophenes which have long trichomes and slimy sheath material. Such movement is also evident in Dr. Mercedes R. Edwards', Dr. Donald Berns', and Dr. Martin Kessel's laboratory cultures of *Calothrix parietina* of the Cambridge culture collection no. 1482/3 (PH), where the colloidal matter is dispersed.

A terminal cell, at one end or at each end of a trichome, may lay down a thickened wall in close proximity to the outer wall and cross wall, retaining its plasmodesmatic connection with the adjacent cell. At the beginning of this process, enlargement and change of shape may be effected so that the resultant thick-walled cell, a "heterocyst," acquires a configuration quite different from that of its former thin-walled state (Figs. 30, 35). A plasmodesma is marked by a collar of wall material about it. The wall may become yellow or brown; and the color of the protoplast usually, but not always, changes gradually to hyaline or yellow. Electronmicrographs and descriptions of the structure and development of heterocysts are presented by Lang (1965, 1971), Lang & Fay (1971), and Fay & Lang (1971). Excretion of colloidal material through the wall stops, and the wall becomes glued to whatever consolidated sheathing matter may be present about it (Figs. 13, 30). Where a terminal heterocyst dies, the cell next to it usually develops forthwith into a heterocyst. If an adjacent cell develops into a mature spore, the terminal heterocyst dies when the common plasmodesma is severed (see the discussion of spores below). Heterocysts may develop, singly or seriately, from intercalary cells of a trichome, each connected by plasmodesmata with adjacent cells. These heterocysts are commonly cylindrical or barrel-shaped where developed within firm restraining sheaths, and more ovoid where the sheath material is gelatinous or dispersed—the same may be said of spores (see below).

Heterocysts are produced under certain types of external conditions (Lemmermann, 1900; Fogg, 1944, 1949). Specimens lacking them have often been disposed in species of imputedly autonomous genera even though, as the environment changes, the same plants later develop them.

Heterocysts are suspected of being reproductive bodies (Canabaeus, 1929; Wolk, 1965), germinating in certain instances, perhaps where the walls have not completely matured; and of being seats of nitrogen fixation (Stewart, Haystead & Pearson, 1968). Their obvious function rests in the fact that, because excretion of colloidal material gradually ceases as they thicken, the walls become anchored to any contiguous solidified sheath material (Drouet, 1951, p. 154) and thus prevent further movement of the trichomes.

In spore formation, an undifferentiated cell may enlarge in diameter and/or length and form a thick wall about the protoplast, as in the maturation of a heterocyst, except that here the plasmodesmatic connections with adjacent cells are eventually cut of by the wall. This cell is a "spore" (Figs. 25, 29, 31–33, 36, 38, 53, 61–64), its protoplasm isolated from the rest of the trichome. As the new wall thickens, and commonly develops yellow or brown pigments, excretion of colloidal material through the wall ceases; and any firm sheath material hitherto accumulated outside the spore becomes affixed to the new wall and in many instances appears to become part of the latter (Figs. 29, 31, 38, 53). In some ecophenes, spores occupy special positions in relation to heterocysts. In others, there seems to be no rule as to their positions. Spores of species treated in this paper appear to develop only in plants which occupy aquatic habitats.

Spores are obviously reproductive cells; they germinate in appropriate environments (Figs. 38, 53, 55, 56), and the germlings are new trichomes. Germination may take place even while a spore is enlarging and before its wall has fully thickened. Spores produced in permanent bodies of water, in my experience, die soon after they are dried. Those of temporary bodies of water, in general, germinate readily after being dried.

Excreted Colloidal Matter

As noted above, under favorable conditions a cell excretes through its wall hyaline colloidal material which may disperse at once in the surrounding water or become changed through external influences into a gel or solid (Drouet, 1968, p. 7). Dispersed colloids and gelatinous accumulations of such substances in the vicinity of trichomes are common in aquatic habitats, whereas discrete and compact aggregations are found mostly in locations subject to alternate wetting and drying. In aerial situations, the colloids become discrete while they are desiccating. Drying of herbarium specimens and fixing of liquid samples cause the sheath material to become optically visible, even if it were originally colloidal or gelatinous. The same happens in various media and reagents used in mounting these algae for observation under the microscope.

Solidified material frequently accretes into a more or less cylindrical sheath completely surrounding a trichome (Fig. 8). As time goes on this material becomes thicker and often laminated by deposition and transformation of colloids from within; and at least parts of it develop yellow or brown pigments, especially where exposed to drying in strong sunlight, or where high concentrations of salts are present in the water, or where it is infested with fungi, bacteria, *Schizothrix calcicola* (Ag.) Gom., or *Spirulina subsalsa* Oerst. Discrete sheaths are thinnest, and most viscous or gelatinous, over the outer walls of terminal cells (Fig. 3) where, during elongation, the tips of trichomes push against them, often rupturing them and/or destroying the terminal cells (Fig. 2). Where one or more cells at a tip of a trichome are destroyed, the space remaining within the tip of the sheath sometimes becomes filled with new colloidal material produced by living cells behind the dead cells.

Within sheaths of prodigiously elongating trichomes, we may safely assume that the excreted colloidal material not only contributes to the thickening of contiguous sheath-segments but is pushed forward to the trichomatal tip and contributes to the accumulation of sheath material there. As occasional or seasonal external factors encourage elongation of trichomes and excretion of colloidal material, the tips of discrete sheaths sometimes appear to be composed of series of funnel-shaped layers (Fig. 14).

In an undifferentiated trichome, or in parts of a trichome between or beyond anchoring cells, the healthy vegetative cells, as noted above, generally divide and elongate in unison; concurrently they excrete much colloidal sheath material. During the processes of growth, most of the cells comprising the progeny of the primordial cells which produced the colloids now transformed into the outer layers of the sheath may, at the time of observation, be far removed from the site of their progenitors. The discrete sheath material, as we see it at any one time, is thus the transmuted product of excretion by numerous cells (many of which have died, or have become thick-walled, or have passed completely out of the sheath material) of several or many cell-generations.

In aquatic habitats, where fragments of trichomes come to rest on a substrate, often at least part of the excreted colloidal material congeals or solidifies and becomes attached to the substrate. Subsequent division and elongation of all cells of such a fragment, with one or both tips orienting toward the source of light, are commonly observed. The lengthening of a trichome with only one tip growing toward a light source may result in an attachment of sheath material which appears to have been produced by a single basal cell (Figs. 47, 70), often a heterocyst; whereas historically this attaching sheath originated from material produced largely or entirely by all or most of the cells of the primordial trichome-fragment.

Growth of Plants

Where a tip of a sheath is broken, an unanchored trichome may move in a rotary motion partially out of, and back and forth into, the opening of the sheath. Such movement is facilitated by the presence of freshly excreted colloidal material within the sheath. Under conditions exceptionally favorable for growth and movement, as in shallow water on hot sunshiny days, segments of trichomes pass out of open sheaths (Figs. 6, 54) in large numbers. Where a trichome is in part anchored to a sheath by heterocysts and/or spores, only that part between an anchoring cell and the open tip of the sheath breaks into segments which pass out into the surrounding water. Trichomes and sheaths are often broken by the browsing of animals and by other mechanical means. Also, trichomes confined within sheaths fragment during elongation, as described below.

Where the surrounding sheath material is firm, portions of a trichome between heterocysts (or, in species of Calothrix, between a heterocyst and an attenuated tip) are restrained in growth by lack of available space for elongation. As external conditions encourage cell division and elongation, the vegetative parts of a trichome are in many plants forced into compressed spirals (Fig. 57) within the sheath material. Frequently during elongation, cells of the middle part of a trichomatal segment burst through a side of the sheath in U-shaped fashion (Figs. 12, 23). The trichome commonly breaks at a point of greatest tension in the U, and the free ends of the two resultant trichomes proceed to elongate (Figs. 1, 16); the new terminal cell of each trichome regenerates into a shape characteristic of the species; and colloidal material is excreted from the exposed parts of the protruding trichomes. Often two adjacent undifferentiated cells, or those on either side of a heterocyst, break apart; and the two new trichomes, in elongating, push through a side of the sheath in similar fashion. Sometimes two or more fragments of trichomes break through a sheath in close proximity to each other. In some ecophenes, a trichome, forming a U which bursts through a side of the sheath, does not break but grows out as a loop which discharges colloidal material that congeals or solidifies about it (Figs. 23, 26); and the trichome continues to elongate indefinitely. Occasionally one finds two or more trichomes elongating side by side within the lumen of a common sheath (Figs. 27, 50).

Where the supply of water is plentiful, a break often occurs between a heterocyst and an undifferentiated cell, or it occurs between two vegetative cells, one of which quickly matures into a heterocyst (Figs. 5, 7); and only the single undifferentiated trichomatal tip bursts through a side of the sheath. Plants which contain gelatinous sheath material exhibit a preponderance of this last-mentioned type of trichomatal multiplication and growth. In aerial plants with very thin sheaths, growth of fragments of trichomes with resultant single-branching of sheaths is often encountered.

The structure of a plant, as noted above, is determined by such kinds of multiplication and growth of trichomes and such methods of branching of excreted material, although this branching may be obscured or not at all evident in very soft gelatinous plants. With the deposition of silt, calcium carbonate, calcium sulfate, and/or silica in and about the sheath material, penetration of light to the basal trichomes, or parts of trichomes, is frequently inhibited. These trichomes, or parts of them, shrink in diameter or become torulose, and eventually die and disintegrate; while the trichomes (or parts of them) above elongate and push outward toward the light. Empty sheaths below may remain intact for many years before they are ultimately destroyed by fungi, bacteria, and small animals; they have remained intact in a culture of my collection of *Scytonema Hofmannii* from Key West, no. *14965,* from April 1969 until April 1972 (PH).

As sheath material becomes thick, firm, and dark-colored, the solidification of colloidal material from within causes the trichomes, perhaps because of pressure and/or dehydration, to shrink in diameter and the cells to become longer and narrower (Fig. 19). This also happens sometimes where trichomes, shrunken somewhat in drying, revive when wetted. Where parts of sheaths toward the ends of trichomes are infested with fungi or become dark-brown in high concentrations of dissolved salts (especially where the water supply dries up), the inclosed trichomes become much constricted at the cross walls, and colloidal material fills the constrictions. These segments of trichome and adjacent sheath material have been distinguished by the name "hormocysts" by Borzi (1914) and Geitler (1932), who ascribe reproductive functions to them. One may hazard the generalization, suggested by my many failures to revive such plants in culture, that the thicker and more opaque a sheath and the more constricted an inclosed trichome, the less likely is the trichome to survive conditions unfavorable for growth.

Many ecophenes of species of these Nostocaceae can exist in a desiccated condition during long periods of the year; and when rains wet them, the trichomes begin the processes of growth again. Those surrounded by thin discrete sheath material appear to initiate cell division very quickly after a long dry season, or after prolonged storage in the herbarium. As among the Oscillatoriaceae, many cells and/or trichomes may die during periods of desiccation or other catastrophe; but those which do not die may grow very rapidly, after being rewetted, until they are again almost dry; at which point fungi grow rapidly in the sheath material (Fig. 4) until the habitat dries out completely.

Other ecophenes of these species originate in parasitization of trichomes by fungi (which together in some cases proceed to become true lichens) and more often in fungal infestation of sheath materials. These processes are usually accompanied by death of at least some cells, by fragmentation (frequently into few- or single-celled segments) of the trichomes, by diminution or cessation of excretion of colloidal materials by the cells, and by degradation of the sheaths. Bacteria and certain algae also inhabit sheath material and contribute to its degradation or augmentation.

Certain greenhouses, spring-fed rock-faces, hot springs, and tropical bodies of water are habitats in which these algae grow continuously, with minimal degradation, for long periods of time. But all environments change, however subtly; and eventually the processes of degradation prevail.

Collection, Preservation, and Observation of Specimens

Ideally, one should study plants of these Nostocaceae under the microscope at the sites where they are alive and growing. Many specimens will be collected already in a dry condition. Where they are wet, the plants can be laid on paper (newspaper is excellent), topside up, and brought to the laboratory for examination. It is important that plenty of air be available to these algae at all times during conveyance. Part of each sample should be dried immediately in the open air and labeled for storage in the herbarium. Since many remain alive for an indefinite period after drying, there may be opportunity to put these collections into laboratory culture at a later date.

Dried algae, or any dried herbarium specimen, can be wetted in minor part with water; a small amount of it can be lifted to a slide with scalpel or forceps, mounted in water or an aqueous solution of a household detergent (containing an alkyl benzene sulfonate), pressed down lightly under a cover slip, and examined under the microscope. As suggested to me by Dr. J. B. Hansen, there is concern that the residue of the detergent may disturb the results of future observers employing other chemical and physical techniques. We may be similarly concerned, now and in the future, about the effects of all chemical reagents used on herbarium specimens. Sand and dirt can be removed from the mount by manipulation with forceps or needles. Many specimens contain deposits of calcium carbonate, which can be dissolved in dilute aqueous solutions of nitric or hydrochloric acid (HNO₃ or HCl); Prud'homme van Reine & van den Hoek (1966) recommend EDTA (ethylenediaminetetraacetic acid) for this purpose with marine organisms. Shrunken cells can usually be restored to their original sizes and shapes in the detergent solution or in chlor-zinc-iodide; sometimes prolonged soaking in one of these solutions is necessary. A dilute aqueous solution of sodium or potassium hydroxide (NaOH or KOH) dissolves the white polymerized formaldehyde from material stored for long periods in liquid preservatives.

Living planktonic blue-green algae, the cells of which often burst while drying, can first be killed (Drouet, 1968, p. 10) by the addition of 2% by volume of commercial formalin. The sample can then be poured in thin layers to dry on slides of mica, clear plastic, or glass. Such slides may readily be studied under the microscope after one places a drop of water and a cover slip on the dried algae.

Classification

The synonymy of each taxon is presented below in the same fashion as was done in the revision of the Oscillatoriaceae (Drouet, 1968). In the present paper, over 3,000 published names are arranged to indicate the precise relationship of each to its Type, specially for convenience to investigators in systematics of Nostocaceae. The basionyms, each with its appropriate Type noted, are disposed in chronological order, beginning with the first published. A basionym and all new or transferred names dependent on a single Type are listed, also in chronological order, in a single paragraph. Since, according to the *International Code of Botanical Nomenclature* (1972), the nomenclature of this group begins with the publication of Bornet & Flahault's "Revision" in 1886–88, validating references are given for a majority of names introduced prior to January 1st of those years, with the original earlier references inclosed in brackets. Time was not available to me for a search of the entire body of botanical literature since 1886 to find validating references for every older name, but it is taken for granted that such references can be found for almost all names listed.

Presumed synonyms which have been interpreted on the bases of descriptions (including illustrations) as Types, where no specimens were available for my study, are listed separately in the synonymy of each taxon treated below, as was done by Drouet & Daily (1956) and Drouet (1968). It is extremely hazardous, and patently unfair to the authors, to employ, for genuine blue-green algae, names based solely upon artifacts (descriptions and illustrations).

15

The abbreviation "fa." is employed throughout the text for the infraspecific taxon *forma*. This is done to avoid confusion with the numerous alphabetical designations of unnamed infraspecific categories employed by nineteenth century phycological authors.

Type specimens listed below are, insofar as possible, the original material, or at least part of it, upon which an author based his descriptions. In the absence of such material, another specimen so named by the author or by an amicable contemporary of the author, or one labeled thus by Bornet or Flahault, or one conforming to the description from the original locality has in some instances been designated as the Type. Only for a few basionyms, upon which nomenclature of prime importance to the classification depends, have other specimens, corresponding accurately with the descriptions, been selected as Types.

Where Type specimens, or descriptions serving as Types, have been interpreted as being beyond the circumscription of this group of Nostocaceae, an alphabetical list of basionyms (to which is added other appropriate nomenclature), each in a separate paragraph, with the respective Types and my interpretations of them, is appended here as "Nomina Excludenda."

Specimens selected for citation as representative of the various species constitute only a small portion of the some 20,000 studied during this investigation. They are arranged in geographical order, beginning with Europe, then Africa, the Atlantic islands, North America north of the Rio Grande, West Indies, Latin America, the Pacific islands, Antarctica, eastern Asia, East Indies, Australia, middle Asia, the islands of the Indian Ocean, and western Asia. Within these regions, specimens are listed from current political subdivisions in north-south tiers from east to west. Names of collectors and their collection numbers are italicized. Locations of specimens are indicated by the abbreviations of herbaria assigned by Lanjouw & Stafleu (1964) or by the unabbreviated names of collections omitted from their index. Specimens in my personal herbarium are noted by the designation: D.

Names selected for the three widely distributed species are those which Bornet & Flahault (1886–88) used for the commonest ecophenes of each.

Key to Genera of Nostocaceae With Cylindrical Trichomes

Terminal vegetative cells of the trichomes at first hemispherical, becoming almost spherical ..*Scytonema*, p. 19
Terminal vegetative cells of the trichomes at first hemispherical, becoming blunt-conical or cylindrical with rotund tips ..*Calothrix*, p. 95
Terminal vegetative cells of the trichomes at first hemispherical, becoming more or less acute-conical ..*Raphidiopsis*, p. 197

Scytonema Agardh

Scytonema Agardh [Disp. Alg. Suec., p. 38. 1812] ex Bornet & Flahault, Ann. Sci. Nat. VII. Bot. 5: 85. 1887. [*Percursaria* Bonnemaison, Journ. de Phys. 94: 188. 1822.] [*Scytonema* Ser. *Myochrotes* Bornet & Flahault, Mém. Soc. Nat. Sci. Nat. & Math. Cherbourg 25: 216. 1885.] *Scytonema* Sect. *Myochrotes* Bornet & Flahault, Ann. Sci. Nat. VII. Bot. 5: 88. 1887. —Type species: *Conferva myochrous* Dillw.

Dillwynella Bory de Saint-Vincent [Dict. Class. d'Hist. Nat. 5: 507. 1824] ex Geitler in Engler & Prantl, Natürl. Pflanzenfam. 1b: 147. 1942. *Plectonema* Thuret [Ann. Sci. Nat. VI. Bot. 1: 375. 1875] ex Gomont, Ann. Sci. Nat. VII. Bot. 16: 96. 1892. —Type species: *Conferva mirabilis* Dillw.

[*Calothrix* Trib. *Distortae* Agardh, Syst. Algar., p. 71. 1824.] —Type species: *Conferva distorta* Müll.

Inoconia Libert [Mém. Soc. Linn. Paris 5: 402. 1827] ex Bornet & Flahault, ibid. 5: 88. 1887. —Type species: *I. Michelii* Lib.

Petalonema Berkeley [Glean. of Brit. Alg., p. 23. 1832] ex Bornet & Flahault, ibid. 5: 85. 1887. *Arthrosiphon* Kützing [Phyc. Germ., p. 177. 1845] ex Bornet & Flahault, loc. cit. 1887. *Scytonema* Sect. *Petalonema* Borzi [N. Giorn. Bot. Ital. 11(4): 372. 1879] ex Bornet & Flahault, ibid. 5: 88. 1887. [*Scytonema* Ser. *Petalonema* Bornet & Flahault, Mém. Soc. Nat. Sci. Nat. & Math. Cherbourg 25: 217. 1885.]—Type species: *Oscillatoria alata* Carm.

Sclerothrix Kützing [Alg. Aq. Dulc. Dec. 2: 17. 1833] ex Bornet & Flahault, Ann. Sci. Nat. VII. Bot. 5: 118. 1887. *Hypheothrix* Kützing [Phyc. Gener., p. 229. 1843] ex Bornet & Flahault, ibid. 5: 85. 1887. *Schizothrix* Subgen. *Hypheothrix* Kützing [ex Gomont, Journ. de Bot. 4(20): 352. 1890] ex Gomont, Ann. Sci. Nat. VII. Bot. 15: 295. 1892. *Lyngbya* Sect. *Lyngbya* (as "Eulyngbya") Subsect. *Hypheothrix* Hansgirg, Prodr. Algenfl. Böhm. 2: 86. 1892. —Type species: *Sclerothrix Callitrichae* Kütz.

19

Drilosiphon Kützing [Phyc. Gener., p. 214. 1843] ex Bornet & Flahault, ibid. 5: 85. 1887. —Type species: *D. muscicola* Kütz.

[*Scytonema* Sect. *Synchaeta* Kützing, ibid., p. 216. 1843.] —Type species: *Scytonema incrustans* Kütz.

Tolypothrix Kützing [ibid., p. 227. 1843] ex Bornet & Flahault, ibid. 5: 118. 1887. —Type species: *T. muscicola* Kütz.

Arthronema Hassall [Hist. Brit. Freshw. Alg. 1: 238. 1845] ex Forti, Syll. Myxoph., p. 558. 1907. —Type species: *Scytonema cirrhosum* Carm.

Petronema Thwaites [in Smith & Sowerby, Suppl. Engl. Bot. 4: 2959. 1849] ex Bornet & Flahault, Ann. Sci. Nat. VII. Bot. 5: 107. 1887. —Type species: *P. fruticulosum* Thw.

[*Rhizonema* Thwaites in Smith & Sowerby, ibid. 4: 2954. 1849.] —Type species: *Calothrix interrupta* Carm.

Desmonema Berkeley & Thwaites [in Smith & Sowerby, ibid. 4: 2958. 1849] ex Bornet & Flahault, ibid. 5: 126. 1887. —Type species: *Microcoleus Dillwynii* Harv.

Hormocoleum De Brébisson *pro synon.* [in Kützing, Sp. Algar., p. 312. 1849] ex Forti, Syll. Myxoph., p. 503. 1907. —Type species: *Calothrix stuposa* Kütz.

Diplocolon Nägeli [in Itzigsohn, Verh. K. Leop.-Carol. Akad. Naturf. 18: 160. 1855; in Rabenhorst, Alg. Sachs. 47–48: 468. 1855] ex Bornet & Flahault, ibid. 5: 129. 1887. —Type species: *D. Heppii* Näg.

Chrysostigma Kirchner [Krypt.-Fl. Schles. 2(1): 238. 1878] ex Bornet & Flahault, ibid. 5: 85. 1887. —Type species. *Lyngbya cincinnata* Kütz.

Coleodesmium Borzi [N. Giorn. Bot. Ital. 11(4): 355. 1879] ex Bornet & Flahault, ibid. 5: 126. 1887. —Type species: *Thorea Wrangelii* Ag.

Gloeochlamys Schmidle in Simmer, Allg. Bot. Zeitschr. 1899(12): 192. 1899. —Type species: *G. Simmeri* Schmidle.

Campylonema Schmidle (corrected from "Camptylonema" by Schmidle in Allg. Bot. Zeitschr. 1900: 233. 1900), Hedwigia 39: 181. 1900. *Schmidleinema* J. de Toni, Noter. di Noter. Algol. 8: [5]. 1936. —Type species: *Stigonema indicum* Schmidle.

Plectonema B *Pseudophormidium* Forti, Syll. Myxoph., p. 493. 1907. —Type species: *Plectonema phormidioides* Hansg.

Handeliella Skuja in Handel-Mazzetti, Symb. Sinic. 1: 30. 1937. —Type species: *H. Stockmayeri* Skuja.

The original material of the type species of the following genera has not been available to me for study:

Coleospermum Kirchner [Krypt.-Fl. Schles. 2(1): 239. 1878] ex Bornet & Flahault, Ann. Sci. Nat. VII. Bot. 5: 83. 1887. —Type species: *C. Goeppertianum* Kirchn.

Sacconema Borzi [N. Giorn. Bot. Ital. 14(4): 298. 1882] ex Bornet & Flahaut, ibid. 3: 381. 1886.—Type species: *S. rupestre* Borzi.

Filarszkya Forti, Syll. Myxoph., p. 258. 1907. —Type species: *Lyngbya saxicola* Fil.

Hormobolus Borzi, Boll. di Studi & Inform. R. Giard. Colon. di Palermo 1: 93. 1914. —Type species: *H. tectorum* Borzi.

Leptobasis Elenkin, Izv. Imp. Bot. Sada Petra Velik. 15(1): 5. 1915. *Fortiea* J. de Toni, Noter. di Nomencl. Algol. 8: [3]. 1936. —Type species: *Leptobasis caucasica* Elenk.

Diplonema Borzi, N. Giorn. Bot. Ital., N. S. 23(4): 70. 1916. *Borzinema* J. de Toni, ibid. 8: [2]. 1936. —Type species: *Diplonema rupicola* Borzi.

Croatella Ercegovic, Acta Bot. Inst. Bot. R. Univ. Zagreb. 1: 91. 1925. —Type species: *C. lithophila* Erceg.

Pseudoscytonema Elenkin, Monogr. Alg. Cyanoph., Pars Spec. 2: 1805. 1949. —Type species: *Plectonema malayense* Bisw.

Symphyonemopsis Tiwari & Mitra, Phykos 7: 186. 1969. —Type species: *S. katniensis* Tiw. & Mitra.

Trichomata cylindrica vel plus minusve torulosa, septata, ambitu recta vel curvantia vel spiralia, ad extrema tumescentia aut apices saepe per aliquot cellulas attenuantia, cellulis terminalibus vegetativis hemisphaericis demum fere sphaericis, heterocystis plus minusve sphaericis vel cylindricis terminalibus vel intercalaribus, sporis plus minusve cylindricis. Materia vaginalis dispersa vel mucosa vel discreta. Planta trichomata nuda vel in muco vel in vaginis plus minusve cylindraceis discretis saepe ramosis comprehens.

Trichomes cylindrical or more or less torulose, septate, straight or curving or spiraled, the ends somewhat swollen or often attenuating through several cells, the terminal vegetative cells hemispherical becoming almost spherical, heterocysts terminal or intercalary and more or less spherical to cylindrical, spores more or less cylindrical. Sheath material dispersed or mucous or discrete. Plant consisting of naked trichomes or trichomes in mucus or in discrete, more or less cylindrical, often branched sheaths.

One species:

Scytonema Hofmannii Agardh

Byssus aeruginosa Weis, Pl. Crypt. Fl. Göttingens., p. 18. 1770. *Lepra aeruginosa* Weber in Wiggers, Primit. Fl. Holsat., p. 97. 1780. *Lepraria aeruginosa* Smith & Sowerby, Engl. Bot. 31: 2182. 1810. *Pulveraria aeruginosa* Rabenhorst, Deutschl. Krypt.-Fl. 2(1): 4. 1845. —TYPE specimen from England: "B. lanuginosa aeruginosa. Filices emortuae stipulis adnascitur, comm. Thom. Cole," in herb. Dillenius (OXF).

Conferva distorta O. F. Müller, Fl. Dan. 5(14): 6. 1782. *Oscillatoria distorta* Agardh [Disp. Alg. Suec., p. 37. 1812] ex Gomont, Ann. Sci. Nat. VII. Bot. 16: 242. 1892. *Elisa distorta* S. F. Gray, Nat. Arr. Brit. Pl. 1: 283. 1821. *Calothrix distorta* Agardh [Syst. Algar., p. 72. 1824] ex Bornet & Flahault, Ann. Sci. Nat. VII. Bot. 5: 119. 1887. [*Lyngbya distorta* Agardh ex Fries, Fl. Scanica, p. 334. 1835.] *Tolypothrix distorta* Kützing [Phyc. Gener., p. 228. 1843] ex Bornet & Flahault, loc. cit. 1887. —TYPE specimen from Denmark: "Conferva distorta fl. dan. In fossis. Septbr. 1813 misit Hornemann" (C).

Lichen velutinus Acharius, Lichenogr. Suec. Prodr., p. 218. 1798. *Cornicularia velutina* Acharius *pro synon.,* loc. cit. 1798. *Collema velutinum* Acharius, Syn. Meth. Lich., p. 329. 1814. *Parmelia velutina* Wallroth, Fl. Crypt. Germ 1: 552. 1831. *Scytonema byssoideum* var. *corticale* Montagne [in de la Sagra, Hist. Phys., Polit. & Nat. de l'Isle de Cuba, Bot., Pl. Cell., p. 10. 1838] ex Bornet & Flahault, ibid. 5: 94. 1887. —TYPE specimen, here designated, presumably from Acharius: "*Collema velutinum* Ach.," in herb. Agardh no. 11235 (LD).

Conferva myochrous Dillwyn, Brit. Conf., pl. 19. 1802. *Scytonema myochrous* Agardh [Disp. Alg. Suec., p. 38. 1812] ex Bornet & Flahault, Ann. Sci. Nat. VII. Bot. 5: 104. 1887. *S. myochrous* fa. *Aegagropila* Ljungquist *pro synon.* ex Kossinskaia in Elenkin, Monogr. Alg. Cyanoph., Pars Spec. 1: 929. 1938. —TYPE specimen from Wales: Snowdon, Jul. 1802, ex Dillwyn in herb. Agardh (LD).

Conferva mirabilis Dillwyn, Brit. Conf., pl. 96. 1808. *Elisa genuflexa* S. F. Gray, Nat. Arr. Brit. Pl. 1: 283. 1821. *E. mirabilis* S. F. Gray, loc. cit. 1821. *Calothrix mirabilis* Agardh [Syst. Algar., p. 72. 1824] ex Bornet & Flahault, ibid. 3: 371. 1886. *Conferva canescens* Agardh *pro synon.,* loc. cit. 1824. [*Dillwynella serpentina* Bory de Saint-Vincent, Dict. Class. d'Hist. Nat. 5: 507. 1824.] [*Oscillatoriella mirabilis* Gaillon, Mém. Soc. d'Emul. d'Abbeville 1833: 472. 1833.] *Scytonema mirabile* Dillwyn [ex Fries, Fl. Scan., p. 333. 1835] ex Bornet, Bull. Soc. Bot. France 36: 155. 1889. [*Oscillatoria mirabilis* Corda in de Carro, Almanach de Carlsbad 6: 213. 1836.] [*Calothrix muscicola* Rabenhorst, Deutschl. Krypt.-Fl. 2(2): 83. 1847.] *Plectonema mirabile* Thuret [Ann. Sci. Nat. VI. Bot. 1: 379. 1875] ex Gomont, Ann. Sci. Nat. VII. Bot. 16: 99. 1892. *Calothrix rupicola* Crouan *pro synon.* in Roumeguère, Dupray & Mougeot, Alg. de France no. 316. 1883–89. *Dillwynella mirabilis* Kuntze, Rev. Gen. Pl. 2: 892. 1891. *Scytonema myochrous* var. *tenue* De Notaris *pro synon.* ex Forti, Syll. Myxoph., p. 519. 1907. *S. figuratum* fa. *Aegagropila* Ljungqvist (as "aegagropilum") *pro synon.* ex Kossinskaja in Elenkin, Monogr. Alg. Cyanoph., Pars Spec. 1: 923. 1938. *S. mirabile* fa. *Aegagropila* Ljungqvist (as "aegagropilum") *pro synon.* ex Kossinskaja in Elenkin, ibid. 1: 925. 1938. —TYPE specimen from Wales: "mirabilis. sp. orig. Dillwyn.," in herb. Agardh (LD).

Trichophorus lanatus Desvaux [Journ. de Bot. 2: 309. 1809] ex Forti, ibid., p. 542. 1907. *Calothrix lanata* Agardh [Syst. Algar., p. 72. 1824] ex Bornet &

Flahault, Ann. Sci. Nat. VII. Bot. 5: 120. 1887. *Oscillatoria lanata* Corda [in de Carro, Almanach de Carlsbad 6: 213. 1836] ex Forti, ibid., p. 499. 1907. *Tolypothrix lanata* Agardh [ex Kützing, Sp. Algar., p. 314. 1849] ex Bornet & Flahault, ibid. 5: 119. 1887. *Calothrix involvens* var. *vadorum* Areschoug *pro synon.* ex Bornet & Flahault, ibid. 3: 371. 1886. *Tolypothrix coactilis* fa. *fuscescens* Areschoug *pro synon.* ex Bornet & Flahault, ibid. 5: 121. 1887. *Scytonema lanatum* Dalla Torre & Sarnthein (as "lanata"), Alg. Tirol, Vorarlb. & Liechtenst., p. 129. 1901. *Tolypothrix tenuis* fa. *lanata* Kossinskaia in Elenkin, ibid. 1: 954. 1938. —TYPE specimen from France: "Trichophorus lanatus Desv. Cl. Desvaux dedit" (PC).

Scytonema Hoffman-Bangii Agardh [Disp. Alg. Suec., p. 39. 1812] ex Bornet & Flahault, Ann. Sci. Nat. VII. Bot. 5: 98. 1887. *S. Hofmannii* Agardh (as "Hoffmannii") [Syn. Alg. Scand., p. 117. 1817] ex Bornet & Flahault, ibid. 5: 97. 1887. *S. Bangii* Lyngbye [in Hornemann, Fl. Dan. 9(27): 8. 1818] ex Bornet & Flahault, ibid. 5: 109. 1887. *Symphyosiphon Bangii* Kützing [Phyc. Gener., p. 218. 1843] ex Forti, Syll. Myxoph., p. 539. 1907. *S. Hofmannii* Kützing [Sp. Algar., p. 323. 1849] ex Gomont, Ann. Sci. Nat. VII. Bot. 15: 306. 1892. [*Symploca Hofmannii* Crouan (as "Hoffmannii"), Fl. Finistère, p. 115. 1867.] *Schizosiphon paradoxus* De Brébisson *pro synon.* in Roumeguère, Dupray & Mougeot, Alg. de France no. 542. 1883–89. *S. intricatus* A. Braun *pro synon.* ex Bornet & Flahault, ibid. 5: 98. 1887. *Scytonema turicense* var. *muscicola* Hepp *pro synon.* [ex Wittrock & Nordstedt, Alg. Aq. Dulc. Exs. 16: 765. 1886] ex Bornet & Flahault, loc. cit. 1887. *S. intricatum* Hennings, Phyk. March. 1: 35. 1893.—TYPE specimen, from Sweden: "Scytonema Hoffman-Bangii mihi," in herb. Agardh (LD).

Scytonema byssoideum Agardh [Disp. Alg. Suec., p. 39. 1812] ex Bornet & Flahault, Ann. Sci. Nat. VII. Bot. 5: 104. 1887. *Hassallia byssoidea* Hassall [Hist. Brit. Freshw. Alg. 1: 233. 1845] ex Bornet & Flahault, ibid. 5: 116. 1887. [*Symploca byssoidea* Crouan, Fl. Finistère, p. 115. 1867.] *Hapalosiphon byssoideus* Kirchner [Krypt.-Fl. Schles. 2(1): 231. 1878] ex Bornet & Flahault, loc. cit. 1887. *Hassallia byssoidea* fa. *lignicola* Bornet & Flahault, ibid. 5: 117. 1887. *Tolypothrix byssoidea* Kirchner in Engler & Prantl, Natürl. Pflanzenfam. 1(1A): 80. 1900. —TYPE specimen from Sweden: Jäder, 1812, in herb. Agardh (LD). Fig. 8.

Thorea Wrangelii Agardh, Disp. Alg. Suec., p. 40. 1812. *Oscillatoria Wrangelii* Agardh [Algar. Dec. 2: 28. 1813] ex Gomont, Ann. Sci. Nat. VII. Bot. 16: 266. 1892. *Calothrix Wrangelii* Agardh [Syst. Algar., p. 71. 1824] ex Bornet & Flahault, Ann. Sci. Nat. VII. Bot. 5: 127. 1887. [*Scytonema Wrangelii* Agardh ex Fries, Fl. Scan., p. 333. 1835.] *Coleodesmium Wrangelii* Borzi [N. Giorn. Bot. Ital. 11(4): 356. 1879] ex Bornet & Flahault, loc. cit. 1887. *Desmonema Wrangelii* Bornet & Flahault [Mém. Soc. Nat. Sci. Nat. & Math. Cherbourg 25: 220. 1885] ex Bornet & Flahault, Ann. Sci. Nat. VII. Bot. 5: 127. 1887. *Dillwynella*

Wrangelii Kuntze, Rev. Gen. Pl. 2: 892. 1891. —TYPE specimen from Sweden: Ramahyttan, in herb. Agardh (LD). Fig. 27.

Conferva cyanea Smith & Sowerby, Engl. Bot. 36: 2578. 1814. *Humida cyanea* S. F. Gray, Nat. Arr. Brit. Pl. 1: 282. 1821. *Oscillatoria cyanea* Agardh [Syst. Algar., p. 68. 1824] ex Gomont, Ann. Sci. Nat. VII. Bot. 16: 242. 1892. [*Oscillatoriella cyanea* Gaillon, Mém. Soc. d'Emul. Abbeville 1833: 472. 1833] —TYPE specimen, here designated in the absence of the original material, from England: Thornhaugh, Northamptonshire, *Berkeley,* in herb. Agardh (LD).

Oscillatoria crispa Agardh [Syn. Alg. Scand., p. 108. 1817] ex Gomont, loc. cit. 1892. *Lyngbya crispa* Agardh [Syst. Algar., p. 74. 1824] ex Gomont, ibid. 16: 153. 1892. [*L. aeruginosa* β *crispa* Rabenhorst, Deutschl. Krypt.-Fl. 2(2): 82. 1847.] *Scytonema crispum* Bornet, Bull. Soc. Bot. France 36: 157. 1889. *Aulosira Sydowii* Lemmermann *pro synon.* ex Frémy in Keissler, Ann. Naturh. Mus. Wien 1932–33: 211. 1933. —TYPE specimen from near Stockholm, Sweden: Haga, in herb. Agardh (LD).

Oscillatoria canescens Agardh [Syn. Alg. Scand., p. 111. 1817] ex Gomont, Ann. Sci. Nat. VII. Bot. 16: 237. 1892. —TYPE specimen from Sweden: Lund, Jun. 1816, in herb. Agardh (LD).

Scytonema myochrous δ *simplex* Agardh [ibid., p. 114. 1817] ex Bornet & Flahault, Ann. Sci. Nat. VII. Bot. 5: 101. 1887. [*S. seriatum* β *simplex* S. F. Gray, Nat. Arr. Brit. Pl., p. 286. 1821.] *S. myochrous* var. *simplex* Lyngbye [ex Bornet & Thuret, Notes Algol. 2: 150. 1880] ex Bornet, Bull. Soc. Bot. France 36: 152. 1889. —TYPE specimen from Sweden: Ramshyttan, in herb. Agardh (LD).

Scytonema penicillatum Agardh [ibid., p. 116. 1817] ex Bornet & Flahault, ibid. 5: 123. 1887. *Tolypothrix penicillata* Thuret [Ann. Sci. Nat. VI. Bot. 1: 380. 1875] ex Bornet & Flahault, loc. cit. 1887. *T. distorta* var. *penicillata* Lemmermann, Krypt.-Fl. Mark Brandenb. 3: 218. 1907. *T. distorta* fa. *penicillata* Kossinskaja in Elenkin, Monogr. Alg. Cyanoph., Pars Spec. 1: 958. 1938. —TYPE specimen from Sweden: Nåtrabyå, in herb. Agardh (LD).

Conferva intexta Pollini, Bibliot. Ital. 7: 418. 1817. [*Scytonema thermale* var. *intextum* Meneghini, Consp. Algol. Eugan., p. 331. 1837.] *S. intextum* Trevisan [Prosp. Fl. Eugan., p. 55. 1842] ex Bornet & Flahault, ibid. 5: 102. 1887. —TYPE specimen, here designated in the absence of the original material, from the original locality in Italy: "Scytonema intextum. In thermis Euganeis," in herb. Meneghini (FI).

Scytonema figuratum Agardh [Syst. Algar., p. 38. 1824] ex Bornet & Flahault, Ann. Sci. Nat. VII. Bot. 5: 101. 1887. —TYPE specimen from Bougainville island: "Conferva B. [*Gaudichaud*] no. 1000, Rawak," in herb. Agardh (LD). Fig. 20.

Scytonema crustaceum Agardh [ibid., p. 39. 1824] ex Bornet & Flahault, ibid. 5: 107. 1887. *Sirosiphon crustaceus* Rabenhorst [Alg. Eur. 133–134: 1334. 1862] ex Bornet & Flahault, ibid. 5: 66. 1887. *Stigonema crustaceum* Kirchner [Krypt.-Fl.

Schles. 2(1): 230. 1878] ex Bornet & Flahault, ibid. 5: 73. 1887. *Calothrix crustacea* Wolle (as "crustaceum"), Fresh-w. Alg. U. S., p. 239. 1887. *Petalonema crustaceum* Kirchner in Engler & Prantl, Natürl. Pflanzenfam. 1(1A): 79. 1900. —TYPE specimen from Stockholm, Sweden, in herb. Agardh (LD). Fig. 17, 23.

Scytonema torridum Agardh [ibid., p. 40. 1824] ex Bornet & Flahault, ibid. 5: 95. 1887. *Byssus aterrima* Bory de Saint-Vincent (as "aterrimus") ex Agardh *pro synon.*, loc. cit. 1824. —TYPE specimen from Réunion, Mascarene islands: sur de gros rochers des torrents de l'ile de Bourbon, in herb. Agardh (LD).

Scytonema Sowerbyanum Agardh [ibid., p. 41. 1824] ex Bornet & Flahault (as "Soverbyanum"), Ann. Sci. Nat. VII. Bot. 5: 112. 1887. —TYPE specimen from England: "Conferva mirabilis misit Hooker," in herb. Agardh (LD).

Calothrix distorta β *flaccida* Agardh [Syst. Algar., p. 72. 1824] ex Bornet & Flahault, ibid. 3: 371. 1886. *Tolypothrix flaccida* Kützing [Phyc. Gener., p. 228. 1843] ex Bornet & Flahault, ibid. 4: 121. 1887. [*T. lanata* b *flaccida* Kirchner, Krypt.-Fl. Schles. 2(1): 228. 1878.] *T. lanata* var. *flaccida* Bornet & Flahault ex Forti, Syll. Myxoph., p. 542. 1907. —TYPE specimen from near Stockholm, Sweden: Traneberg, 30 Jul. 1824, in herb. Agardh (LD).

Calothrix lanata β *fuscescens* Agardh [loc. cit. 1824] ex Bornet, Bull. Soc. Bot. France 36: 156. 1889. *C. fuscescens* De Brébisson & Godey [Mém. Soc. Acad. Sci. Arts & Belles-Lettr. Falaise 1835: 24. 1836] ex Bornet & Flahault, Ann. Sci. Nat. VII. Bot. 3: 371. 1886. —TYPE here designated, in the absence of the original material, filed thus in the Agardh herbarium: "Oscillatoria distorta. Eau douce, Paris, dedit Desvaux" (LD).

Oscillatoria alata Carmichael [in Greville, Scott. Crypt. Fl. 4: 222. 1826] ex Gomont, Ann. Sci. Nat. VII. Bot. 16: 241. 1892. *Petalonema alatum* Berkeley [Gleanings of Brit. Alg., p. 23. 1832] ex Bornet & Flahault, ibid. 5: 110. 1887. *Arthrosiphon Grevillei* Kützing [Phyc. Germ., p. 177. 1845] ex Bornet & Flahault, loc. cit. 1887. *A. alatus* Rabenhorst [Fl. Eur. Algar. 2: 265. 1865] ex Bornet & Flahault, loc. cit. 1887. *Scytonema alatum* Borzi [N. Giorn. Bot. Ital. 11(4): 373. 1879] ex Bornet & Flahault, loc. cit. 1887. *S. crustaceum* var. *alatum* Hansgirg, Prodr. Algenfl. Böhmen 2: 32. 1892. —TYPE specimen from Argyll, Scotland: Appin, *Carmichael*, in herb. Agardh (LD). Fig. 14.

Oscillatoria torpens Bory de Saint-Vincent (as "Oscillaria") [Dict. Class. d'Hist. Nat. 12: 476. 1827] ex Gomont, ibid. 16: 240. 1892. —TYPE specimen from Vaucluse, France: coulée un rocher à travers lequel l'eau filtrait, *Requien*, in herb. Bory (PC).

Oscillatoria lanosa Bory de Saint-Vincent (as "Oscillaria") [ibid. 12: 477. 1827] ex Gomont, ibid. 16: 243. 1892. [*Plectonema lanosum* Bornet & Thuret, Notes Algol. 2: 137. 1880.] —TYPE specimen from Ile-de-France, France: contre les parois du canal de la grande rivière, *Bory*, in herb. Bornet—Thuret (PC).

Oscillatoria pannosa Bory de Saint-Vincent (as "Oscillaria") [ibid. 12: 478. 1827] ex Gomont, ibid. 16: 244. 1892. *Phormidium pannosum* Kützing [Phyc. Gener., p. 195. 1843] ex Gomont, ibid. 16: 173. 1892. [*Microcoleus pannosus* Rabenhorst, Deutschl. Krypt.-Fl. 2(2): 78. 1847.] —TYPE specimen from Hérault, France: e font. Perou prop. Montempessulanum, in herb. Bory (PC).

Inoconia Michelii Libert (as "Micheli") [Mém. Soc. Linn. Paris 5: 403. 1827] ex Bornet & Flahault, Ann. Sci. Nat. VII. Bot. 5: 95. 1887. *Scytonema cinereum* a *Michelii* Rabenhorst (as "Micheli") [Fl. Eur. Algar. 2: 247. 1865] ex Bornet & Flahault, ibid. 5: 96. 1887. —TYPE specimen from Liége, Belgium: près de Malmédy, in Libert, Plantae Cryptogamicae in Arduenna no. 96 (FH).

[*Oscillatoria rupestris* Greville, Scott. Crypt. Fl. 5: 246. 1827.] *O. limbata* Greville [ibid., Synops., p. 40. 1828] ex Gomont, Ann. Sci. Nat. VII. Bot. 16: 238. 1892. *Phormidium rivulare β rupestre* Kützing [Phyc. Gener., p. 195. 1843] ex Gomont, ibid. 16: 176. 1892. *Hassallia limbata* Hassall [Hist. Brit. Freshw. Alg. 1: 234. 1845] ex Bornet & Flahault, ibid. 5: 117. 1887. [*Scytonema limbatum* Kützing, Tab. Phyc. 2: 7. 1850–52.] *Phormidium rivulare* var. *rupestre* Kützing [ex Mougeot & Roumeguère in Louis, Le Départ. des Vosges 2: 583. 1887] ex Elenkin, Monogr. Alg. Cyanoph., Pars Spec. 2: 1490. 1949. *Tolypothrix limbata* Forti, Syll. Myxoph., p. 556. 1907. —TYPE specimen from Scotland: "O. rupestris Grev. Ad saxas humidas. Misit Greville," in herb. Agardh (LD).

Sclerothrix Callitrichae Kützing [Alg. Aq. Dulc. Dec. 2: 17. 1833] ex Bornet & Flahault, Ann. Sci. Nat. VII. Bot. 5: 122. 1887. [*Oscillatoria Kuetzingii* Corda in de Carro, Almanach de Carlsbad 6: 213. 1836.] *Hypheothrix Callitrichae* Kützing [Phyc. Gener., p. 229. 1843] ex Bornet & Flahault, loc. cit. 1887. *Scytonema Callitrichae* Rabenhorst [Fl. Eur. Algar. 2: 260. 1865] ex Bornet & Flahault, ibid. 5: 123. 1887. —TYPE specimen from Sachsen, Germany: an Callitrichen, Teich bei Hirschbach zwischen Sühl und Schleusingen, 1830, in herb. Kützing (L).

Conferva muscosa Beggiato, D. Terme Eugan. Mem., p. 51. 1833. [*Scytonema thermale* var. *muscosum* Meneghini, Consp. Algol. Eugan., p. 331. 1837.] [*S. intextum* var. *muscosum* Trevisan, Prosp. d. Fl. Eugan., p. 55. 1842.] [*S. thermale* var. *muscorum* Meneghini ex Hohenbühel-Heufler, Abh. k. Zool.-bot. Ges. Wien 21: 315. 1871.] *Scytonema muscosum* Bornet & Thuret [Notes Algol. 2: 152. 1880] ex Bornet & Flahault, Ann. Sci. Nat. VII. Bot. 5: 108. 1887. —TYPE specimen, here designated in the absence of material collected by Beggiato, from the Euganean springs, Italy: "Conferva muscosa Begg." in herb. Meneghini (FI).

Scytonema cirrhosum Carmichael [in Harvey in Hooker, Engl. Fl. 5 (Brit. Fl. 2): 366. 1833] ex Bornet & Flahault, ibid. 5: 127. 1887. *Arthronema cirrhosum* Hassall [Hist. Brit. Freshw. Alg. 1: 238. 1845] ex Forti, Syll. Myxoph., p. 558, 559. 1907. [*Desmonema cirrhosum* Berkeley & Thwaites, Engl. Bot. Suppl., pl. 2958. 1849.] *Calothrix cirrhosa* Bonhomme [Notes s. Quelq. Alg. d'Eau Douce 1:

4. 1858] ex Forti, loc. cit. 1907. [*Tolypothrix cirrhosa* Cooke, Brit. Fresh-w. Alg., p. 270. 1884.] —TYPE specimen from Argyll, Scotland: Appin, *Carmichael*, ex Harvey in herb. Agardh (LD).

Scytonema contextum Carmichael [in Harvey in Hooker, loc. cit. 1833] ex Bornet & Flahault, ibid. 5: 101. 1887. *S. myochrous* i *contextum* Rabenhorst [Fl. Eur. Algar. 2: 255. 1865] ex Bornet & Flahault, ibid. 5: 103. 1887. —TYPE specimen from Ireland: Killarney, *W. H. Harvey*, in herb. Agardh (LD).

Calothrix interrupta Carmichael [in Harvey in Hooker, ibid., p. 368. 1833] ex Bornet & Flahault, Ann. Sci. Nat. VII. Bot. 3: 371. 1886. *Stigonema interruptum* Hassall [Hist. Brit. Freshw. Alg. 1: 299. 1845] ex Bornet & Flahault, ibid. 5: 79. 1887. [*Symphyosiphon interruptus* Kützing, Bot. Zeit. 5(12): 197. 1847.] [*Rhizonema interruptum* Thwaites in Berkeley & Thwaites, Suppl. Engl. Bot., pl. 2954. 1849.] *Scytonema interruptum* Thwaites [ex Cooke, Brit. Fresh-w. Alg., p. 266. 1884] ex Bornet & Flahault, ibid. 5: 113. 1887. —TYPE specimen, presumably from Argyll, Scotland: "Calothrix interrupta from Carmichael," in herb. Hassall (BM).

Oscillatoria lucifuga Harvey [in Hooker, Engl. Fl. 5(Brit. Fl. 2): 373. 1833] ex Gomont, Ann. Sci. Nat. VII. Bot. 16: 243. 1892. [*Calothrix lucifuga* Carmichael *pro synon.* in Harvey in Hooker, loc. cit. 1833.] *Symploca lucifuga* De Brébisson [in Kützing, Sp. Algar., p. 271. 1849] ex Gomont, ibid. 16: 118. 1892. —TYPE specimen from Argyll, Scotland: Appin, *Carmichael* (K in BM).

Scytonema Kneiffii Wallroth [Fl. Crypt. German. 2: 56. 1833] ex Bornet & Flahault, Ann. Sci. Nat. VII. Bot. 5: 121. 1887. [*S. globosum* Kneiff *pro synon.* in Wallroth, ibid. 2: 57. 1833.] *Tolypothrix Kneiffii* Kützing [Sp. Algar., p. 314. 1849] ex Bornet & Flahault, loc. cit. 1887. *T. aegagropila* b *Kneiffii* Kützing [ex Rabenhorst, Fl. Eur. Algar. 2: 274. 1865] ex Bornet & Flahault, ibid. 5: 122. 1887. *Calothrix aegagropila* var. *Kneiffii* Kirchner ex Forti, Syll. Myxoph., p. 543. 1907. —TYPE specimen, presumably from Baden, Germany: "Scytonema Kneiffii, S. globosum. *Kneiff*," in herb. Meneghini (FI).

[*Oscillatoria Juergensii* Corda in de Carro, Almanach de Carlsbad 6: 212. 1836.] —TYPE specimen from Jever, Germany: in lacu Wiedlermeer, in Jürgens, Algae Aquaticae, Dec. 11, no. 5, as "Conferva distorta" (FH).

Scytonema thermale Kützing [Alg. Aq. Dulc. Germ. Dec. 14: 139. 1836] ex Bornet & Flahault, ibid. 5: 102. 1887. *S. myochrous* var. *tenuius* Kützing [ex Bornet & Thuret *pro synon.*, Not. Algol. 2: 150. 1880] ex Bornet & Flahault, ibid. 5: 103. 1887. —TYPE specimen from the Euganean springs, Italy: Abano, in herb. Kützing (L).

Scytonema cinereum Meneghini [Consp. Algol. Eugan., p. 331. 1837] ex Bornet & Flahault, Ann. Sci. Nat. VII. Bot. 5: 95. 1887. *S. cinereum* c *cinereum* Rabenhorst [Fl. Eur. Algar. 2: 248. 1865] ex Bornet & Flahault, ibid. 5: 96. 1887. —TYPE specimen from the Euganean springs, Italy: in umbrosis Euganeorum, *Meneghini* (PC).

[*Scytonema thermale* var. *hemisphaericum* Meneghini, loc. cit. 1837.] *Byssus hemisphaerica* Vandelli (as "hemisphaericus") *pro synon.* in Meneghini, loc. cit. 1837. [*Scytonema intextum* var. *hemisphaericum* Trevisan, Prosp. d. Fl. Eugan., p. 55. 1842.] —TYPE specimen from Italy: Therm. Eugan., in herb. Meneghini (FI).

Scytonema rubrum Montagne [Ann. Sci. Nat. II. Bot. 8: 349. 1837] ex Bornet & Flahault, ibid. 5: 112. 1887. —TYPE specimen from West Indies: Cuba, ad folias, *Ramon de la Sagra,* in herb. Montagne (PC).

Scytonema julianum Meneghini [Giorn. Toscana Sci. Med., Fis. & Nat. 1: 188. 1840] ex Bornet & Flahault, Ann. Sci. Nat. VII. Bot. 5: 98. 1887. *Drilosiphon julianus* Kützing [Bot. Zeit. 5(12): 197. 1847] ex Bornet & Flahault, loc. cit. 1887. *Scytonema cinereum* b *julianum* Rabenhorst [Fl. Eur. Algar. 2: 248. 1865] ex Bornet & Flahault, loc. cit. 1887. *S. cinereum* var. *julianum* Rabenhorst [ex Zeller, Journ. Asiatic Soc. Bengal 42(2:3): 181. 1873] ex Desikachary, Cyanoph., p. 472. 1959. *S. Hofmannii* var. *julianum* Bornet & Thuret (as "juliana") [Not. Algol. 2: 149. 1880] ex Forti, Syll. Myxoph., p. 514. 1907. *Symphyosiphon julianus* Kützing ex Forti, loc. cit. 1907. —TYPE specimen from Toscana, Italy: Therm. Julian., *Meneghini,* in herb. Kützing (L).

[*Calothrix atroviridis* Harvey, Man. Brit. Alg., p. 159. 1841.] *C. atrovirens* Harvey [ex Rabenhorst, Fl. Eur. Algar. 2: 271. 1865] ex Forti, ibid., p. 490. 1907. —TYPE specimen, here designated, from Wales: Swansea, *Ralfs,* in herb. Hassall (BM).

Microcoleus Dillwynii Harvey [Man. Brit. Alg., p. 169. 1841] ex Gomont, Ann Sci. Nat. VII. Bot. 15: 362. 1892. *Calothrix Dillwynii* Kützing [Phyc. Germ., p. 182. 1845] ex Bornet & Flahault, Ann. Sci. Nat. VII. Bot. 5: 127. 1887. *Tolypothrix Dillwynii* Hassall [Hist. Brit. Freshw. Alg. 1: 242. 1845] ex Bornet & Flahault, loc. cit. 1887. [*Scytonema Dillwynii* Harvey & Ralfs *pro synon* in Hassall, loc. cit. 1845.] *Desmonema Dillwynii* Berkeley & Thwaites [Suppl. Engl. Bot., pl. 2958. 1849] ex Bornet & Flahault, loc. cit. 1887. —TYPE specimen, here designated in the absence of the original material, from Wales: Dolgelly, *J. Ralfs 18* (BM).

Scytonema chthonoplastes Liebman [Forhandl. Skandinav. Naturforsk. 2: 339. 1841] ex Bornet & Flahault, ibid. 5: 111. 1887. —TYPE specimen from Iceland: Reikum, 100–102° (C).

[*Scytonema intextum* var. *fasciculatum* Trevisan, Prosp. d. Fl. Eugan., p. 55. 1842.] [*S. thermale* var. *fasciculatum* Meneghini *pro synon.* in Trevisan, loc. cit. 1842.] [*Symphyosiphon velutinus* β *Meneghinianus* Kützing, Sp. Algar., p. 323. 1849.] *Scytonema velutinum* var. *Meneghinianum* Kützing (as "Meneghiniana") [ex Hohenbühel-Heufler, Abh. Zool.-bot. Ges. Wien 21: 316. 1871] ex Forti, Syll. Myxoph., p. 527. 1907. [*S. fasciculatum* Meneghini *pro synon.* ex Hohenbühel-Heufler, loc. cit. 1871.] —TYPE specimen from the Euganean springs, Italy: Abano, *Meneghini,* in herb. Kützing (L).

[*Scytonema intextum* var. *fibrosum* Trevisan, loc. cit. 1842.] [*S. thermale* var. *fibrosum* Meneghini *pro synon.* ex Trevisan, loc. cit. 1842.] —TYPE specimen from the Euganean springs, Italy: *Naccari 15*, 1832, as "Symphyosiphon" in herb. Meneghini (FI).

Scytonema aerugineo-cinereum Kützing [Phyc. Gener., p. 214. 1843] ex Bornet & Flahault, Ann. Sci. Nat. VII. Bot. 5: 110. 1887. [*S. cinereum* e *aerugineo-cinereum* Rabenhorst, Fl. Eur. Algar. 2:248. 1865.] *S. cinereum* c *aerugineo-caeruleum* Rabenhorst ex Bornet & Flahault, ibid. 5: 111. 1887. *S. cinereum* var. *aerugineo-cinereum* Rabenhorst ex Forti, Syll. Myxoph., p. 532. 1907. —TYPE specimen from Thüringen, Germany: Hainleite, auf Maschelkalk, mixed with *Anacystis montana* (Lightf.) Dr. & Daily, in herb. Kützing (L).

Drilosiphon muscicola Kützing [loc. cit. 1843] ex Bornet & Flahault, ibid. 5: 98. 1887. *Scytonema Drilosiphon* Elenkin & Poliansky, Bot. Mat. Inst. Sporov. Rast. Glavn. Bot. Sada Petrograd 1(12): 189. 1922. —TYPE specimen from Italy: in speluncis Montis Spaccatae, Triest, in herb. Kützing (L).

Scytonema thermale var. *decumbens* Kützing [Phyc. Gener., p. 215. 1843] ex Forti, Syll. Myxoph., p. 518. 1907. —TYPE specimen from the Euganean springs, Italy: Abano, in herb. Kützing (L).

Scytonema allochroum Kützing [loc. cit. 1843] ex Bornet & Flahault, Ann. Sci. Nat. VII. Bot. 5: 110. 1887. *Tolypothrix allochroa* Borzi [N. Giorn. Bot. Ital. 11(4): 371. 1879] ex Bornet & Flahault, ibid. 5: 124. 1887. —TYPE specimen from Italy: Triest, Monte Spaccato in torrentibus, in herb. Kützing (L).

Scytonema gracillimum Kützing [loc. cit. 1843] ex Bornet & Flahault, ibid. 5: 102. 1887. —TYPE specimen from Sachsen, Germany: Sachswerfen, in gypsac. stillicidis, in herb. Kützing (L).

Scytonema castaneum Kützing [loc. cit. 1843] ex Bornet & Flahault, ibid. 5: 96. 1887. *S. fasciculatum* b *castaneum* Rabenhorst [Fl. Eur. Algar. 2: 257. 1865] ex Bornet & Flahault, loc. cit. 1887. —TYPE specimen from Calvados, France: Falaise, in herb. Kützing (L).

Scytonema turfosum Kützing [Phyc. Gener., p. 216. 1843] ex Bornet & Flahault, ibid. 5: 104. 1887. [*S. nigrum* Meneghini *pro synon.* ex Hohenbühel-Heufler, Abh. k. Zool.-bot. Ges. Wien 21: 315. 1871.] —TYPE specimen from Sachsen, Germany: Steigerthal bei Nordhausen, 1839, in herb. Kützing (L).

Scytonema pachysiphon Kützing [loc. cit. 1843] ex Bornet & Flahault, Ann. Sci. Nat. VII. Bot. 5: 106. 1887. [*S. secundatum* Kützing *pro synon.*, Sp. Algar., p. 309. 1849.] *S. myochrous* e *rivulare* Rabenhorst [ibid. 2: 255. 1865] ex Bornet & Flahault, ibid. 5: 107. 1887. *S. myochrous* var. *rivulare* Rabenhorst ex Hansgirg, Beih. z. Bot. Centralbl. 18(2): 489. 1905. —TYPE specimen from Italy: Triest, Cataract am Monte Spaccato, in herb. Kützing (L).

Scytonema naideum Kützing [Phyc. Gener., p. 216. 1843] ex Bornet & Flahault (as "naiadeum"), ibid. 5: 111. 1887. —TYPE specimen from Switzerland: "Schweiz, Original!", in herb. Kützing (L).

Scytonema chlorophaeum Kützing [loc. cit. 1843] ex Bornet & Flahault, ibid. 5: 104. 1887. *S. myochrous* g *chlorophaeum* Rabenhorst [Fl. Eur. Algar. 2: 255. 1865] ex Bornet & Flahault, ibid. 5: 105. 1887. —TYPE specimen from Scotland: Scotiae, *Greville,* in herb. Kützing (L).

Scytonema incrustans Kützing [loc. cit. 1843] ex Bornet & Flahault, Ann. Sci. Nat. VII. Bot. 5: 107. 1887. [*S. clavatum* var. *crassum* Nägeli *pro synon.* ex Bornet & Thuret, Not. Algol. 2: 152. 1880.] [*Symphyosiphon incrustans* Kirchner, Jahresh. Ver. Vaterl. Naturk. Württemb. 36: 194. 1880.] *Sytonema crustaceum* β *incrustans* Bornet & Flahault [Mém. Soc. Nat. Sci. Nat. & Math. Cherbourg 25: 217. 1885] ex Bornet & Flahault, Revision, Index, p. 15. 1888. *S. crustaceum* var. *incrustans* Bornet & Flahault, Ann. Sci. Nat. VII. Bot. 5: 107. 1887. *Petalonema crustaceum* var. *incrustans* Bornet & Flahault ex Migula, Krypt.-Fl. 2(1): 131. 1907. *Scytonema crustaceum* fa. *incrustans* Kossinskaia in Elenkin, Monogr. Alg. Cyanoph., Pars Spec. 1: 934. 1938. —TYPE specimen from Italy: Triest, Monte Spaccato, Apr. 1835, in herb. Kützing (L). Fig. 26.

Scytonema tomentosum Kützing [Phyc. Gener., p. 217. 1843] ex Bornet & Flahault, ibid. 5: 105. 1887. *S. myochrous* var. *tomentosum* Hansgirg, Beih. z. Bot. Centralbl. 18(2): 489. 1905. —TYPE specimen from Sachsen, Germany: Nordhausen, in herb. Kützing (L).

Scytonema fasciculatum Kützing [loc. cit. 1843] ex Bornet & Flahault, ibid. 5: 96. 1887. —TYPE specimen from Germany: Harz, *Kunze,* in herb. Kützing (L).

Scytonema helveticum Kützing [loc. cit. 1843] ex Bornet & Flahault, ibid. 5: 105. 1887. *S. alpinum* Meneghini *pro synon.* [ex Kützing, Sp. Algar., p. 308. 1849] ex Bornet & Flahault, loc. cit. 1887. *S. myochrous* f *helveticum* Rabenhorst [Fl. Eur. Algar. 2: 255. 1865] ex Bornet & Flahault, loc. cit. 1887. —TYPE specimen from Switzerland: Staubbach im Lauterbrunner Thale, Aug. 1835, in herb. Kützing (L).

Symphyosiphon spongiosus Kützing [Phyc. Gener., p. 218. 1843] ex Forti, Syll. Myxoph., p. 522. 1907. *Scytonema spongiosum* Rabenhorst [Deutschl. Krypt.-Fl. 2(2): 86. 1847] ex Bornet & Flahault, loc. cit. 1887. —TYPE specimen from the Euganean springs, Italy: Abano, in herb. Kützing (L).

Symphyosiphon caespitulus Kützing [loc. cit. 1843] ex Forti, ibid., p. 514. 1907. [*Scytonema caespitulum* Rabenhorst, loc. cit. 1847.] —TYPE specimen from Sachsen, Germany: Nordhausen, Sept. 1841, in herb. Kützing (L).

Symphyosiphon velutinus Kützing [loc. cit. 1843] ex Bornet & Flahault, Ann. Sci. Nat. VII. Bot. 5: 108. 1887. *Scytonema velutinum* Kützing *pro synon.* [ibid., p. 219. 1843] ex Bornet & Flahault, loc. cit. 1887. *Petalonema velutinum* Migula, Krypt.-Fl. 2(1): 131. 1907. —TYPE specimen from the Euganean springs, Italy: Abano, in herb. Kützing (L). Fig. 24.

Lyngbya cincinnata Kützing [Phyc. Gener., p. 226. 1843] ex Gomont, Ann. Sci. Nat. VII. Bot. 16: 153. 1892. *Scytonema cincinnatum* Thuret [Ann. Sci. Nat. VI. Bot. 1: 380. 1875] ex Bornet & Flahault, ibid. 5: 89. 1887. *Chrysostigma cincin-*

natum Kirchner [Krypt.-Fl. Schles. 2(1): 238. 1878] ex Bornet & Flahault, loc. cit. 1887. *Calothrix zeylanica* Berkeley *pro synon.* ex Bornet & Flahault, ibid. 3: 372. 1886. —TYPE specimen from Germany: in stehenden Gewässer bei Halle, in herb. Kützing (L).

Lyngbya cincinnata β annosa Kützing [loc. cit. 1843] ex Gomont, ibid. 16: 150. 1892. —TYPE specimen from Thüringen, Germany: Schleusingen, in herb. Kützing (L).

Tolypothrix muscicola Kützing [ibid., p. 227. 1843] ex Bornet & Flahault, Ann. Sci. Nat. VII. Bot. 5: 121. 1887. [*T. aegagropila* f *muscicola* Kirchner, ibid. 2(1): 228. 1878.] *T. aegagropila* var. *muscicola* Kützing [ex Cooke, Brit. Fresh-w. Alg., p. 269. 1884] ex Forti, Syll. Myxoph., p. 544. 1907. *Calothrix aegagropila* var. *muscicola* Kirchner ex Forti, ibid., p. 543. 1907. —TYPE specimen from Sachsen, Germany: Weissenfels, in herb. Kützing (L).

Tolypothrix pygmaea Kützing [Phyc. Gener., p. 227. 1843] ex Bornet & Flahault, ibid. 5: 122. 1887. [*Calothrix muscicola* b *pygmaea* Rabenhorst, Deutschl. Krypt.-Fl. 2(2): 84. 1847.] [*Tolypothrix aegagropila* e *pygmaea* Kirchner, Krypt.-Fl. Schles. 2(1): 228. 1878.] *T. aegagropila* var. *pygmaea* Kützing [ex Cooke, loc. cit. 1884] ex Forti, ibid., p. 546. 1907. *T. tenuis* var. *pygmaea* Hansgirg, Prodr. Algenfl. Böhm. 2: 37. 1892. —TYPE specimen from Sachsen, Germany: Merseburg, natans, in herb. Kützing (L).

Tolypothrix tenuis Kützing [Phyc. Gener., p. 228. 1843] ex Bornet & Flahault, loc. cit. 1887. *T. lanata* fa. *tenuis* Cedergren, Ark. f. Bot. 25A(4): 15. 1933. *T. lanata* fa. *aegagropila* Ljungqvist *pro synon.* ex Kossinskaia in Elenkin, Monogr. Alg. Cyanoph., Pars Spec. 1: 953. 1938. —TYPE specimen from Thüringen, Germany: Kloster Vestra bei Schleusingen, in herb. Kützing (L). Fig. 7

Tolypothrix thermalis Kützing [loc. cit. 1843] ex Bornet & Flahault, Ann. Sci. Nat. VII. Bot. 5: 125. 1887. *Scytonema chloroides* Kützing [Sp. Algar., p. 304. 1849] ex Bornet & Flahault, ibid. 5: 113. 1887. *S. thermale* b *chloroides* Rabenhorst [Fl. Eur. Algar. 2: 250. 1865] ex Bornet & Flahault, ibid. 5: 112. 1887. *S. thermale* var. *chloroides* Rabenhorst ex Forti, Syll. Myxoph., p. 521. 1907. —TYPE specimen from the Euganean springs, Italy: Abano, *Meneghini,* in herb. Kützing (L).

Tolypothrix pulchra Kützing [Phyc. Gener., p. 228. 1843] ex Bornet & Flahault, ibid. 5: 121. 1887. [*T. aegagropila* c *pulchra* Rabenhorst, Fl. Eur. Algar. 2: 274. 1865.] *T. aegagropila* var. *pulchra* Kützing *pro synon.* ex Wolle, Fresh-w. Alg. U. S., p. 264. 1887. *Calothrix aegagropila* var. *pulchra* Kirchner ex Forti, ibid., p. 543. 1907. —TYPE specimen from Thüringen, Germany: auf dem Bruchteiche bei Tennstädt, Feb. 1832, in herb. Kützing (L).

Tolypothrix coactilis Kützing [Phyc. Gener., p. 228. 1843] ex Bornet & Flahault, loc. cit. 1887. *Calothrix coactilis* Rabenhorst [Deutschl. Krypt.-Fl. 2(2): 84. 1847] ex Bornet & Flahault, loc. cit. 1887. [*Tolypothrix aegagropila* d *coactilis* Kirchner, Krypt.-Fl. Schles. 2(1): 228. 1878.] *Calothrix aegagropila* var.

coactilis Kirchner ex Forti, Syll. Myxoph., p. 543. 1907. —TYPE specimen from Sachsen, Germany: Nordhausen, in herb. Kützing (L).

Scytonema cyaneum Meneghini [Giorn. Bot. Ital. 1(1): 304. 1844] ex Bornet & Flahault, Ann. Sci. Nat. VII. Bot. 5: 111. 1887. —TYPE specimen from Dalmatia, Yugoslavia: ad muscos inundatos, *Stalio 1410*, in herb. Meneghini (FI).

Scytonema tenue Kützing [Phyc. Germ., p. 176. 1845] ex Bornet & Flahault, ibid. 5: 102. 1887. [*Symphyosiphon tenuis* Kirchner, ibid. 2(1): 226. 1878.] —TYPE specimen from Switzerland: Bern, Jolimont, in herb. Kützing (L).

Symphyosiphon caespitulus β hercynicus Kützing [ibid., p. 177. 1845] ex Forti, Syll. Myxoph., p. 514. 1907. [*S. hercynicus* Kützing in Roemer, Alg. Deutschl., p. 37. 1845.]—TYPE specimen from the Harz mountains, Germany: *Römer 101*, in herb. Kützing (L).

Symphyosiphon Contarenii Kützing [Phyc. Germ., p. 177. 1845] ex Forti, ibid., p. 535. 1907. *Scytonema Contarenii* Rabenhorst [Deutschl. Krypt.-Fl. 2(2): 86. 1847] ex Bornet & Flahault, Ann. Sci. Nat. VII. Bot. 5: 111. 1887. —TYPE specimen from Italy: Friuli, *Meneghini,* in herb. Kützing (L).

Symphyosiphon crustaceus Kützing [loc. cit. 1845] ex Forti, ibid., p. 525. 1907. —TYPE specimen from Sachsen, Germany: Nordhausen, Jul. 1843, in herb. Kützing (L).

Calothrix olivacea Hooker & Harvey [London Journ. of Bot. 4: 296. 1845] ex Bornet & Flahault, ibid. 3: 371. 1886. *Dichothrix olivacea* Bornet & Flahault [Mém. Soc. Nat. Sci. Nat. & Math. Cherbourg 25: 204. 1885] ex Bornet & Flahault, Ann. Sci. Nat. VII. Bot. 3:371. 1886. —TYPE specimen from Kerguelen island, Indian Ocean: no. 640, in herb. Harvey (TCD). Fig. 16.

Scytonema piligerum De Notaris [Prosp. d. Fl. Ligust., p. 74. 1846] ex Forti, Syll. Myxoph., p. 536. 1907. —TYPE specimen from Genova, Italy: ne' siti bagnati da continuo stillicidio lunghesso l'acquedotto, in herb. Ardissone (PAD).

Sytonema gracile Kützing [Bot. Zeit. 5(12): 196. 1847] ex Bornet & Flahault, ibid. 5: 105. 1887. [*Tolypothrix gracilis* Borzi, N. Giorn. Bot. Ital. 11(4): 371. 1879.] —TYPE specimen from Calvados, France: marais calcaire, *De Brébisson 163,* mixed with *Stigonema* sp., in herb. Kützing (L).

Scytonema clavatum Kützing [loc. cit. 1847] ex Bornet & Flahault, Ann. Sci. Nat. VII. Bot. 5: 107. 1887. —TYPE specimen from Calvados, France: rochers granitiques, Falaise, *De Brébisson 177,* in herb. Kützing (L).

Scytonema flagelliferum Kützing [ibid. 5(12): 197. 1847] ex Bornet & Flahault, ibid. 5: 127. 1887. —TYPE specimen from France: Montaud près Marseille, *Lenormand,* in herb. Kützing (L).

[*Arthrosiphon Grevillei β nodosus* Kützing, loc. cit. 1847.] [*A. Grevillei* var. *nodosus* Kützing (as "nodosa"), Tab. Phyc. 2: 8. 1850–52.] [*A. alatus* b *nodosus* Kützing ex Rabenhorst, Fl. Eur. Algar. 2: 266. 1865.] *Scytonema alatum* var. *nodosum*

Kützing ex Forti, Syll. Myxoph., p. 528. 1907. *S. alatum* fa. *nodosum* Kossinskaia in Elenkin, Monogr. Alg. Cyanoph., Pars Spec. 1: 945. 1938.—TYPE specimen from Calvados, France: Falaise, *Lenormand,* in herb. Kützing (L).

[*Arthrosiphon Grevillei* γ *plicatulus* Kützing, Bot. Zeit. 5(12): 197. 1847.] *A. alatus* var. *plicatulus* Kützing [ex Rabenhorst, Alg. Eur. 218–220: 2183. 1870] ex Bornet & Flahault, Ann. Sci. Nat. VII. Bot. 5: 110. 1887. —TYPE specimen from Calvados, France: Falaise, in herb. Kützing (L).

Symploca Flotowiana Kützing [Bot. Zeit. 5(13): 219. 1847] ex Gomont, Ann. Sci. Nat. VII. Bot. 16: 117. 1892. *S. minuta* var. *Flotowiana* Hansgirg, Prodr. Algenfl. Böhm. 2: 82. 1892. —TYPE specimen from Silesia, Poland: Eichberg bei Hirschberg, *Flotow 9b,* Oct. 1836, in herb. Kützing (L).

Symploca Flotowiana β *tenuior* Kützing [loc. cit. 1847] ex Gomont, loc. cit. 1892.—TYPE specimen from Silesia, Poland: Kauffnez, Hitzelberg, *Flotow 8,* 1839, in herb. Kützing (L).

Symphyosiphon javanicus Kützing [Flora, N. R. 5(2:48): 772. 1847] ex Forti, Syll. Myxoph., p. 507. 1907. *Scytonema javanicum* Bornet & Thuret [Not. Algol. 2: 148. 1880] ex Bornet & Flahault, Ann. Sci. Nat. VII. Bot. 5: 95. 1887. *S. Hofmannii* b *javanicum* Hansgirg, ibid. 2: 34. 1892. *S. Hofmannii* subsp. *javanicum* Hansgirg ex Forti, loc. cit. 1907. —TYPE specimen from Indonesia: Java, *Zollinger 1355,* in herb. Kützing (L).

Tolypothrix subsalsa Zanardini [Not. Intorno alle Cellul. Mar. d. Lag. & Lit. Venezia, p. 79. 1847] ex Bornet & Flahault, ibid. 5: 125. 1887. —TYPE specimen from Italy: in vadib. subsalsis circa Venetiam, *Contarini,* in herb. Zanardini (Mus. Storia Nat., Venezia).

Symploca scytonemacea β *major* Kützing [Sp. Algar., p. 272. 1849] ex Gomont, Ann. Sci. Nat. VII. Bot. 16: 118. 1892. *S. Flotowiana* b *major* Rabenhorst [Fl. Eur. Algar. 2: 155. 1865] ex Gomont, ibid. 16: 117. 1892. —TYPE specimen from Calvados, France: Vire, in herb. Kützing (L).

[*Scytonema parietinum* Meneghini *pro synon. S. tenuis* γ in Kützing, ibid., p. 303. 1849.] —TYPE specimen from Italy: Patavii, *Meneghini,* in herb. Kützing (L).

Scytonema nigrescens Kützing [loc. cit. 1849] ex Bornet & Flahault, Ann. Sci. Nat. VII. Bot. 5: 112. 1887. —TYPE specimen from Switzerland: Zürich, *Nägeli 268,* in herb. Kützing (L).

[*Scytonema maculiforme* Meneghini *pro synon. S. tenuis* β. in Kützing, loc. cit. 1849.]—TYPE specimen from Italy: Patavii, *Meneghini,* in herb. Kützing (L).

[*Scytonema crispum* Meneghini *pro synon. S. tenuis* δ in Kützing, loc. cit. 1849.] —TYPE specimen from Italy: Patavii, *Meneghini,* in herb. Kützing (L).

Scytonema Kuetzingianum Nägeli (as "Kützingianum") [in Kützing, loc. cit. 1849] ex Bornet & Flahault, ibid. 5: 96. 1887. *S. cinereum* f *Kuetzingianum* Rabenhorst [Fl. Eur. Algar. 2: 248. 1865] ex Bornet & Flahault, loc. cit. 1887. —TYPE specimen from Switzerland: Zürich, Sihlwald, Felsen, *Nägeli 407,* in herb. Kützing (L).

Scytonema Meneghinianum Kützing [Sp. Algar., p. 304. 1849] ex Bornet & Flahault, Ann. Sci. Nat. VII. Bot. 5: 111. 1887. *S. allochroum* b *Meneghinianum* Rabenhorst [Fl. Eur. Algar. 2: 256. 1865] ex Bornet & Flahault, ibid. 5: 110. 1887. *S. allochroum* var. *Meneghinianum* Rabenhorst ex Forti, Syll. Myxoph., p. 531. 1907. —TYPE specimen from Italy: Treviso, *Meneghini,* in herb. Kützing (L).

Scytonema elegans Kützing [loc. cit. 1849] ex Bornet & Flahault, ibid. 5: 111. 1887. [*S. fasciculatum* c *elegans* Rabenhorst, Fl. Eur. Algar. 2: 257. 1865.] —TYPE specimen from the Euganean springs, Italy: Abano, in herb. Kützing (L).

[*Scytonema allochroum* β *rhenanum* Kützing, loc. cit. 1849.] *S. allochroum* var. *rhenanum* Kützing ex Forti, loc. cit. 1907. —TYPE specimen from Switzerland: Rheinfall, *Nägeli 175,* in herb. Kützing (L).

Scytonema Panicii Montagne (as "Panici") [in Kützing, Sp. Algar., p. 305. 1849] ex Bornet & Flahault, Ann. Sci. Nat. VII. Bot. 5: 102. 1887. —TYPE specimen from Brazil: ad radices Panici inter muscos Bahiae, *Gaudichaud,* in herb. Kützing (L).

Scytonema coactile Montagne [in Kützing, loc. cit. 1849] ex Bornet & Flahault, ibid. 5: 90. 1887. —TYPE specimen from Guadeloupe, West Indies: *Perrottet,* in herb. Kützing (L).

Scytonema chrysochlorum Kützing [loc. cit. 1849] ex Bornet & Flahault, ibid. 5: 102. 1887. —TYPE specimen from the Euganean springs, Italy: *Meneghini,* in herb. Kützing (L).

Scytonema aureum Meneghini [in Kützing, ibid., p. 306. 1849] ex Bornet & Flahault, ibid. 5: 111. 1887. *Tolypothrix aurea* Borzi [N. Giorn. Bot. Ital. 11(4): 372. 1879] ex Bornet & Flahault, ibid. 5: 125. 1887. —TYPE specimen from Toscana, Italy: in aquariis horti Pisani, *Meneghini,* in herb. Kützing (L).

Scytonema turicense Nägeli [in Kützing, Sp. Algar., p. 306. 1849] ex Bornet & Flahault, Ann. Sci. Nat. VII. Bot. 5: 123. 1887. —TYPE specimen from Switzerland: Zürich, Ausfluss an Steinen, *Nägeli 173,* in herb. Kützing (L)

Scytonema turicense β *rigidum* Kützing [loc. cit. 1849] ex Bornet & Flahault, ibid. 5: 112. 1887. —TYPE specimen from Switzerland: Rheinfall, an Felsen, in herb. Kützing (L).

Scytonema gracillimum β *curvatum* Kützing [loc. cit. 1849] ex Bornet & Flahault, ibid. 5: 102. 1887. *S. gracillimum* fa. *terrestre* Rabenhorst (as "terrestris") [Fl. Eur. Algar. 2; 254. 1865] ex Bornet & Flahault, loc. cit. 1887. —TYPE specimen, here designated, the only specimen annotated thus in the Kützing herbarium, from Switzerland: Zürich, Küssnach, auf feuchten sandigen Felsen, *Nägeli 417* (L).

Scytonema gracillimum γ *obscurum* Kützing [loc. cit. 1849] ex Forti, Syll. Myxoph., p. 725. 1907. *S. gracillimum* var. *obscurum* Rabenhorst [Alg. Eur. 109–110: 1097. 1861] ex Bornet & Flahault, ibid. 5: 105. 1887. —TYPE specimen from Calvados, France: Falaise, *De Brébisson 193*, in herb. Kützing (L).

Scytonema decumbens Kützing [Sp. Algar., p. 307. 1849] ex Bornet & Flahault, Ann. Sci. Nat. VII. Bot. 5: 68. 1887. *S. myochrous* h *decumbens* Rabenhorst [Fl. Eur. Algar. 2: 255. 1865] ex Bornet & Flahault, loc. cit. 1187.—TYPE specimen from Switzerland: Zürich, an feuchten Felsen, *Nägeli 227*, in herb. Kützing (L).

Scytonema tolypothrichoides Kützing [loc. cit. 1849] ex Bornet & Flahault, ibid. 5: 100. 1887. *S. mirabile* var. *tolypothrichoides* Lobik, Izv. Imp. Bot. Sada Petra Velik. 25(1): 43. 1915. —TYPE specimen from Calvados, France: Falaise, *De Brébisson 527*, in herb. Kützing (L).

Scytonema varium Kützing [loc. cit. 1849] ex Bornet & Flahault, ibid. 5: 97. 1887. —TYPE specimen from Indonesia: Java, in herb. Kützing (L).

Scytonema Leprieurii Kützing [Sp. Algar., p. 307. 1849] ex Bornet & Flahault, Ann. Sci. Nat. VII. Bot. 5: 103. 1887. [*S. figuratum* β *Leprieurii* Bornet & Flahault, Mém. Soc. Nat. Sci. Nat. & Math. Cherbourg 25: 217. 1885.] *S. figuratum* var. *Leprieurii* Bornet & Flahault, Ann. Sci. Nat. VII. Bot. 5: 103. 1887. *S. mirabile* var. *Leprieurii* Collins, Holden & Setchell, Phyc. Bor.-Amer. 21: 1014. 1903. *S. mirabile* fa. *Leprieurii* Kossinskaia in Elenkin, Mongr. Alg. Cyanoph., Pars Spec. 1: 925. 1938. —TYPE specimen from French Guyana: Cayenne, *Leprieur*, in herb. Kützing (L).

Scytonema calothrichoides Kützing [loc. cit. 1849] ex Bornet & Flahault (as "calotrichoides"), ibid. 5: 102. 1887. —TYPE specimen from Calvados, France: in ericetis humidis, Falaise, *De Brébisson 195*, in herb. Kützing (L).

Scytonema flexuosum Meneghini [in Kützing, ibid., p. 308. 1849] ex Bornet & Flahault, ibid. 5: 105. 1887. [*S. Orsinianum* Meneghini *pro synon.* in Kützing, loc. cit. 1849.] *S. myochrous* c *flexuosum* Rabenhorst [Fl. Eur. Algar. 2: 254. 1865] ex Bornet & Flahault, loc. cit. 1887. —TYPE specimen from Italy: Monte Corno, *Meneghini*, in herb. Kützing (L).

Scytonema flexuosum β *gallicum* Kützing [loc. cit. 1849] ex Bornet & Flahault, loc. cit. 1887. —TYPE specimen from Calvados, France: in paludibus, Falaise, *De Brébisson 196*, in herb. Kützing (L).

Scytonema coalitum Nägeli [in Kützing, Sp. Algar., p. 308. 1849] ex Bornet & Flahault, Ann. Sci. Nat. VII. Bot. 5: 111. 1887. *S. myochrous* d *coalitum* Rabenhorst [loc. cit. 1865] ex Bornet & Flahault, loc. cit. 1887. *S. myochrous* var. *coalitum* Rabenhorst [ex Crouan in Mazé & Schramm, Essai Class. Alg. Guadeloupe, ed. 2, p. 34. 1870–77] ex Forti, Syll. Myxoph., p. 524. 1907. —TYPE specimen from Switzerland: Rheinfall, an Felsen, *Nägeli 171*, in herb. Kützing (L).

Scytonema dimorphum Kützing [loc. cit. 1849] ex Bornet & Flahault, ibid. 5: 102. 1887. *S. myochrous* b *dimorphum* Rabenhorst [Fl. Eur. Algar. 2: 254. 1865] ex Bornet & Flahault, loc. cit. 1887. —TYPE specimen from Switzerland: an feuchten Felsen beim Rheinfall bei Schaffhausen, *A. Braun 16,* Jul. 1847, in herb. Kützing (L).

[*Scytonema myochrous β tenue* Kützing, ibid., p. 309. 1849.] —TYPE specimen from Switzerland: Zürich, auf feuchtem Sand, *Nägeli 386,* in herb. Kützing (L).

Scytonema Heerianum Nägeli [in Kützing, loc. cit. 1849] ex Bornet & Flahault, Ann. Sci. Nat. VII. Bot. 5: 105. 1887. —TYPE specimen from Switzerland: Zürich, *Nägeli 265,* in herb. Kützing (L).

Calothrix pulchra Kützing [Sp. Algar., p. 311. 1849] ex Bornet & Flahault, ibid. 3: 370. 1886. —TYPE specimen from Calvados, France: rochers d'une cascade, Falaise, *De Brébisson 501,* in herb. Kützing (L).

Calothrix radiosa Kützing [loc. cit. 1849] ex Bornet & Flahault, ibid. 3: 372. 1886. *Dillwynella radiosa* Kuntze, Rev. Gen. Pl. 2: 892. 1891. —TYPE specimen from Manche, France: rochers d'une cascade, Mortain, *De Brébisson 515,* in herb. Kützing (L).

Calothrix cespitosa Kützing [ibid., p. 312. 1849] ex Bornet & Flahault (as "caespitosa"), ibid. 3: 370. 1886. *Dillwynella cespitosa* Kuntze (as "caespitosa"), loc. cit. 1891. —TYPE specimen from Baden, Germany: Titisee, *A. Braun 15,* Sept. 1847, in herb. Kützing (L).

Calothrix Leineri A. Braun [in Kützing, loc. cit. 1849] ex Bornet & Flahault, ibid. 3: 371. 1886. —TYPE specimen from Baden, Germany: Constanz, *Leiner,* in herb. Kützing (L).

Calothrix stuposa Kützing [loc. cit. 1849] ex Bornet & Flahault, ibid. 5: 92. 1887. *Hormocoleum stuposum* De Brébisson *pro synon.* [in Kützing, loc. cit. 1849] ex Forti, Syll. Myxoph., p. 503. 1907. *Scytonema stuposum* Bornet & Thuret [Not. Algol. 2: 146. 1880] ex Bornet & Flahault, loc. cit. 1887. [*S. faliconis* J. Agardh *pro synon.* ex Bornet & Thuret, loc. cit. 1880.] *Dillwynella stuposa* Kuntze, loc. cit. 1891. —TYPE specimen from Manche, France: sur l'*Hypnum alopecurum,* Cascade de Mortain, *De Brébisson 428,* in herb. Kützing (L).

Tolypothrix intricata Nägeli [in Kützing, Sp. Algar., p. 314. 1849] ex Bornet & Flahault, Ann. Sci. Nat. VII. Bot. 5: 125. 1887. —TYPE specimen from Switzerland: Zürich, in e. kleinen Teiche an Würzeln, *Nägeli 264,* in herb. Kützing (L).

Tolypothrix Naegelii Kützing [loc. cit. 1849] ex Bornet & Flahault, ibid. 5: 124. 1887. —TYPE specimen from Switzerland: Zürich, am Brettern im See, *Nägeli 80,* in herb. Kützing (L).

Tolypothrix bicolor Kützing [loc. cit. 1849] ex Bornet & Flahault, ibid. 5: 119. 1887. *T. aegagropila* d *bicolor* Rabenhorst [Fl. Eur. Algar. 2: 274. 1865] ex Bornet & Flahault, ibid. 5: 121. 1887. —TYPE specimen from Bouches-du-Rhône, France: stagnis pr. St.-Chamas, *Lenormand,* in herb. Kützing (L).

Symphyosiphon vaporarius Nägeli [in Kützing, ibid., p. 323. 1849] ex Forti, Syll. Myxoph., p. 526. 1907. *Scytonema vaporarium* Nägeli (as "vaporarius") [ex

Rabenhorst, ibid. 2: 263. 1865] ex Bornet & Flahault, ibid. 5: 108. 1887. —TYPE specimen from Italy: Ischia, *Nägeli 92*, in herb. Kützing (L).

Symphyosiphon hirtulus Kützing [loc. cit. 1849] ex Forti, ibid., p. 531. 1907. *Scytonema hirtulum* Kützing (as "hirtulus") [ex Rabenhorst, ibid. 2: 265. 1865] ex Bornet & Flahault, Ann. Sci. Nat. VII. Bot. 5: 111. 1887. —TYPE specimen from Sachsen, Germany: Petersdorf, Nordhausen, in herb. Kützing (L).

Scytonema lignicola Nägeli [in Kützing, Sp. Algar., p. 894. 1849] ex Bornet & Flahault, loc. cit. 1887. —TYPE specimen from Switzerland: Zürich, an feuchten Felsen, *Nägeli 597*, in herb. Kützing (L).

Scytonema crassum Nägeli [in Kützing, loc. cit. 1849] ex Bornet & Flahault, ibid. 5: 109. 1887. *Petalonema crassum* Migula, Krypt.-Fl. 2(1): 131. 1907. —TYPE specimen from Switzerland: Zürich, Erlenbach, an feuchten Felsen, *Nägeli 410*, in herb. Kützing (L).

Arthrosiphon densus A. Braun [in Kützing, loc. cit. 1849] ex Bornet & Flahault, loc. cit. 1887. *Scytonema densum* Bornet & Thuret [Not. Algol. 2: 152. 1880] ex Bornet & Flahault, loc. cit. 1887. *Petalonema densum* A. Braun ex Migula, ibid. 2(1): 132. 1907. —TYPE specimen from Neuchâtel, Switzerland: St.-Aubin, auf überrieselten Felsen bei der Grotte aux Filles, *Braun 107*, Sept. 1848, in herb. Kützing (L). Fig. 3.

Calothrix indica Montagne [Ann. Sci. Nat. III. Bot. 12: 287. 1849] ex Bornet & Flahault, Ann. Sci. Nat. VII. Bot. 3: 370. 1886. —TYPE specimen from India: in Scapam, Assam, *Berkeley*, in herb. Montagne (PC).

Petronema fruticulosum Thwaites [in Berkeley & Thwaites, Suppl. Engl. Bot. 4: 2959. 1849] ex Bornet & Flahault, ibid. 5: 107. 1887. —TYPE specimen from England: St. Vincent's Rocks near Bristol, Feb. 1848 (BM).

[*Lyngbya bicolor* A. Braun in Mougeot & Nestler, Stirp. Crypt. Vogeso-rhen. 13: 1287. 1850.] *L. discolor* A. Braun ex Gomont, Ann. Sci. Nat. VII. Bot. 16: 153. 1892. —TYPE specimen from Baden, Germany: in fossis pratorum Brisgoviae, in Mougeot & Nestler, Stirpes Cryptogamae Vogeso-rhenanae no. 1287 (FH).

[*Arthrosiphon Grevillei* var. *cirrhosiphon* Kützing, Tab. Phyc. 2: 8. 1850–52.] [*A. alatus* c *cirrhosiphon* Kützing ex Rabenhorst, Fl. Eur. Algar. 2: 266. 1865.] *Scytonema alatum* var. *cirrhosiphon* Kützing (as "cirrosiphon") ex Forti, Syll. Myxoph., p. 529. 1907. *S. alatum* fa. *cirrhosiphon* Kossinskaia in Elenkin, Monogr. Alg. Cyanoph., Pars Spec. 1: 944. 1938. —TYPE specimen with that of γ *plicatulus* from Calvados, France: Falaise, *Lenormand*, in herb. Kützing (L).

Scytonema tectorum Itzigsohn [in Rabenhorst, Alg. Sachs. 27–28: 263. 1853] ex Bornet & Flahault, Ann. Sci. Nat. VII. Bot. 5: 107. 1887. —TYPE specimen from Brandenburg, Germany: auf einem Lattendache, Neudamm, *Itzigsohn*, Dec. 1852, in Rabenhorst, Algen Sachsens no. 263 (PH).

Scytonema salisburgense Rabenhorst [Hedwigia 1(4): 16. 1853] ex Bornet & Flahault, ibid. 5: 105. 1887. *S. gracile* fa. *crassius* Rabenhorst (as "crassior") [Fl.

38 REVISION OF THE NOSTOCACEAE

Eur. Algar. 2: 251. 1865] ex Bornet & Flahault, loc. cit. 1887. —TYPE specimen from Austria: bei Salzburg, *Sauter,* in Rabenhorst, Algen Sachsens no. 117 (PH).

Sirosiphon Bouteillei De Brébisson & Desmazières [in Desmazières, Pl. Crypt. de France, sér. 2, 3: 140. 1854] ex Bornet & Flahault, ibid. 5: 116. 1887. *Hapalosiphon Bouteillei* Borzi [N. Giorn. Bot. Ital. 11(4): 384. 1879] ex Bornet & Flahault, loc. cit. 1887. *Stigonema Bouteillei* Cooke (as "Bouteillii") [Brit. Fresh-w. Alg., p. 271. 1884] ex Forti, Syll. Myxoph., p. 552. 1907. *Hassallia Bouteillei* Bornet & Flahault [Mém. Soc. Nat. Sci. Nat. & Math. Cherbourg 25: 218. 1885] ex Bornet & Flahault, Ann. Sci. Nat. VII. Bot. 5: 116. 1887. *Tolypothrix Bouteillei* Lemmermann, Krypt.-Fl. Mark Brandenb. 3: 219. 1907; Forti, loc. cit. 1907. —TYPE specimen from Seine-et-Oise, France: in rupibus cretaceis, Magny, *L.-N. Bouteille,* Dec. 1852 (PC).

Scytonema truncicola Rabenhorst [Alg. Sachs. 35–36: 352. 1854; Hedwigia 1(9): 47. 1854] ex Bornet & Flahault, loc. cit. 1887. *Tolypothrix truncicola* Thuret [Ann. Sci. Nat. VI. Bot. 1: 380. 1875] ex Bornet & Flahault, loc. cit. 1887. —TYPE specimen from Piemonte, Italy: an der Rinde von Ulmus, bei Ver-celli, *Cesati,* in Rabenhorst, Algen Sachsens no. 352 (FH).

Tolypothrix Bulnheimii Rabenhorst [Alg. Sachs. 39–40: 393. 1854] ex Bornet & Flahault, ibid. 5: 121. 1887. —TYPE specimen from Sachsen, Germany: Leipzig, in Tümpeln der Harth, *O. Bulnheim,* Jun. 1854, in Rabenhorst, Algen Sachsens no. 393 (F).

Scytonema polymorphum Nägeli & Wartmann [in Jack, Leiner & Stizenberger, Krypt. Badens 8: 342. 1854] ex Bornet & Flahault, ibid. 5: 112. 1887. —TYPE specimen from Baden, Germany: beim Hirschensprung im Höllenthal, *C. Cramer,* Sept. 1854, in Jack, Leiner & Stizenberger, Kryptogamen Badens no. 342 (W).

Diplocolon Heppii Nägeli [in Itzigsohn, Verh. k. Leop.-Carol. Akad. Naturf. 18: 160. 1855; in Itzigsohn in Rabenhorst, Alg. Sachs. 47–48: 468. 1855] ex Bornet & Flahault, ibid. 5: 129. 1887. *Scytonema Heppii* Wolle, Fresh-w. Alg. U. S., p. 260. 1887. —TYPE specimen from Aargau, Switzerland: Jurakalkfelsen der Lägern, oberhalb Wittingen bei Baden am Stein, *Hepp,* Jun. 1852, in Rabenhorst, Algen Sachsens no. 468 (FH). Fig. 13.

Symphyosiphon involvens A. Braun [Hedwigia 1856: 105. 1856] ex Forti, Syll. Myxoph., p. 527. 1907. *Scytonema involvens* Rabenhorst [Fl. Eur. Algar. 2: 262. 1865] ex Bornet & Flahault, Ann. Sci. Nat. VII. Bot. 5: 108. 1887. *Petalonema involvens* A. Braun ex Migula, Krypt.-Fl. 2(1): 131. 1907. —TYPE specimen from Brandenburg, Germany: Grunewald bei Berlin in Torfgräben, *A. Braun,* Mar. 1856 (L). Fig. 9.

Scytonema pellucidum Cramer [in Rabenhorst, Alg. Sachs. 55–56: 542. 1856] ex Bornet & Flahault, ibid. 5: 102. 1887. —TYPE specimen from Switzerland: auf feuchten Felsen am Fusse des Rigi, *C. Cramer,* Sept. 1856, in Rabenhorst, Algen Sachsens no. 542 (NY).

Tolypothrix andina Montagne [Ann. Sci. Nat. IV. Bot. 6: 182. 1856] ex Bornet & Flahault, ibid. 5: 122. 1887. —TYPE specimen from Lake Chuquiaguillo, prov. La Paz, Bolivia: *Weddell,* in herb. Montagne (PC).

Scytonema vasconicum Lespinasse & Montagne [in Montagne, ibid. 6: 185. 1856] ex Bornet & Flahault, ibid. 5: 105. 1887. —TYPE specimen from Lot-et-Garonne, France: secus flumen Lotus ad rupes, *Lespinasse,* in herb. Montagne (PC).

Scytonema vinosum Montagne [loc. cit. 1856] ex Bornet & Flahault, Ann. Sci. Nat. VII. Bot. 5: 95. 1887. —TYPE specimen from French Guyana: Cayenne, *Leprieur 1370,* in herb. Bornet–Thuret (PC).

Tolypothrix pulchra fa. *tenuior* Suringar [Obs. Phycol., p. 41. 1857] ex Forti, Syll. Myxoph., p. 545. 1907. *T. lanata* fa. *tenuior* Forti, loc. cit. 1907. —TYPE specimen from the Netherlands: Giekerk, *W. F. R. Suringar 90b,* Jul. 1854 (L).

Ephebella Hegetschweileri Itzigsohn in Rabenhorst, Alg. Sachs. 59–60: 598. 1857. *Scytonema Hegetschweileri* Rabenhorst [Fl. Eur. Algar. 2: 252. 1865] ex Bornet & Flahault, ibid. 5: 113. 1887. —TYPE specimen from Graubünden, Switzerland: an Granitfelsen auf der Albula (Engadin), *P. Hepp,* Aug. 1855, in Rabenhorst, Algen Sachsens no. 598 (PH).

Calothrix pilosa Harvey [Ner. Bor.-Amer. 3: 106. 1858] ex Bornet & Flahault, ibid. 3: 363. 1886. *Schizosiphon pilosus* Crouan [in Mazé & Schramm, Essai Class. Alg. Guadeloupe, ed. 2, p. 31. 1870–77] ex Bornet & Flahault, ibid. 3: 359. 1886. *Tildenia pilosa* Poliansky, Izv. Glavn. Bot. Sada SSSR 27(3): 327. 1928. *Setchelliella pilosa* J. de Toni, Noter. di Nomencl. Algol. 8: [6]. 1936. —TYPE specimen from Florida: Key West, *W. H. Harvey* (K in BM). Fig. 18.

Scytonema incrustans var. *fuscum* Rabenhorst [Alg. Sachs. 67–68: 670. 1858] ex Bornet & Flahault, ibid. 5: 108. 1887. *S. incrustans* fa. *crassius* Rabenhorst (as "crassior") [Fl. Eur. Algar. 2: 264. 1865] ex Bornet & Flahault, loc. cit. 1887. —TYPE specimen from Bayern, Germany: bei Untersontheim (Schwäbisch Hall), *Kemmler,* Aug. 1857, in Rabenhorst, Algen Sachsens no. 670 (F).

Tolypothrix Wartmanniana Rabenhorst [Alg. Sachs. 77–78: 769. 1858] ex Bornet & Flahault, ibid. 5: 122. 1887. *T. tenuis* var. *Wartmanniana* Hansgirg, Prodr. Algenfl. Böhm. 2: 37. 1892. —TYPE specimen from Switzerland: an Wänden eines Wassersammlers bei Peter und Paul unweit St. Gallen, *Wartmann,* Apr. 1858, in Rabenhorst, Algen Sachsens no. 769 (F).

Symphyosiphon guyanensis Montagne [Ann. Sci. Nat. IV. Bot. 12: 171. 1859] ex Forti, Syll. Myxoph., p. 506. 1907. *Scytonema guyanense* Bornet & Flahault [Mém. Soc. Nat. Sci. Nat. & Math. Cherbourg 25: 215. 1885] ex Bornet & Flahault, Ann. Sci. Nat. VII. Bot. 5: 94. 1887. —TYPE specimen from French Guyana: tiges de palétuviers dans la mer, Cayenne, *Leprieur 542,* in herb. Montagne (PC).

Lyngbya effusa Harvey [Proc. Amer. Acad. 4: 335. 1859] ex Gomont, Ann. Sci. Nat. VII. Bot. 16: 153. 1892. —TYPE specimen from Japan: Loo Choo islands, *C. Wright* (US).

Scytonema asphaltii Fiorini-Mazzanti (as "asphalti") [Atti Accad. Pontif. N. Lincei Roma 13(4): 260. 1860] ex Forti, ibid., p. 539. 1907. —TYPE specimen from Roma, Italy: a Filettino, *Fiorini* (PAD).

Tolypothrix gracilis Rabenhorst [Alg. Eur. 105–106: 1052. 1861] ex Bornet & Flahault, ibid. 5: 123. 1887. [*T. flaccida* fa. *gracilis* Rabenhorst, Fl. Eur. Algar. 2: 277. 1865.] —TYPE specimen from Jutland, Denmark: in Wassergräben bei Björnsholm, *T. Jensen,* in Rabenhorst, Algen Europas no. 1052 (PH).

Tolypothrix flavo-viridis Kützing [Ostern-Progr. Realsch. Nordhausen 1862–63: 8. 1863] ex Bornet & Flahault, Ann. Sci. Nat. VII. Bot. 5: 101. 1887. *T. flavo-virens* Kützing [ex Rabenhorst, Fl. Eur. Algar. 2: 282. 1865] ex Bornet & Flahault, loc. cit. 1887. *Scytonema flavo-viride* Bornet & Flahault [Mém. Soc. Nat. Sci. Nat. & Math. Cherbourg 25: 216. 1885] ex Bornet & Flahault, Ann. Sci. Nat. VII. Bot. 5: 101. 1887. —TYPE specimen from Mexico: in Sümpfen bei Vera Cruz, *F. Müller,* in herb. Kützing (L).

Tolypothrix mexicana Kützing [Ostern-Progr. Realsch. Nordhausen 1862–63: 8. 1863] ex Bornet & Flahault, ibid. 5: 125. 1887. —TYPE specimen from Orizaba, Mexico: Río Blanco, *F. Müller,* Aug. 1853, in herb. Kützing (L).

Scytonema bormiense Brügger [Jahresber. Naturf. Ges. Graubündens, N. F. 8: 265. 1863] ex Bornet & Flahault, ibid. 5: 102. 1887. [*S. burmiense* Brügger in Wartmann & Schenk, Schw. Krypt. no. 241. 1863; Hedwigia 3(4): 58. 1864.] —TYPE specimen from Lombardia, Italy: bei den Bädern von Bormio, *C. Brügger,* Sept. 1862, in Wartmann & Schenk, Schweizerische Kryptogamen no. 241 (MIN).

Hydrocoleum calotrichoides Grunow [in Rabenhorst, Fl. Eur. Algar. 2: 152. 1865] ex Gomont, Ann. Sci. Nat. VII. Bot. 15: 346. 1892. —TYPE specimen from Steiermark, Austria: Bach, Schladming, *A. Grunow,* Sept. 1860 (W).

Scytonema Naegelii Kützing [in Rabenhorst, ibid. 2: 252. 1865] ex Bornet & Flahault, Ann. Sci. Nat. VII. Bot. 5: 124. 1887. —TYPE specimen from Switzerland: Zürichsee, an Brettern in der Badeanstalt, *Nägeli 176,* Aug. 1847 (ZT).

Scytonema calothrichoides fa. *natans* Rabenhorst [ibid. 2: 253. 1865] ex Wolle, Fresh-w. Alg. U. S., p. 251. 1887. —TYPE specimen from Brandenburg, Germany: Berlin, in Torfbrüchen bei Grunewald, *A. de Bary,* Aug. 1852, in Rabenhorst, Algen Sachsens no. 248 (F).

Scytonema gracillimum fa. *crassius* Rabenhorst (as "crassior") [ibid. 2: 254. 1865] ex Bornet & Flahault, ibid. 5: 102. 1887. —TYPE specimen from Bayern, Germany: bei Eichstadt, *F. Arnold,* as *S. tenue* in Rabenhorst, Algen Sachsens no. 652 (PH).

[*Scytonema penicillatum* b *muscicola* Hepp in Rabenhorst, Fl. Eur. Algar. 2: 256. 1865.] *S. turicense* var. *muscicola* Hepp ex Bornet & Flahault, ibid. 5: 98. 1887. —TYPE specimen from Switzerland: oberhalb des Schiessstandes Neuenburg, *P. Hepp,* in Rabenhorst, Algen Sachsens no. 695 (F).

Scytonema truncicola b *saxicola* Grunow [in Rabenhorst, Fl. Eur. Algar. 2: 257. 1865] ex Bornet & Flahault, Ann. Sci. Nat. VII. Bot. 5: 112. 1887. —TYPE specimen from Tirol, Austria: Klobenstein, an schattigen Felsen, *Hausmann,* 1859 (W).

Scytonema incrustans fa. *bryophilum* Rabenhorst (as "bryophila") [ibid. 2: 264. 1865] ex Bornet & Flahault, ibid. 5: 111. 1887. *S. crustaceum* fa. *bryophilum* Rabenhorst (as "bryophila") ex Forti, Syll. Myxoph., p. 526. 1907. —TYPE specimen, here designated, from Baden, Germany: auf *Hypnum Halleri,* Münsterthal, *A. Braun,* Aug. 1858 (L).

[*Arthrosiphon alatus* d *tinctus* A. Braun in Rabenhorst, Fl. Eur. Algar. 2: 266. 1865.] [*A. Grevillei* var. *tinctus* A. Braun (as "tinctum") in Rabenhorst, Alg. Eur. 185–186: 1843. 1866.] *Scytonema alatum* var. *tinctum* A. Braun (as "tinctus") ex Forti, ibid., p. 529. 1907. *S. alatum* fa. *tinctum* Kossinskaia in Elenkin, Monogr. Alg. Cyanoph., Pars Spec. 1: 945. 1938. —TYPE specimen from Switzerland: in den Grotten bei St. Aubin am Neuenburger See, *A. Braun,* Sept. 1865, in Rabenhorst, Algen Europas no. 1843 (F).

[*Arthrosiphon alatus* e *inconspicuus* A. Braun in Rabenhorst, Fl. Eur. Algar. 2: 266. 1865.] *Scytonema alatum* var. *inconspicuum* A. Braun (as "inconspicuus") ex Forti, loc. cit. 1907. *S. alatum* fa. *inconspicuum* Kossinskaia in Elenkin, loc. cit. 1938. —TYPE specimen from Switzerland, with that of *Arthrosiphon alatus* d *tinctus* A. Br. above (F).

[*Arthrosiphon densus* b *strictus* A. Braun in Rabenhorst, loc. cit. 1865.] *Scytonema densum* var. *strictum* A. Braun ex Forti, ibid., p. 527. 1907. —TYPE specimen, with that of *Arthrosiphon densus* a *densus* above, from Neuchâtel, Switzerland: St.-Aubin, *Braun 107,* Sept. 1848, in herb. Kützing (L).

Tolypothrix tenuis fa. *bryophila* Rabenhorst [Fl. Eur. Algar. 2: 273. 1865] ex Wolle, Fresh-w. Alg. U. S., p. 265. 1887. —TYPE specimen from Sachsen, Germany: an *Hypnum fluitans,* bei Weissenfels in Gräben, as "Calothrix mirabilis" in Kützing, Algarum Aquae Dulcis Germanicarum Dec. 1: 6 (F).

Tolypothrix tenuis fa. *pallescens* Rabenhorst [ibid. 2: 274. 1865] ex Forti, Syll. Myxoph., p. 547. 1907. *T. lanata* var. *tenuior* Hilse *pro synon.* [in Rabenhorst, loc. cit. 1865] ex Bornet & Flahault, Ann. Sci. Nat. VII. Bot. 5: 123. 1887. —TYPE specimen from Silesia, Poland: Peterwitz bei Strehlen, *Hilse,* Mai 1860, in Rabenhorst, Algen Europas no. 1033 (PH).

[*Tolypothrix pygmaea* fa. *crassior* Rabenhorst, Fl. Eur. Algar. 2: 275. 1865.] —TYPE specimen from Sachsen, Germany: in der Torfgrube am Bienitz bei Leipzig, *Bulnheim,* Apr. 1859, in Rabenhorst, Algen Sachsens no. 973 (F).

[*Tolypothrix distorta* fa. *tenuior* Rabenhorst, Fl. Eur. Algar. 2: 276. 1865.] —TYPE specimen from Schleswig, Germany: in Torfgruben bei Flensburg, *R. Häcker,* Apr. 1859, in Rabenhorst, Algen Sachsens no. 824 (PH).

[*Tolypothrix flaccida* fa. *tenuior* Rabenhorst, Fl. Eur. Algar. 2: 277. 1865.] —TYPE specimen, here designated, from Hessen, Germany: Hengster bei

Offenbach am Main, *A. de Bary,* Aug. 1853, in Rabenhorst, Algen Sachsens no. 311 (F).

Scytonema lichenicola Kunze [in Rabenhorst, Fl. Eur. Algar. 2: 281. 1865] ex Bornet & Flahault, Ann. Sci. Nat. VII. Bot. 5: 95. 1887. —TYPE specimen from the Madeira islands: *Holl,* in herb. Bornet–Thuret (PC).

Tolypothrix Selaginellae Rabenhorst [ibid. 2: 282. 1865] ex Bornet & Flahault, ibid. 5: 98. 1887. —TYPE specimen from the West Indies: auf St. Kitts in Sumpfe, *Breutel,* in herb. Kützing (L).

Lyngbya decipiens Crouan [in Schramm & Mazé, Essai Class. Alg. Guadeloupe, p. 32. 1865] ex Gomont, Ann. Sci. Nat. VII. Bot. 16: 150. 1892. *Scytonema submarinum* Crouan [in Mazé & Schramm, Essai Class. Alg. Guadeloupe, ed. 2, p. 33. 1870–77] ex Bornet & Flahault, ibid. 3: 363. 1886. —TYPE specimen from Guadeloupe, West Indies: Vieux-Fort, anse de la Petite Fontaine, *Mazé & Schramm 144* , in herb. Bornet-Thuret (PC).

Tolypothrix guadelupensis Crouan [in Schramm & Mazé, Essai Class. Alg. Guadeloupe, p. 32. 1865] ex Bornet & Flahault, ibid. 5: 91. 1887.—TYPE specimen from Guadeloupe, West Indies: Moule, mare de l'habitation Caillebot, *Mazé & Schramm 256,* Aug. 1861, in herb. Bornet-Thuret (PC).

Symphyosiphon Wimmeri Hilse [Jahres-ber. Schles. Ges. Vaterl. Cultur 1864: 95. 1865; in Rabenhorst, Alg. Eur. 177–178: 1775. 1865] ex Hansgirg, Prodr. Algenfl. Böhm. 2: 38. 1892. *Scytonema Wimmeri* Hilse [ex Rabenhorst, Fl. Eur. Algar. 2: 263. 1865] ex Bornet & Flahault, Ann. Sci. Nat. VII. Bot. 5: 114. 1887. *Tolypothrix Wimmeri* Kirchner [Krypt.-Fl. Schles. 2(1): 228. 1878] ex Bornet & Flahault, ibid. 5: 121. 1887. *T. lanata* var. *Wimmeri* Hansgirg, loc. cit. 1892. —TYPE specimen from Silesia, Poland: auf feuchter Erde unweit Siemsdorf bei Breslau, *Hilse,* Nov. 1864, in Rabenhorst, Alg. Eur. no. 1775 (F).

[*Calothrix mirabilis* var. *rupicola* Crouan, Fl. Finistère, p. 118. 1867.] *Plectonema Tomasinianum* var. *rupicola* Crouan ex Forti, Syll. Myxoph., p. 492. 1907. —TYPE specimen from Finistère, France: Fontaine de Coatodon, Aug. 1843, in herb. Crouan (Lab. Biol. Mar., Concarneau).

Calothrix synplocoides Reinsch [Algenfl. Mittl. Th. Franken, p. 51. 1867; in Rabenhorst, Alg. Eur. 192–193: 1923. 1867] ex Bornet & Flahault, ibid. 3: 372. 1886. *Scytonema Hofmannii* var. *synplocoides* Bornet & Thuret (as "symplocoides") [Not. Algol. 2: 149. 1880] ex Bornet & Flahault, ibid. 5: 99. 1887. *S. synplocoides* Hieronymus (as "symplocoides") in De Wildeman, Ann. Jard. Bot. Buitenzorg, Suppl. 1: 46. 1897. *S. Hofmannii* fa. *synplocoides* Kossinskaia (as "symplocoides") in Elenkin, Monogr. Alg. Cyanoph., Pars Spec. 1: 913. 1938. —TYPE specimen from Franken, Bayern, Germany: in scopulo aqua destillante humido, in sylva Sebaldiana, *P. Reinsch,* in Rabenhorst, Algen Europas no. 1923 (FH).

Scytonema cataracta Wood [Proc. Amer. Philos. Soc. 11: 129. 1869] ex Wolle, Fresh-w. Alg. U. S., p. 252. 1887. *S. cataractae* Wood ex Bornet & Flahault,

Ann. Sci. Nat. VII. Bot. 5: 105. 1887. —TYPE specimen from New York: Niagara river, *H. C. Wood* (PH).

Scytonema cortex Wood [Proc. Amer. Philos. Soc. 11: 130. 1869] ex Bornet & Flahault, ibid. 5: 98. 1887. —TYPE specimen from South Carolina: on bark of *Platanus occidentalis, H. W. Ravenel 47* (FH).

Scytonema Ravenellii Wood [loc. cit. 1869] ex Bornet & Flahault, ibid. 5: 94. 1887. *S. cortex* fa. *Ravenellii* Wood ex Wolle, ibid., p. 257. 1887. —TYPE specimen from South Carolina: ramul. Celtidis, Society Hill, *Ravenel,* in herb. Princeton University (NY).

Sirosiphon scytonematoides Wood [Proc. Amer. Philos. Soc. 11: 134. 1869] ex Bornet & Flahault, ibid. 5: 63. 1887. —TYPE specimen from South Carolina: bark of trees, Aiken, *Ravenel 183* (FH).

Scytonema Vieillardii Martens [Proc. Asiatic Soc. Bengal 1870: 258. 1870] ex Bornet & Flahault, ibid. 5: 112. 1887. —TYPE specimen from India: Calcutta, *S. Kurz 2675* (L).

Scytonema parietinum Crouan [in Mazé & Schramm, Essai Class. Alg. Guadeloupe, ed. 2, p. 33. 1870–77] ex Bornet & Flahault, Ann. Sci. Nat. VII. Bot. 5: 96. 1887. —TYPE specimen from Guadeloupe, West Indies: *Mazé & Schramm 960,* Sept. 1869, in herb. Bornet-Thuret (PC).

Scytonema cyanescens Crouan [in Mazé & Schramm, ibid., p. 34. 1870–77] ex Bornet & Flahault, ibid. 5: 92. 1887. —TYPE specimen from Guadeloupe, West Indies: Dolé, Bassin de la Digue, *Mazé & Schramm 182,* Mai 1864, in herb. Bornet–Thuret (PC).

Scytonema elegans var. *antillarum* Crouan [in Mazé & Schramm, ibid., p. 35. 1870–77] ex Bornet & Flahault, ibid. 5: 90. 1887. —TYPE specimen from Guadeloupe, West Indies: Morne de la Grande-Découverte, *Mazé & Schramm 150,* Aug. 1863, in herb. Bornet–Thuret (PC).

Scytonema coactile var. *radians* Crouan [in Mazé & Schramm, loc. cit. 1870–77] ex Bornet & Flahault, loc. cit. 1887. —TYPE specimen from Guadeloupe, West Indies: Pointe-à-Pitre, mare Chauvel, *Mazé & Schramm 1160,* Feb. 1870, in herb. Bornet–Thuret (PC).

Calothrix conferta Crouan [in Mazé & Schramm, ibid., p. 36. 1870–77] ex Bornet & Flahault, ibid. 3: 371. 1886. —TYPE specimen from Guadeloupe, West Indies: Matouba, Rivière Saint-Louis, *Mazé & Schramm 587,* Oct. 1868, in herb. Bornet–Thuret (PC).

Scytonema simplice Wood [Smithson. Contrib. Knowl. 241: 57. 1872] ex Wolle, Fresh-w. Alg. U. S., p. 259. 1887. *S. simplex* Wood ex Bornet & Flahault, Ann. Sci. Nat. VII. Bot. 5: 112. 1887. —TYPE specimen, here designated, from Pennsylvania: on wet rocks, *F. Wolle,* 1875, in herb. Bornet–Thuret (PC).

Scytonema Austinii Wood [Smithson. Contrib. Knowl. 241: 58. 1872] ex Bornet & Flahault, ibid. 5: 111. 1887. *Symphyosiphon Austinii* Wood [ex Wolle, ibid., p. 261. 1887] ex Stokes, Analyt. Keys Gen. & Sp. Fresh-w. Alg., p. 65.

44

REVISION OF THE NOSTOCACEAE

1893. —TYPE specimen from Little Falls, New Jersey: "same as sent to Dr. Wood," *C. F. Austin,* in herb. Wolle (D).

[*Symphyosiphon multistratus* Zanardini, Phyc. Indic. Pugill., p. 27. 1872.] —TYPE specimen, the sheaths infested with fungi, from Ceylon: *E. Beccari* (L).

Tolypothrix flexuosa Zanardini [ibid., p. 28. 1872] ex Bornet & Flahault, ibid. 5: 125. 1887. —TYPE specimen from Sarawak, Malaysia: Marop, *Beccari 6,* May 1867, in herb. Beccari (FI).

Calothrix maculiformis Zanardini (as "maculaeformis") [ibid., p. 29. 1872] ex Bornet & Flahault, ibid. 3: 370. 1886. —TYPE specimen from Sarawak, Malaysia: ruscelli di Mte. Mattang, *Beccari 33,* 1866, in herb. Beccari (FI).

Scytonema fulvum Zeller [Journ. Asiatic Soc. Bengal 42(2:3): 182. 1873] ex Bornet & Flahault, Ann. Sci. Nat. VII. Bot. 5: 113. 1887. —TYPE specimen from Burma: Pegu, *S. Kurz 3146* (L).

Scytonema Kurzianum Zeller [loc. cit. 1873] ex Bornet & Flahault, ibid. 5: 98. 1887. —TYPE specimen from Burma: an Baumrinde im Yomah-Gebirge, *S. Kurz,* in Rabenhorst, Algen Europas no. 2343 (NY).

Scytonema murale Zeller [loc. cit. 1873] ex Bornet & Flahault, ibid. 5: 96. 1887. —TYPE specimen from Burma: an den Mauern des Gerichtshauses zu Rangoon, *S. Kurz,* in Rabenhorst, Algen Europas no. 2344 (NY).

Scytonema subclavatum Zeller [ibid. 42(2:3): 183. 1873] ex Bornet & Flahault, ibid. 5: 112. 1887. —TYPE specimen from Burma: Pegu, Henzada, *S. Kurz 3168* (L).

Scytonema olivaceum Zeller [loc. cit. 1873] ex Bornet & Flahault, loc. cit. 1887. —TYPE specimen from Burma: Pegu, Yomah, *S. Kurz 3235* (L).

Scytonema parvulum Zeller [loc. cit. 1873] ex Bornet & Flahault, loc. cit. 1887. —TYPE specimen from Burma: Pegu, *Kurz 3156* (L).

Dictyonema membranaceum var. *guadelupense* Rabenhorst, Alg. Eur. 236–237: 2361. 1873. —TYPE specimen from Guadeloupe, West Indies: auf *Mastigobryum vincentinum* L., *l'Herminier,* in Rabenhorst, Algen Europas no. 2361 (PH).

[*Calothrix calibaea* Rabenhorst, ibid. 236–237. 2362. 1873.] [*Scytonema caribaeum* Bornet & Thuret, Not. Algol. 2: 146. 1880.] *Calothrix caribaea* Rabenhorst *pro synon.* [ex Bornet & Thuret, loc. cit. 1880] ex Bornet & Flahault, ibid. 3: 371. 1886. —TYPE specimen from Guadeloupe, West Indies: auf *Dumortiera hirsuta* R. Br., *l'Herminier,* in Rabenhorst, Algen Europas no. 2362 (PH).

Symphyosiphon Wollei Bornet [in Wolle, Bull. Torrey Bot. Club 6(26): 139. 1877] ex Forti, Syll. Myxoph., p. 506. 1907. —TYPE specimen from Pennsylvania: Bethlehem, eau douce, *F. Wolle,* in herb. Bornet—Thuret (PC).

Scytonema Brandegei Wolle [ibid. 6(35): 184. 1877] ex Bornet & Flahault, Ann. Sci. Nat. VII. Bot. 5: 111. 1887. —TYPE specimen from Colorado: wet rocks, *T. S. Brandegee,* in herb. Wolle (D).

Tolypothrix rupestris Wolle [ibid. 6(35): 185. 1877] ex Wolle, Fresh-w. Alg. U. S., p. 265. 1887. —TYPE specimen from Pennsylvania: Delaware Water Gap, Jul. 1878, in herb. Wolle (D).

Tolypothrix glacialis Dickie [Journ. Linn. Soc. Bot. 17: 8. 1878] ex Forti, ibid., p. 556. 1907. —TYPE specimen from Arctic Canada: edge of Glacier lake, Cape Baird, 81° 31′ N, *Feilden, Moss & Hart* (BM).

Scytonema pulvinatum Nordstedt [Minneskr. K. Fysiogr. Sållsk. Lund 1878(7): 6. 1878] ex Bornet & Flahault, ibid. 5: 94. 1887. —TYPE specimen from Hawaii: Honolulu, *S. Berggren*, 1873 (LD).

Scytonema mirabile Wolle [Bull. Torrey Bot. Club 6(40): 217. 1878] ex Forti, Syll. Myxoph., p. 513. 1907. *S. Wolleanum* Forti, loc. cit. 1907. —TYPE specimen from Florida: Gainesville, *H. W. Ravenel*, in herb. Wolle (D).

Scytonema Welwitschii Rabenhorst [Hedwigia 17(8): 116. 1878] ex Forti, ibid., p. 533. 1907. —TYPE specimen from Angola: Benguela, *F. Welwitsch*, in herb. Kützing (L).

Scytonema cortex var. *bruneum* Wolle [ibid. 6(49): 284. 1879] ex Drouet, Field Mus. Bot. Ser. 20(2): 37. 1939. *S. cortex* fa. *bruneum* Wolle (as "brunea"), Fresh-w. Alg. U. S., p. 258. 1887. *S. Hofmannii* fa. *bruneum* Wolle (as "brunnea") ex Forti, ibid., p. 515. 1907. —TYPE specimen from South Carolina: cortice Cephalanthi, *H. W. Ravenel 65* (D).

Scytonema cortex var. *corrugatum* Wolle [Bull. Torrey Bot. Club 6(49): 284. 1879] ex Forti, ibid., p. 516. 1907. *S. cortex* fa. *corrugatum* Wolle (as "corrugata"), Fresh-w. Alg. U. S., p. 257. 1887. —TYPE specimen, the sheaths infested with fungi, from Florida: *J. D. Smith*, Mar. 1878 (D).

Tolypothrix Ravenelii Wolle [Bull. Torrey Bot. Club 6(49): 285. 1879] ex Wolle, Fresh-w. Alg. U. S., p. 265. 1887. —TYPE specimen from Florida: on sandstone rocks, Gainesville, *H. W. Ravenel* (BM).

Microchaete tenera Thuret [in Bornet & Thuret, Not. Algol. 2: 128. 1880] ex Bornet & Flahault, Ann. Sci. Nat. VII. Bot. 5: 84. 1887. *Coleospermum tenerum* Elenkin, Izv. Imp. Bot. Sada Petra Velik. 25(1): 20. 1915. *Fremyella tenera* J. de Toni, Noter. di Nomencl. Algol. 8: [4]. 1936. —TYPE specimen, containing fragments of trichomes escaped from the sheaths, from France: étang de Vaugrenier près d'Antibes, *G. Thuret*, Mar. 1875, in herb. Bornet—Thuret (PC).

Scytonema Clevei Grunow [ex Bornet & Thuret, ibid. 2: 146. 1880] ex Bornet & Flahault, ibid. 3: 363. 1886. —TYPE specimen from the West Indies: Tortola, *Cleve*, in herb. Bornet—Thuret (PC).

Scytonema Millei Bornet & Thuret [ibid. 2: 147. 1880] ex Bornet & Flahault, ibid 5: 93. 1887. —TYPE specimen from French Guyana: sur la face des rochers de la partie nord, ville de Cayenne, près la mer, *Mille*, 1834, in herb. Bornet—Thuret (PC).

Scytonema gracile var. *tolypotrichoides* Wolle [in Wittrock & Nordstedt, Alg. Aq.

46

Dulc. Exs. 8: 389. 1880] ex Bornet & Flahault, ibid. 5: 111. 1887. —TYPE specimen from New Jersey: Morris pond, *F. Wolle*, Jul. 1879, in Wittrock & Nordstedt, Algae Exsiccatae no. 389 (PH).

Tolypothrix bombycina Wolle [Bull. Torrey Bot. Club 7: 44. 1880] ex Drouet, Field Mus. Bot. Ser. 20(2): 33. 1939. —TYPE specimen from New Jersey: Hopatcong lake, *Wolle*, 1879 (D).

Plectonema Kirchneri Cooke [Grevillea 11(58): 75. 1882] ex Gomont, Ann. Sci. Nat. VII. Bot. 16: 103. 1892. —TYPE specimen from England: in piscina horti botanici Kewensis, *M. C. Cooke*, in Wittrock & Nordstedt, Algae Exsiccatae no. 883 (LD).

[*Lyngbya scabrosa* Dickie, Proc. R. Soc. Edinburgh 11(110): 456. 1882.] —TYPE specimen from Socotra: *B. Balfour*, 1880 (BM).

Scytonema aerugineum Lespinasse [Actes Soc. Linn. Bordeaux 36: 197. 1882] ex Bornet & Flahault, Ann. Sci. Nat. VII. Bot. 5: 121. 1887. —TYPE specimen from France: au jardin bot. de Bordeaux, dans des pots à fleurs où M. Durieu cultivait le *Riella gallica* rapporté d'Agde (mares de Roquehaute), *Lespinasse*, Sept. 1870, in herb. Bornet—Thuret (PC).

[*Arthrosiphon Grevillei* fa. *americanus* Lespinasse (as "americana"), ibid. 36: 198. 1882.] —TYPE specimen from New York: Niagara, *Bailey*, in herb. Kützing (L).

Scytonema coactile var. *brasiliense* Nordstedt [in Wittrock & Nordstedt, Alg. Aq. Dulc. Exs. 10: 488. 1882] ex Wittrock & Nordstedt, ibid. 21: 52. 1889. —TYPE specimen from São Paulo, Brazil: Olaria do Faustino prope Pirassununga, *A. Löfgren 162*, Feb. 1880, in Wittrock & Nordstedt, Algae Exsiccatae no. 488 (LD).

Coleospermum Goeppertianum var. *minus* Hansgirg (as "minor") [Österr. Bot. Zeitschr. 33: 224. 1883] ex Drouet, Ann. Naturh. Mus. Wien 61: 46. 1957. *Microchaete tenera* var. *minor* Hansgirg, Prodr. Algenfl. Böhm. 2: 55. 1892. *Fremyella tenera* var. *minor* J. de Toni, Diagn. Alg. Nov. 1(4): 346. 1938. —TYPE specimen from Czechoslovakia: bei Prag, *A. Hansgirg*, Oct. 1883 (W).

Scytonema Hansgirgianum Richter [Hedwigia 23(5): 67. 1884; in Wittrock & Nordstedt, Alg. Aq. Dulc. Exs. 14: 674. 1884] ex Bornet & Flahault, Ann. Sci. Nat. VII. Bot. 5: 98. 1887. *S. Hofmannii* var. *Hansgirgianum* Richter [ex Hansgirg in Kerner, Schedae ad Fl. Exs. Austro-Hungar. 4: 112. 1886] ex Hansgirg, Prodr. Algenfl. Böhm. 2: 33. 1892. *S. Hofmannii* fa. *Hansgirgianum* Kossinskaia in Elenkin, Monogr. Alg. Cyanoph., Pars Spec. 1: 913. 1938. —TYPE specimen, in part infested with fungi, from Sachsen, Germany: ad parietes caldarii in Anger prope Lipsiam, *P. Richter*, Aug. 1883, in Wittrock & Nordstedt, Algae Exsiccatae no. 674 (FH).

Scytonema polycystum Bornet & Flahault [Mém. Soc. Nat. Sci. Nat. & Math. Cherbourg 25: 214. 1885] ex Bornet & Flahault, Ann. Sci. Nat. VII. Bot. 5: 90. 1887. —TYPE specimen from New Caledonia: prope Nouméa, *A. Grunow*, Nov. 1884, in herb. Bornet—Thuret (PC). Fig. 11.

Scytonema Arcangelii Bornet & Flahault [Mém. Soc. Nat. Sci. Nat. & Math. Cherbourg 25: 215. 1885] ex Bornet & Flahault, Ann. Sci. Nat. VII. Bot. 5: 92. 1887. —TYPE specimen from Italy: nelle conche delle stufe dell 'orto botanico fiorentino, *G. Arcangeli,* Mar. 1878, in Erbario Crittogamico Italiano, ser. 2, no. 785 (PC).

Tolypothrix limbata Thuret [ex Bornet & Flahault, Mém. Soc. Nat. Sci. Nat. & Math. Cherbourg 25: 219. 1885] ex Bornet & Flahault, Ann. Sci. Nat. VII. Bot. 5: 124. 1887. —TYPE specimen from France: Montpellier, *C. Flahault 704,* Oct. 1883, in herb. Bornet—Thuret (PC). Fig. 5.

Aulosira implexa Bornet & Flahault [Bull. Soc. Bot. France 32: 120. 1885] ex Bornet & Flahault, Ann. Sci. Nat. VII. Bot. 7: 257. 1888. *Nodularia implexa* Bourrelly, Alg. d'Eau Douce 3: 418. 1970. —TYPE specimen from Uruguay: eaux stagnantes, marais des environs de Montevido, *Arechavaleta,* Mar. 1884, in herb. Bornet—Thuret (PC). Fig. 25.

Hassallia byssoidea fa. *saxicola* Grunow in Bornet & Flahault, ibid. 5: 117. 1887. *Scytonema cortex* fa. *saxicola* Grunow ex Wolle, Fresh-w. Alg. U. S., p. 355. 1887. *Tolypothrix byssoidea* fa. *saxicola* Grunow ex Forti, Syll. Myxoph., p. 552. 1907. —TYPE specimen from France: collines de Biot près d'Antibes, *E. Bornet,* Apr. 1872, in herb. Bornet—Thuret (PC).

Microchaete striatula Hy, Journ. de Bot. 1(13): 197. 1887. *Leptobasis striatula* Elenkin, Izv. Imp. Bot. Sada Petra Velik. 25(1): 11. 1915. *Fortiea striatula* J. de Toni, Noter. di Nomencl. Algol. 8: [3]. 1936. *Fremyella striatula* Drouet, Field Mus. Bot. Ser. 20(6): 130. 1942. —TYPE specimen from Anjou, France: mares tourbeuses, Juigné-sur-Loire, *F. Hy,* Jun. 1887, in herb. Bornet—Thuret (PC).

Plectonema phormidioides Hansgirg, Physiol. & Algol. Stud., p. 108. 1887; Österr. Bot. Zeitschr. 37: 121. 1887. —TYPE specimen from Bohemia, Czechoslovakia: Siehdichfür nächst Neuwelt im Riesengebirge, *A. Hansgirg,* Jul. 1886 (W).

Symphyosiphon Bornetianus Wolle (as "Bornetianum"), Fresh-w. Alg. U. S., p. 261. 1887.—TYPE specimen from South Carolina: clay cliffs, *H. W. Ravenel 218* (FH).

Scytonema burdigalense Durieu de Maisonneuve & Roumeguère in Mougeot, Dupray & Roumeguère, Alg. de France, no. 424. 1888. —TYPE specimen from Gironde, France: carrière abandonnée, Cambes, *C. Roumeguère,* Jun. 1888 (BM).

Tolypothrix penicillata var. *tenuis* Hansgirg, Sitzungsber. K. Böhm. Ges. Wiss., Math.-nat. Cl. 1890(1): 14. 1890. —TYPE specimen from Slovenia, Yugoslavia: bei Pisino, *A. Hansgirg,* Aug. 1889 (W).

Scytonema obscurum var. *terrestre* Hansgirg, Prodr. Algenfl. Böhmen 2: 36. 1892. *S. obscurum* fa. *terrestre* Hansgirg ex Elenkin, Monogr. Alg. Cyanoph., Pars Spec. 1: 902. 1938. —TYPE specimen from Czechoslovakia: in den Schanzgräben hinter dem Kornthore, Prag, *A. Hansgirg,* Nov. 1883 (W).

Tolypothrix distorta var. *symplocoides* Hansgirg, Prodr. Algenfl. Böhm. 2: 39.

1892. *T. distorta* fa. *symplocoides* Kossinskaia in Elenkin, ibid. 1: 959. 1938. —TYPE specimen from Prague, Czechoslovakia: in einem Warmhause des botan. Gartens am Smichow, *A. Hansgirg* (W).

Plectonema Tomasinianum var. *cincinnatum* Hansgirg, ibid. 2: 40. 1892. *P. Tomasinianum* fa. *cincinnatum* Poliansky in Hollerbach, Kossinskaia & Poliansky, Opred. Presnov. Vodor. SSSR 2: 595. 1953. —TYPE specimen from Bohemia, Czechoslovakia: in einer Mühlschleuse bei Eisenbrod, *A. Hansgirg,* Jul. 1885 (W).

Scytonema amplum W. & G. S. West, Journ. Linn. Soc. Bot. 30: 270. 1895. *S. mirabile* var. *amplum* Playfair, Proc. Linn. Soc. N. S. Wales 37(3): 533. 1913. —TYPE specimen, with that of *Symploca cuspidata* W. & G. S. West [= *Schizothrix Friesii* (Ag.) Gom.], from Dominica, West Indies: on trees, summit of Trois Pitons, *W. R. Elliott 903,* 1892 (UC).

Hassallia usambarensis Hieronymus in Engler, Pflanzenw. Ost-Afr. C: 10. 1895. —TYPE specimen from Tanzania: Usambara, Waldgebüsch bei Doda, *Holst 3021,* Jun. 1893 (BM).

Scytonema Holstii Hieronymus in Engler, loc. cit. 1895. —TYPE specimen from Tanzania: unter Rasen von Selaginella, Usambara, *Holst 831* (BM).

Tolypothrix polymorpha Lemmermann, Forschungsber. Biol. Sta. Plön 4: 184. 1896. *T. tenuis* fa. *polymorpha* Kossinskaia in Elenkin, Monogr. Alg. Cyanoph., Pars Spec. 1: 954. 1938. —TYPE specimen from Holstein, Germany: Tümpel in der Nähe der Parnassus, Plöner-Gebiet, *E. Lemmermann,* Jul. 1895 (BREM).

Tolypothrix fasciculata Gomont, Bull. Soc. Bot. France 43: 381. 1896. —TYPE specimen from Haute-Auvergne, France: rochers secs, Route de Thiézac à St.-Jacques, *M. Gomont 40,* Aug. 1894, in herb. Gomont (PC).

Calothrix breviarticulata W. & G. S. West, Journ. of Bot. 35: 240. 1897. —TYPE specimen from Angola: Pungo Andongo, in rivulis, *F. Welwitsch 105,* Mar. 1857 (UC).

Scytonema cincinnatum var. *aethiopicum* W. & G. S. West, ibid. 35: 264. 1897. *S. crispum* var. *aethiopicum* W. & G. S. West ex Forti, Syll. Myxoph., p. 500. 1907. —TYPE specimen, filed with *Mougeotia irregularis* W. & G. S. West, from Angola: Pungo Andongo, in pascuis prope Catete, *Welwitsch 111,* May 1857 (BM).

Scytonema myochrous var. *chorographicum* W. & G. S. West, ibid. 35: 265. 1897. *S. chorographicum* Welwitsch ex W. & G. S. West *pro synon.,* ibid. 35: 303. 1897. *S. myochrous* fa. *chorographicum* West (as "chorographica") ex Frémy, Rev. Algol. 1(1): 47. 1924. —TYPE specimen from Angola: Pungo Andongo, *Welwitsch 6,* Feb. 1857 (UC).

Scytonema insigne W. & G. S. West, ibid. 35: 266. 1897. —TYPE specimen from Angola: Golungo Alto, ad rupes madidas, *Welwitsch 5,* May 1856 (UC).

Tolypothrix phyllophila W. & G. S. West, ibid. 35: 267. 1897. —TYPE specimen from Angola: Golungo Alto, ad folia in sylvis de Alto Queta, *Welwitsch,* Dec. 1855 (BM).

Tolypothrix arenophila W. & G. S. West, loc. cit. 1897.—TYPE specimen from Angola: Pungo Andongo, ad terram, *Welwitsch 151,* Jan. 1857 (UC).

Tolypothrix crassa W. & G. S. West, loc. cit. 1897. —TYPE specimen from Angola: Pungo Andongo, ad flum. Cuanza, *Welwitsch 11,* Mar. 1857 (UC).

Scytonema dubium De Wildeman, Ann. Jard. Bot. Buitenzorg, Suppl. 1: 43. 1897. *S. De-Wildemanii* J. de Toni, Noter. di Nomencl. Algol. 8: [6]. 1936. —TYPE specimen from Java, Indonesia: Jardin Botanique, Buitenzorg, *J. Massart 993* (L).

Scytonema foliicola De Wildeman (as "foliicolum"), ibid. 1: 44. 1897. —TYPE specimen from Java, Indonesia: Gorges de Tjaipoes, *Massart 817* (L).

Tolypothrix Setchellii Collins, Erythea 5(9): 96. 1897.—TYPE specimen from Connecticut: on shells, Twin lakes, Salisbury, *W. A. Setchell & I. Holden,* Aug. 1897, in Collins, Holden & Setchell, Phycotheca Boreali-Americana no. 310, in herb. Collins (NY).

Gloeochlamys Simmeri Schmidle in Simmer, Allg. Bot. Zeitschr. 1899(12): 192. 1899. —TYPE specimen from Kärnten, Austria: an der Westseite des Kleinen Knoten, *H. Simmer,* Jun. 1898 (W).

Scytonema Simmeri Schmidle in Simmer, ibid. 1899(12): 193. 1899. *Petalonema Simmeri* Migula, Krypt.-Fl. 2(1): 132. 1907. —TYPE specimen from Kärnten, Austria: bei Irschen, in einer Felsspalte, *Simmer 1106,* May 1898 (W).

Scytonema caldarium Setchell, Erythea 7(5): 48. 1899; in Collins, Holden & Setchell, Phyc. Bor.-Amer. 12: 559. 1899. —TYPE specimen from California: on cliff, Waterman hot spring near San Bernardino, *S. B. Parish,* Apr. 1897 (UC).

Scytonema occidentale Setchell, ibid. 7(5): 49. 1899. —TYPE specimen from California: La Jota creek on Howell mountain near St. Helena, Napa county, *W. A. Setchell 1095,* Nov. 1895 (UC).

Stigonema indicum Schmidle (as "indica"), Allg. Bot. Zeitschr. 1900: 54. 1900. *Camptylonema indicum* Schmidle, Hedwigia 39: 181. 1900. *Campylonema indicum* Schmidle ex Forti, Syll. Myxoph., p. 540. 1907. *Schmidleinema indicum* J. de Toni, Noter. di Nomencl. Algol. 8: [5]. 1936 —TYPE specimen, with that of *Trentepohlia Monilia* fa. *hyalina* Schmidle, from India: ad muros vetustos in silvis palmarum prope Mahim (Bombay), *A. Hansgirg,* in Musei Vindobonensis Kryptogamae Exsiccatae no. 858 (W).

Scytonema junipericola Farlow in Collins, Holden & Setchell, Phyc. Bor.-Amer. 16: 756. 1900. —TYPE specimen from Bermuda: on bark of *Juniperus bermudiana,* Hamilton, *W. G. Farlow,* Jan. 1900 (FH).

Hassallia byssoidea fa. *cylindrica* Tilden, Amer. Alg. 4: 398. 1900. *Tolypothrix byssoidea* fa. *cylindrica* Tilden, Minn. Alg. 1: 233. 1910. —TYPE specimen from British Columbia: on vertical rocks, Baird point, Vancouver island, *J. E. Tilden,* Aug. 1898, in Tilden, American Algae no. 398 (MIN).

Scytonema Steindachneri Krasser in Zahlbrückner, Ann. K. K. Naturh. Hofmus. Wien 15: 173. 1900. —TYPE specimen from Venezia Giulia, Italy: in rupibus irrigatis ad Barcola prope Triest, *F. Krasser,* in Musei Vindobonensis Kryptogamae Exsiccatae no. 422 (W).

Scytonema bruneum Schmidle (as "brunea") in Simmer, Allg. Bot. Zeitschr. 1901 (5): 85. 1901. —TYPE specimen from Kärnten, Austria: in Zwickenberg auf feuchtem Bachsande, *H. Simmer 63,* Aug. 1898 (W).

Scytonema figuratum fa. *minus* Schmidle (as "minor") in Simmer, loc. cit. 1901. *S. minus* Lemmermann (as "minor"), Krypt.-Fl. Mark Brandenb. 3: 313. 1907. *S. mirabile* fa. *minus* Schmidle (as "minor") ex Forti, Syll. Myxoph., p. 520. 1907. —TYPE specimen from Kärnten, Austria: im Gnoppnitzthale auf Thonschieferfelsen, *Simmer,* Nov. 1898 (W).

Scytonema conchophilum Humphrey in Collins, Proc. Amer. Acad. Arts & Sci. 37(9): 241. 1901. —TYPE specimen from Jamaica: in shells and bones, Navy island, *J. E. Humphrey,* Aug. 1897, in Collins, Holden & Setchell, Phycotheca Boreali-Americana no. LII (NY).

Scytonema Schmidtii Gomont, Bot. Tidsskr. 24: 207. 1902. —TYPE specimen from Koh Chang, Cambodia: *Danske Siamexpedition X,* 1899–1900, in herb. Gomont (PC).

Scytonema hawaiianum Tilden (as "hawaiiana"), Amer. Alg. 6: 573. 1902. —TYPE specimen from Hawaii: on rocks, Pahala Plantation beach, *J. E. Tilden,* Jul. 1900, in Tilden, American Algae no. 573 (MIN).

Scytonema hawaiianum var. *terrestre* Tilden (as "terrestris"), ibid. 6: 574. 1902. —TYPE specimen from Hawaii: terrestrial, Laie point, Koolauloa, *Tilden,* June 1900, in Tilden, American Algae no. 574 (MIN).

Microchaete robusta Setchell & Gardner, Univ. Calif. Publ. Bot. 1: 194. 1903. *Fremyella robusta* J. de Toni, Noter. di Nomencl. Algol. 8: [4]. 1936. —TYPE specimen from Washington: ponds, Seattle, *T. C. D. Kincaid 768,* 1902 (UC).

Tolypothrix helicophila Lemmermann, Krypt.-Fl. Mark Brandenb. 3: 219. 1907. —TYPE specimen from Germany: Berlin, Weissensee, *A. Braun* (BREM).

Scytonema azureum Tilden, Amer. Alg. 7(1): 630. 1909. —TYPE specimen from Hawaii: warm spring, Puna, *Tilden,* Jul. 1900, in Tilden, American Algae no. 630 (MIN).

Scytonema saleyerense Weber-van Bosse (as "saleyerensis"), Liste d. Alg. d. Siboga, p. 31. 1913. —TYPE specimen from Indonesia: Saleyer, op den grond bij eer beekje, *A. Weber-van Bosse* (L).

Scytonema samoense Wille, Hedwigia 53: 145. 1913. —TYPE specimen from Upolu, Samoa: *K. Rechinger 3171,* in slide collection (O).

Scytonema coactile var. *minus* Wille (as "minor"), loc. cit. 1913. —TYPE specimen from Upolu, Samoa: *Rechinger 3174,* in slide collection (O).

Tolypothrix distorta var. *samoensis* Wille, loc. cit. 1913. —TYPE specimen from Savaii, Samoa: in ausgetrocknetem Flussbette bei Potasuka, *Rechinger 2969,* Jul. 1905 (O).

Hassallia Rechingeri Wille, loc. cit. 1913. *Tolypothrix Rechingeri* Geitler in Pascher, Süsswasserfl. 12: 259. 1925. —TYPE specimen from Upolu, Samoa: *Rechinger 2726* (O).

Hassallia Rechingeri fa. *saxicola* Wille, loc. cit. 1913. *Tolypothrix Rechingeri* fa. *saxicola* Wille ex Geitler, Rabenh. Krypt.-Fl., ed. 2, 14: 731. 1932. —TYPE specimen from Upolu, Samoa: an Felsen im Wasserfalle Papaseea, *Rechinger 5092* (O).

Tolypothrix conglutinata var. *colorata* Ghose, Journ. Linn. Soc. Bot. 46: 345. 1923.—TYPE specimen from Punjab, India: Simla, on damp rocks etc., Aug. 1919, in coll. Fritsch no. 336 (BM).

Scytonema Zellerianum Brühl & Biswas, Journ. Dept. Sci. Calcutta Univ. 5(Bot., Comment. Algol. 2): 16. 1923. —TYPE specimen from Burma: Pegu, *S. Kurz 3352* (L).

Scytonema Bewsii Fritsch & Rich, Trans. R. Soc. So. Africa 11: 364. 1924. —TYPE specimen from South Africa: mountain stream at Vryheid, *J. W. Bews 18*, Mar. 1915 (BM).

Scytonema splendens Fritsch & Rich, ibid. 11: 366. 1924. —TYPE specimen from South Africa: on moist cliff, Drakensberg, Goodoo Pass, *Bews 109*, Sept. 1915 (BM).

Homoeothrix aequalis Fritsch & Rich, ibid. 11: 372. 1924. —TYPE specimen from South Africa: bricks in greenhouse, Maritzburg, *Bews 21*, Mar. 1915 (BM).

Calothrix parietina var. *africana* Fritsch in Fritsch & Rich, ibid. 11: 375. 1924. *C. parietina* fa. *africana* Poliansky in Elenkin, Monogr. Alg. Cyanoph., Pars Spec. 2: 1076. 1949. —TYPE specimen from Drakensberg, South Africa: *Bews 113*, in slide collection (BM).

Tolypothrix Chungii Gardner, Rhodora 28: 4. 1926. —TYPE specimen from Fukien, China: in a pool near the University of Amoy, *H. H. Chung A71* (FH).

Tolypothrix tenella Gardner, loc. cit. 1926. —TYPE specimen from Fukien, China: mountain stream, Amoy island, *Chung A75* (FH).

Synechococcus intermedius Gardner, Mem. New York Bot. Gard. 7: 3. 1927. —TYPE specimen from Puerto Rico: Caguas, *N. Wille 439b*, 1915 (NY). See Drouet & Daily, Butler Univ. Bot. Stud. 12: 166, fig. 354 (1956).

Lyngbya Martensiana var. *minor* Gardner, ibid. 7: 41. 1927. *L. Martensiana* fa. *minor* Elenkin, Monogr. Alg. Cyanoph., Pars Spec. 2: 1628. 1949.—TYPE specimen from Puerto Rico: on limestone, Arecibo to Hatillo, *Wille 1392b*, Feb. 1915 (NY).

Calothrix conica Gardner, ibid. 7: 66. 1927. —TYPE specimen from Puerto Rico: trees near Coamo Springs, *Wille 1895* (NY).

Scytonema evanescens Gardner, ibid. 7: 71. 1927. —TYPE specimen from Puerto Rico: on limestone, Arecibo to Utuado, *Wille 1482*, Mar. 1915 (NY).

Scytonema capitatum Gardner, ibid. 7: 72. 1927. —TYPE specimen from

Puerto Rico: rocks 10 km. north of Utuado, *Wille 1537a,* Mar. 1915 (NY).

Scytonema longiarticulatum Gardner, ibid. 7: 73. 1927. —TYPE specimen from Puerto Rico: rock at Campo, Maricao, *Wille 1229a,* Feb. 1915 (NY).

Scytonema subgelatinosum Gardner, ibid. 7:74. 1927. —TYPE specimen from Puerto Rico: rocks by water reservoir, Rio Piedras, *Wille 108,* Dec. 1914 (NY).

Scytonema variabile Gardner, loc. cit. 1927. —TYPE specimen from Puerto Rico: on waterpipe near Maricao, *Wille 1149a,* Feb. 1915 (NY).

Scytonema magnum Gardner, ibid. 7: 75. 1927. —TYPE specimen from Puerto Rico: rocks in a brook 5 km. north of Utuado, *Wille 1609,* Mar. 1915 (NY).

Scytonema punctatum Gardner, loc. cit. 1927. —TYPE specimen from Puerto Rico: on rocks, Utuado to Adjuntas, *Wille 1658,* Mar. 1915 (NY).

Scytonema pulchellum Gardner, ibid. 7: 76. 1927. —TYPE specimen from Puerto Rico: rocks near Utuado, *Wille 1574,* Mar. 1915 (NY).

Scytonema Millei var. *majus* Gardner (as "Milleri"), loc. cit. 1927. —TYPE specimen from Puerto Rico: rocks near Maricao, *Wille 1166,* Feb. 1915 (NY).

Scytonema catenulum Gardner, ibid. 7: 77. 1927. *S. catenatum* Gardner *pro synon.,* ibid. 7: 128. 1927. —TYPE specimen from Puerto Rico: rocks 10 km. north of Utuado, *Wille 1556,* Mar. 1915 (NY).

Scytonema mirabile var. *majus* Gardner, ibid. 7: 78. 1927. —TYPE specimen from Puerto Rico: rocks 5 km. north of Utuado, *Wille 1617,* Mar. 1915 (NY).

Scytonema lyngbyoides Gardner, loc. cit. 1927. —TYPE specimen from Puerto Rico: rocks in a brook 5 km. north of Utuado, *Wille 1607,* Mar. 1915 (NY).

Scytonema ocellatum var. *constrictum* Gardner, ibid. 7: 79. 1927. —TYPE specimen from Puerto Rico: rocks north of Maricao, *Wille 1253,* Feb. 1915 (NY).

Scytonema ocellatum var. *majus* Gardner, loc. cit. 1927. —TYPE specimen from Puerto Rico: limestone at Hacienda, Lake Tortuguero, *Wille 867,* Feb. 1915 (NY).

Scytonema ocellatum var. *purpureum* Gardner, loc. cit. 1927. —TYPE specimen from Puerto Rico: soil, Coamo Springs, *Wille 292c,* Jan. 1915 (NY).

Scytonema guyanense var. *minus* Gardner, loc. cit. 1927. —TYPE specimen from Puerto Rico: rocks, Arecibo to Utuado, *Wille 1455,* Mar. 1915 (NY).

Scytonema tenellum Gardner, ibid. 7: 80. 1927. —TYPE specimen from Puerto Rico: rock north of Sabana Grande, *Wille 936a,* Feb. 1915 (NY).

Scytonema spirulinoides Gardner, loc. cit. 1927. —TYPE specimen from Puerto Rico: rocks near San Lorenzo, *Wille 534,* Jan. 1915 (NY).

Scytonema multiramosum Gardner, ibid. 7: 81. 1927. —TYPE specimen from Puerto Rico: rocks 10 km. north of Utuado, *Wille 1527,* Mar. 1915 (NY).

Scytonema javanicum var. *distortum* Gardner, loc. cit. 1927. —TYPE specimen from Puerto Rico: trees north of Mayaguez, *Wille 1000,* Feb. 1915 (NY).

Scytonema javanicum var. *pallidum* Gardner, loc. cit. 1927. —TYPE specimen from Puerto Rico: rock near Mayaguez, *Wille 899a,* Feb. 1915 (NY).

Hassallia brevis Gardner (as "Hassalia"), ibid. 7: 82. 1927. *Tolypothrix brevis*

Geitler, Rabenh. Krypt.-Fl., ed. 2, 14: 726. 1932. —TYPE specimen from Puerto Rico: fountain in Fajardo, *Wille 659,* Jan. 1915 (NY).

Hassallia granulata Gardner (as "Hassalia"), loc. cit. 1927. *Tolypothrix granulata* Geitler, ibid. 14: 727. 1932. —TYPE specimen from Puerto Rico: bark near Coamo Springs, *Wille 1913,* Mar. 1915 (NY).

Hassallia heterogenea Gardner (as "Hassalia"), ibid. 7: 83. 1927. —TYPE specimen from Puerto Rico: rocks, Hacienda Holm, Mayaguez, *Wille 1191,* Feb. 1915 (NY).

Hassallia discoidea Gardner (as "Hassalia"), loc. cit. 1927. *Tolypothrix discoidea* Geitler, ibid. 14: 733. 1932. —TYPE specimen from Puerto Rico: rocks 7 km. east of Coamo, *Wille 1870,* Mar. 1915 (NY).

Hassallia rugulosa Gardner (as "Hassalia"), ibid. 7: 84. 1927. *Tolypothrix rugulosa* Geitler, ibid. 14: 727. 1932. —TYPE specimen from Puerto Rico: rocks near San Lorenzo, *Wille 517,* Jan. 1915, the sheaths somewhat infested with fungi (NY).

Hassallia scytonematoides Gardner (as "Hassalia"), loc. cit. 1927. *Tolypothrix scytonematoides* Geitler, ibid. 14: 732. 1932. —TYPE specimen from Puerto Rico: bark, road to Monte Montoro, Maricao, *Wille 1087a,* Feb. 1915 (NY).

Hassallia fragilis Gardner (as "Hassalia"), ibid. 7: 85. 1927. *Tolypothrix fragilis* Geitler, ibid. 14: 724. 1932. —TYPE specimen from Puerto Rico: wood at Hacienda Catalina, Palmer, *Wille 747c,* Jan. 1915 (NY).

Tolypothrix papyracea Gardner, loc. cit. 1927. —TYPE specimen from Puerto Rico: near Manati, *W. C. Earle,* 1925 (NY).

Tolypothrix penicillata var. *brevis* Gardner, loc. cit. 1927. *T. distorta* var. *brevis* Gardner ex Geitler, ibid. 14: 721. 1932. —TYPE specimen from Puerto Rico: water reservoir, Rio Piedras, *Wille 118,* Dec. 1914 (NY).

Tolypothrix amoena Gardner, Mem. New York Bot. Gard. 7: 86. 1927. —TYPE specimen from Puerto Rico: pool west of Experiment Station, Rio Piedras, *Wille 1932,* Mar. 1915 (NY).

Tolypothrix robusta Gardner, ibid. 7: 87. 1927. —TYPE specimen from Puerto Rico: in Laguna Tortuguero, *Wille 827,* Feb. 1915 (NY).

Tolypothrix Willei Gardner, loc. cit. 1927. —TYPE specimen from Puerto Rico: reservoir west of Experiment Station, Rio Piedras, *Wille 209a,* Jan. 1915 (NY).

Scytonema hyalinum Gardner, Univ. Calif. Publ. Bot. 14(1): 7. 1927. —TYPE specimen from Fukien, China: Kushan near Foochow, *H. H. Chung A387,* 1926 (FH).

Scytonema crassum var. *majus* Gardner (as "major"), ibid. 14(1): 8. 1927. —TYPE specimen from Fukien, China: rocks in streamlets, Kushan near Foochow, *Chung A371* (FH).

Tolypothrix consociata Gardner, loc. cit. 1927. —TYPE specimen from Fukien, China: rock, Kushan near Foochow, *Chung a327a,* Jul. 1926 (FH).

Tolypothrix curta Gardner, loc. cit. 1927. —TYPE specimen from Fukien,

China: stone work in hot spring, Huangshun, Foochow, *Chung A425b,* Sept. 1926 (FH).

Tolypothrix tenuis fa. *terrestris* J. B. Petersen, Bot. Icel. 2(2): 306. 1928. —TYPE specimen from Iceland: Hallormstadir, *70(70a)* (C).

Hassallia byssoidea var. *polyclados* Frémy Rev. Algol. 3: 84. 1928. *Tolypothrix byssoidea* var. *polyclados* Frémy ex Geitler, Rabenh. Krypt.-Fl., ed. 2, 14: 729. 1932. *T. byssoidea* fa. *polyclados* Kossinskaia in Elenkin, Monogr. Alg. Cyanoph., Pars Spec. 1: 961. 1938. —TYPE specimen from Manche, France: Saint-Lô, sur les Phyllades, *P. Frémy,* Feb. 1921 (NY).

Aulosira implexa fa. *kerguelensis* Wille, Deutsch. Südpolar-Exped. 8(Bot.): 420. 1928. —TYPE specimen from Schwarzer See, Kerguelen island: "Kerguelen 20," in the slide collection (O).

Microchaete spiralis Ackley, Trans. Amer. Microsc. Soc. 48(3): 302. 1929. *Fremyella spiralis* J. de Toni, Noter. di Nomencl. Algol. 8: [4]. 1936. —TYPE specimen from Michigan: ditch near Germanfask, *A. B. Ackley 153C,* Jul. 1926 (Ackley).

Scytonema guyanense var. *epiphyllum* Gardner, New York Acad. Sci. Sci. Surv. Porto Rico 8(2): 299. 1932. —TYPE specimen from Puerto Rico: leaves, Las Marias, *F. L. Stevens 1175,* Jul. 1915 (NY).

Scytonema Boergesenii Gardner, ibid. 8(2): 301. 1932. —TYPE specimen from Virgin islands, West Indies: Coral bay, St. John, *F. Børgesen 131,* Sept. 1905 (NY).

Petalonema alatum var. *continuum* Nägeli ex Geitler, Rabenh. Krypt.-Fl., ed. 2, 14: 792. 1932. —TYPE specimen from Switzerland: Zürich, an feuchten Felsen, *Nägeli,* 1847 (D).

Scytonema crispum var. *minus* Li, Ohio Journ. Sci. 33(3): 151. 1933. —TYPE specimen from China: on rock with dripping water, Nanking, *Y. C. H. Wang 98,* 1930 (D).

Tolypothrix fusca Schwabe, Bol. Soc. Biol. Concepción 10(2): 117. 1936. *T. distorta* var. *fusca* Schwabe, Verh. Deutsch. Wiss. Ver. Santiago de Chile 3: 127. 1936. —TYPE specimen, here designated, from southern Chile: Puerto Montt, *G. H. Schwabe 2,* 1935 (D).

Tolypothrix metamorpha Skuja in Handel-Mazzetti, Symb. Sinic. 1: 25. 1937. —TYPE specimen from Yünnan, China: prope Yungning, in rivo supra Mudidjin, *H. Handel-Mazzetti 3196,* Jun. 1914 (W).

Scytonema praegnans Skuja in Handel-Mazzetti, ibid. 1: 27. 1937. —TYPE specimen from Szechwan, China: prope Lemoka in Lolo ad rupes fonte calido irrigatas, *Handel-Mazzetti 1620,* Apr. 1914 (W).

Handeliella Stockmayeri Skuja in Handel-Mazzetti, ibid. 1: 30. 1937. —TYPE specimen from Tonkin, North Viet Nam: Laogai, murum quendam siccum obducens, *Handel-Mazzetti 1,* Feb. 1914 (W). Both Skuja (loc. cit.) and Bourrelly (Alg. d'Eau Douce 3: 378. 1970) describe ramification of trichomes in this material, but I have been unable to find such branching in the Type (W) or in duplicates of the Type (D, FH).

Scytonema guyanense var. *marinum* Setchell & Gardner, Proc. Calif. Acad. Sci., ser. 4, 22(2): 72. 1937. —TYPE specimen from Galapagos islands, Ecuador: in tide pools, Narborough island, *J. T. Howell 819,* May 1932 (CAS).

Scytonema burmanicum Skuja, N. Acta R. Soc. Sci. Upsal., ser. 4, 14(5): 34. 1949. —TYPE specimen from Burma: Kemmendine, *L. P. Khanna 433,* Aug. 1935 (D).

Scytonema pseudopunctatum Skuja, ibid. 14(5): 38. 1949. —TYPE specimen from Burma: on Zizyphus, University Estate, Rangoon, *Khanna 359,* May 1935 (D).

Tolypothrix Roberti-Lamii Bourrelly in Bourrelly & Manguin, Alg. d'Eau Douce Guadeloupe, p. 151. 1952. —TYPE specimen from Guadeloupe, West Indies: propriété Marsalle à Pigeon, *P. Allorge 25B,* Mar. 1936 (PC).

Coleodesmium swazilandicum Welsh, Rev. Algol., N. S. 6(3): 231. 1962. —TYPE specimen from Swaziland: Great Usutu river 30 km. south of Mbabane, *J. D. Agnew,* Jul. 1961 (BM).

Scytonema Masonianum Welsh (as "Masoniana"), Revista de Biologia 3(2–4): 227. 1963. —TYPE specimen from Rhodesia: rock pool 65 km. east of Nuanetsi, *A. D. Harrison,* Apr. 1962 (BM).

Fortiea rugulosa Skuja, N. Acta R. Soc. Sci. Upsal., ser. 4, 18(3): 67. 1964. —TYPE specimen from Lapland, Sweden: Sphagnum-Tümpel, Abisko, *H. Skuja* (Skuja).

Coleodesmium Scottianum Welsh, N. Hedwigia 9(1–4): 139. 1965. —TYPE specimen from South West Africa: rapids of Okavango river near Tsheye, *B. J. Cholnoky 174,* Aug. 1962 (BM).

Scytonema bivaginatum Welsh (as "bivaginata"), ibid. 9(1–4): 151. 1965. —TYPE specimen from South West Africa: confluence of Omatako Omuramba and Okavango rivers, *Cholnoky 187,* Aug. 1962 (BM).

Scytonema Seagriefianum, Welsh (as "Seagriefiana"), ibid. 10(1–2): 16. 1965. —TYPE specimen from Cape province, South Africa: Three Sisters Rocks near mouth of Riet river 56 km. from Grahamstown, *S. C. Seagrief,* Jul. 1964 (BM).

Scytonema Twymanianum Welsh (as "Twymaniana"), ibid. 11: 488. 1966. —TYPE specimen from Cape province, South Africa: on *Aloe Bainesii* Dyer, Grahamstown, *E. S. Twyman,* Dec. 1964 (BM).

Fortiea africana Compère, Bull. Jard. Bot. Nat. Belg. 37(2): 153, 247. 1967. —TYPE specimen from Tchad: Lac Tchad, delta du Chari, dans l'eau, *J. Léonard 3806,* Dec. 1964 (BR).

Original specimens have not been available to me for the following names; the original descriptions must serve as Types until the specimens are found:

[*Calothrix olivacea* Biasoletto, Viaggio di S. M. Federico Augusto, Re di Sassonia, per l'Istria, Dalmazia e Montenegro, p. 233. 1841.]

Scytonema hibernicum Hassall [Hist. Brit. Freshw. Alg. 1: 236. 1845] ex Bornet & Flahault, Ann. Sci. Nat. VII. Bot. 5: 105. 1887.

Tolypothrix punctata Hassall [ibid. 1: 240. 1845] ex Bornet & Flahault, ibid. 5: 125. 1887.

Scytonema asperum Cesati [Hedwigia 1(9): 47. 1854; in Rabenhorst, Alg. Sachs. 35–36: 352. 1854] ex Bornet & Flahault, ibid. 5: 110. 1887.

Scytonema rubicundum Itzigsohn [Verh. K. Leop.-Carol. Akad. Naturf. 26(1): 155. 1857] ex Bornet & Flahault, ibid. 5: 112. 1887.

Oscillatoria crenata Fiorini Mazzanti (as "Oscillaria") [Atti Accad. Pontif. N. Lincei (Roma) 16(5): 631. 1863] ex Gomont, Ann. Sci. Nat. VII. Bot. 16: 237. 1892.

[*Scytonema thermale* var. *rhaeticum* Brügger, Jahresber. Naturf. Ver. Graubündens, N. F. 8: 266. 1863.] *S. thermale* d *rhaeticum* Brügger ex Bornet & Flahault, Ann. Sci. Nat. VII. Bot. 5: 112. 1887. *S. mirabile* var. *rhaeticum* Brügger ex Forti, Syll. Myxoph., p. 520. 1907. *S. mirabile* fa. *rhaeticum* Kossinskaia in Elenkin, Monogr. Alg. Cyanoph., Pars Spec. 1: 925. 1938.

Tolypothrix geminata A. Braun [in Rabenhorst, Fl. Eur. Algar. 2: 276. 1865] ex Bornet & Flahault, ibid. 5: 125. 1887.

Scytonema immersum Wood [Smithson. Contrib. Knowl. 241: 59. 1872] ex Bornet & Flahault, ibid. 5: 111. 1887.

Scytonema violascens Zeller [Journ. Asiatic Soc. Bengal 42(2: 3): 183. 1873] ex Bornet & Flahault, ibid. 5: 112. 1887.

Scytonema dendrophilum Grunow [in Reichardt, Sitzungsber. K. Akad. Wiss. Wien, Math.-nat. Cl. 75(1): 556. 1877] ex Forti, Syll. Myxoph., p. 537. 1907.

Lyngbya spongiosa Zanardini [N. Giorn. Bot. Ital. 10: 39. 1878] ex Forti, ibid., p. 276. 1907.

Coleospermum Goeppertianum Kirchner [Krypt.-Fl. Schles. 2(1): 239. 1878] ex Bornet & Flahault, Ann. Sci. Nat. VII. Bot. 5: 84. 1887. *Microchaete Goeppertiana* Kirchner in Engler & Prantl, Natürl. Pflanzenfam. 1(1A): 76. 1900. *Fremyella Goeppertiana* J. de Toni, Noter. di Nomencl. Algol. 8: [4]. 1936.

[*Tolypothrix muscicola* var. *hawaiensis* Nordstedt, Minneskr. K. Fysiogr. Sällsk, Lund 1878 (7): 7. 1878.] *T. muscicola* β *sandvicense* Nordstedt, K. Svensk Vet. Akad. Handl. 22(8): 80. 1888. *T. lanata* var. *hawaiensis* Nordstedt ex Forti, Syll. Myxoph., p. 545. 1907.

Coleodesmium floccosum Borzi [N. Giorn. Bot. Ital. 11(4): 356. 1879] ex Bornet & Flahault, ibid. 5: 128. 1887. *Tolypothrix floccosa* Meneghini *pro synon.* [ex Borzi, loc. cit. 1879] ex Bornet & Flahault, loc. cit. 1887. *Desmonema floccosum* Bornet & Flahault, loc. cit. 1887.

[*Tolypothrix tenuis* β *discolor* Borzi, ibid. 11(4): 371. 1879.]

Tolypothrix conglutinata Borzi [loc. cit. 1879] ex Bornet & Flahault, ibid. 5: 125. 1887.

Scytonema rivulare Borzi [ibid. 11(4): 373. 1879] ex Bornet & Flahault, Ann. Sci. Nat. VII. Bot. 5: 91. 1887.

Scytonema siculum Borzi [ibid. 11(4): 374. 1879] ex Bornet & Flahault, ibid. 5: 96. 1887.

Tolypothrix penicillata var. *gracilis* Nordstedt [Bot. Not. 1882: 50. 1882] ex Forti, Syll. Myxoph., p. 550. 1907.

Sacconema rupestre Borzi [N. Giorn. Bot. Ital. 14(4): 298. 1882] ex Bornet & Flahault, ibid. 3: 381. 1886. *Calothrix gypsophila* fa. *rupestris* Poliansky, Tr. Bot. Inst. Akad. Nauk SSSR, ser. 2, 2: 17. 1935. *C. gypsophila* fa. *rupestris-saccoidea* Poliansky, loc. cit. 1935.

Chroococcus Zopfii Hansgirg, Bot. Centralbl. 22: 351. 1885.

Tolypothrix distorta fa. *magma* Wolle, Fresh-w. Alg. U. S., p. 263. 1887. *T. distorta* fa. *magna* Wolle ex Forti, ibid., p. 542. 1907.

Scytonema papuasicum Borzi, N. Notarisia 1892: 42. 1892.

Tolypothrix woodlarkiana Borzi, ibid. 1892: 43. 1892.

Scytonema Cookei W. West, Journ. R. Microsc. Soc. London 1892: 740. 1892.

Desmonema Wrangelii var. *minus* W. West (as "minor"), loc. cit. 1892. *Coleodesmium Wrangelii* var. *minus* W. West ex J. de Toni, Gen. & Sp. Myxophyc. 1(A–C): 161. 1949.

Scytonema subtile Möbius, Flora 75: 448. 1892.

Microchaete tenera var. *major* Möbius, Abh. Senckenb. Naturf. Ges. 18(3): 343. 1894. *Fremyella tenera* var. *major* J. de Toni, Diagn. Alg. Nov. 1(4): 343. 1938.

Scytonema figuratum var. *samoense* Hieronymus in Schmidle in Reinecke, Bot. Jahrb. 23: 253. 1896. *S. mirabile* var. *samoense* Hieronymus & Schmidle ex Forti, Syll. Myxoph., p. 520. 1907.

Scytonema Hieronymii Schmidle (as "Hieronymi") in Reinecke, ibid. 23: 254. 1896.

Tolypothrix tjipanasensis De Wildeman, Ann. Jard. Bot. Buitenzorg, Suppl. 1: 34. 1897.

Scytonema intermedium De Wildeman, ibid. 1: 45. 1897.

Scytonema tenuissimum Schmidle (as "tenuissima"), Flora 83: 323. 1897. *S. Schmidlei* J. de Toni, Noter. di Nomencl. Algol. 8: [6]. 1936.

Tolypothrix calcarata Schmidle in Simmer, Allg. Bot. Zeitschr. 1899(12): 193. 1899. *Hassallia calcarata* Schmidle ex Forti, ibid., p. 554. 1907.

Tolypothrix chathamensis Lemmermann, Abh. Nat. Ver. Bremen 16: 355. 1899.

Tolypothrix ceylonica Schmidle, Allg. Bot. Zeitschr. 1900(5): 78. 1900. *Hassallia ceylonica* Schmidle, Hedwigia 39: 185. 1900.

Scytonema maculiforme Schmidle (as "maculiformis"), Allg. Bot. Zeitschr. 1900(5): 78. 1900.

Scytonema Hansgirgii Schmidle (as "Hansgirgi"), ibid. 1900(5): 79. 1900

Lyngbya saxicola Filarszky, Hedwigia 39: 140. 1900. *Filarszkya saxicola* Forti, Syll. Myxoph., p. 258. 1907.

Scytonema Bohneri Schmidle, Bot. Jahrb. 30: 60. 1901.

Plectonema Volkensii Schmidle, Hedwigia 40: 343. 1901.

Scytonema Gomontii Gutwiński, Rozpr. Akad. Umiej. Kraków., Wydz. Mat.-przyr., ser. 2, 19: 303. 1902.

Microchaete calothrichoides Hansgirg, Beih. z. Bot. Centralbl. 18(2): 494. 1905. *Fremyella calothrichoides* J. de Toni (as "calotrichoides"), Noter. di Nomencl. Algol. 8: [3]. 1936.

Scytonema javanicum var. *havaiense* Lemmermann, Bot. Jahrb. 34: 624. 1905. *S. javanicum* var. *hawaiiense* Lemmermann ex Forti, ibid., p. 507. 1907.

Microchaete catenata Lemmermann, Bot. Jahrb. 38: 352. 1907. *Fremyella catenata* J. de Toni, ibid. 8: [4]. 1936.

Scytonema amplum fa. *hibernicum* W. West (as "hibernica"), Proc. R. Irish Acad. 31(1: 16): 44. 1912.

Tolypothrix cavernicola Weber-van Bosse, Liste d. Alg. d. Siboga, p. 27. 1913.

Hormobolus tectorum Borzi, Boll. di Studi & Inform. R. Giard. Colon. di Palermo 1: 93. 1914.

Tolypothrix crassa Borzi, ibid. 1: 103. 1914.

Tolypothrix tenuis var. *calida* Elenkin, Kamshatsk. Eksped. vod. P. Riabushinskago, Bot. Otd. 2(Sporov. Rast. 1): 195. 1914.

Microchaete crassa G. S. West, Mém. Soc. Sci. Nat. Neuchâtel 5(2): 1017. 1914. *Leptobasis crassa* Geitler, Beih. z. Bot. Centralbl. 41(2): 275. 1925. *Fortiea crassa* J. de Toni, Noter. di Nomencl. Algol. 8: [3]. 1936.

Scytonema calcicola Kufferath (as "calcicolum"), Ann. de Biol. Lac. 7: 269. 1914.

Leptobasis caucasica Elenkin, Izv. Imp. Bot. Sada Petra Velik. 15(1): 5. 1915. *Fortiea caucasica* J. de Toni, loc. cit. 1936.

Tolypothrix lophopodellophila W. West, Journ. Asiatic Soc. Bengal, N. S. 7: 83. 1915.

Diplonema rupicola Borzi, Studi sulle Mixof., p. 141. 1917. *Borzinema rupicola* J. de Toni, loc. cit. 1936.

Tolypothrix brevicellularis Klugh, Contrib. to Canad. Biol. 1918–20: 182. 1921.

Scytonema Fritchii [sic!] Ghose, Journ. Linn. Soc. Bot. 46: 342. 1923.

Tolypothrix campylonemoides Ghose, ibid. 46: 344. 1923.

Aulosira striata Voronikhin, Not. Syst. Inst. Crypt. Hort. Bot. Petropol. 2(8): 113. 1923.

Tolypothrix Elenkinii Hollerbach, Bot. Mat. Inst. Sporov. Rast. Glavn. Bot. Sada Petrograd 2(12): 183 [misprinted as "175"]. 1923.

Tolypothrix Elenkinii fa. *saccoideo-fruticulosa* Hollerbach, ibid. 2(12): 184 [misprinted as "176"]. 1923.

Scytonema pulchrum Frémy, Rev. Algol. 1(1): 48. 1924. *Petalonema pulchrum* Geitler, Arch. f. Hydrobiol. Suppl.-Bd. 12(6): 453. 1935.

Tolypothrix fragilissima Ercegović, Acta Bot. Inst. Bot. R. Univ. Zagreb. 1: 90. 1925.

Tolypothrix Setchellii var. *epilithica* Ercegović, loc. cit. 1925. *T. epilithica* Geitler, Rabenh. Krypt.-Fl., ed. 2, 14: 739. 1932.

Croatella lithophila Ercegović, ibid. 1: 91. 1925.

Scytonema chiastum Geitler in Pascher, Süsswasserfl. 12: 269. 1925.

Microchaete uberrima Carter, Rec. Bot. Surv. India 9(4): 268. 1926. *Fremyella uberrima* J. de Toni, Noter. di Nomencl. Algol. 8: [4]. 1936.

Microchaete uberrima fa. *minor* Carter, loc. cit. 1926. *Fremyella uberrima* fa. *minor* J. de Toni, Archivio Bot. 20: 2. 1946.

Scytonema alatum fa. *majus* Kossinskaia, Bot. Mat. Inst. Sporov. Rast. Glavn. Bot. Sada SSSR 4(5–6): 71. 1926. *Petalonema alatum* var. *majus* Kossinskaia (as "maius") ex Geitler, Rabenh. Krypt.-Fl., ed. 2, 14: 792. 1932.

Scytonema ocellatum var. *capitatum* Ghose, Journ. Burma Res. Soc. 17(3): 247. 1927.

Tolypothrix inflata Ghose, ibid. 17(3): 254. 1927.

Tolypothrix Saviczii Kossinskaia, Izv. Glavn. Bot. Sada SSSR 27(3): 294. 1928.

Hassallia Magninii Frémy, Rev. Algol. 3: 79. 1928. *Tolypothrix Magninii* Geitler, ibid. 14: 725. 1932.

Tolypothrix tenuis fa. *australica* Möbius ex Wille, Deutsch. Südpolar-Exped. 8(Bot.): 421. 1928.

Scytonema minus var. *Istvanffianum* Kol, Folia Crypt. 1(6): 615. 1928.

Microchaete investiens Frémy, Arch. de Bot. Caen 3(Mém. 2): 283. 1929. *Fremyella investiens* J. de Toni, Noter. di Nomencl. Algol. 8: [4]. 1936.

Tolypothrix Letestui Frémy, ibid. 3(Mém. 2): 288. 1929.

Tolypothrix arboricola Frémy, ibid. 3(Mém. 2): 291. 1929.

Hassallia pulvinata Frémy, ibid. 3(Mém. 2): 294. 1929. *Tolypothrix pulvinata* Geitler, Rabenh. Krypt.-Fl., ed. 2, 14: 721. 1932.

Scytonema Arcangelii fa. *minus* Frémy, ibid. 3(Mém. 2): 302. 1929.

Scytonema Hofmannii fa. *phormidioides* Frémy, ibid. 3(Mém. 2): 315. 1929.

Desmonema Wrangelii fa. *majus* Frémy, ibid. 3(Mém. 2): 327. 1929. *Coleodesmium Wrangelii* fa. *majus* J. de Toni, Gen. & Sp. Myxoph. 1(A–C): 160. 1949.

Scytonema Conardii Kufferath (as "Conardi"), Ann. Crypt. Exot. 2(1): 36 1929.

Scytonema inaequale Kufferath, loc. cit. 1929.

Scytonema costaricense Kufferath, ibid. 2(1): 37. 1929.

Scytonema Echeverriai Kufferath, ibid. 2(1): 38. 1929.

Tolypothrix Haumanii Kufferath (as "Haumani"), ibid. 2(1): 43. 1929.

Plectonema malayense Biswas, Journ. Fed. Malay St. Mus. 14: 411. 1929. *Pseudoscytonema malayense* Elenkin, Monogr. Alg. Cyanoph., Pars Spec. 2: 1805. 1949.

Spelaeopogon Fridericii Budde, Arch. f. Hydrobiol. 20: 456. 1929.

Tolypothrix Werneckei Budde, ibid. 20: 457. 1929.

Scytonema Cuatrecasasii Gonzalez Guerrero, Cavanillesia 3(1–5): 55. 1930.

Scytonema malaviyaense Bharadwaja (as "malaviyaensis"), Rev. Algol. 5(2): 223. 1930.

Tolypothrix Foreaui Frémy, Trav. Crypt. Déd. à L. Mangin, p. 105. 1931.

Scytonema leptobasis Ghose, Journ. Indian Soc. Bot. 10(1): 36. 1932.

Tolypothrix limbata var. *cylindrica* Ghose, loc. cit. 1932.

Leptobasis striatula fa. *laevis* Kossinska, Zhurn. Bio-Bot. Tsikly Vseukrain, Akad. Nauk 1(3–4): 115. 1932. *Fortiea striatula* fa. *laevis* Kossinskaia (as "levis") ex Starmach, Fl. Slodkow. Polski 2: 554. 1966.

Scytonema mirabile fa. *zonatum* Geitler (as "zonata"), Rabenh. Krypt.-Fl., ed. 2, 14: 778. 1932.

Desmonema Wrangelii var. *majus* Geitler (as "maior"), ibid. 14: 800. 1932. *D. Wrangelii* var. *Geitleri* J. de Toni, Archivio Bot. 20: 1. 1946. *Coleodesmium Wrangelii* var. *Gajanum* J. de Toni, Gen. & Sp. Myxoph. 1(A–C): 160. 1949.

Scytonema coactile var. *thermale* Geitler, Arch. f. Hydrobiol. Suppl.-Bd. 12(4): 632. 1933.

Aulosira pseudoramosa Bharadwaja, Ann. of Bot. 47(185): 137. 1933.

Scytonema simplex Bharadwaja, Rev. Algol. 7(1–2): 157. 1934. *S. Bharadwajae* J. de Toni, Archivio Bot. 15(3–4): 291. 1939.

Scytonema Pascheri Bharadwaja, ibid. 7(1–2): 158. 1934.

Scytonema Iyengarii Bharadwaja (as "Iyengari"), ibid. 7(1–2): 159. 1934.

Scytonema saleyerense var. *indicum* Bharadwaja (as "indica"), ibid. 7(1–2): 160. 1934.

Scytonema ocellatum fa. *minus* Bharadwaja (as "minor"), ibid. 7(1–2): 161. 1934.

Scytonema dilatatum Bharadwaja, loc. cit. 1934.

Scytonema dilatatum fa. *majus* Bharadwaja (as "major"), ibid. 7(1–2): 163. 1934.

Scytonema guyanense var. *proliferum* Bharadwaja (as "prolifera"), ibid. 7(1–2): 164. 1934.

Scytonema pseudoguyanense Bharadwaja, loc. cit. 1934.

Scytonema Hofmannii var. *crassum* Bharadwaja (as "crassa"), ibid. 7(1–2): 166. 1934.

Scytonema multiramosum var. *ceylonicum* Bharadwaja (as "ceylonica"), loc. cit. 1934.

Scytonema pseudohofmannii Bharadwaja (as "pseudohofmanni"), ibid. 7(1–2): 167. 1934.

Scytonema mirabile fa. *minus* Bharadwaja (as "minor"), ibid. 7(1–2): 170. 1934.

Scytonema tolypothrichoides fa. *terrestre* Bharadwaja (as "terrestris"), loc. cit. 1934.

Scytonema Geitleri Bharadwaja, ibid. 7(1–2): 172. 1934.

Scytonema Geitleri fa. *tenue* Bharadwaja (as "tenuis"), loc. cit. 1934.

Tolypothrix nodosa Bharadwaja, ibid. 7(1–2): 175. 1934.

Tolypothrix magna Bharadwaja, ibid. 7(1–2): 177. 1934.

Scytonema Chengii Wang, Contrib. Biol. Lab. Sci. Soc. China 9(2): 117. 1934.

Scytonema splendens fa. *variabile* Wang, ibid. 9(2): 118. 1934.

Tolypothrix cucullata Jaag, Österr. Bot. Zeitschr. 83(4): 288. 1934.

Campylonema Danilovii Hollerbach (as "Camptylonema"), Tr. Bot. Inst. Akad. Nauk SSSR, ser. 2, 2: 40. 1935. *Schmidleinema Danilovii* J. de Toni, Archivio Bot. 20(1–4): 3. 1946. *Camptylonemopsis Danilovii* Desikachary, Proc. Indian Acad. Sci. 28(2B): 45. 1948.

Microchaete tenera fa. *minor* Hollerbach, ibid. 2: 41. 1935. *Fremyella tenera* fa. *Hollerbachii* J. de Toni, ibid. 20(1–4): 2. 1946.

Desmonema Lievreae Frémy, Bull. Soc. d'Hist. Nat. Afr. Nord 26: 93. 1935. *Coleodesmium Lievreae* Geitler in Engler & Prantl, Natürl. Pflanzenfam., ed. 2, 1b: 156. 1942.

Scytonema Racovitzae De Wildeman, Résult. du Voy. Belgica, Rapp. Sci., Bot., Obs. s. l. Alg., p. 13. 1935.

Aulosira implexa var. *crassa* Dixit, Proc. Indian Acad. Sci. 3(1B): 98. 1936.

Tolypothrix lanata fa. *minor* Dixit, ibid. 3(1B): 100. 1936.

Scytonema caldarium var. *terrestre* Copeland, Ann. New York Acad. Sci. 36: 94. 1936.

Scytonema induratum Copeland, ibid. 36: 96. 1936.

Scytonema planum Copeland, loc. cit. 1936.

Tolypothrix distorta var. *endophytica* Copeland, ibid. 36: 99. 1936.

Scytonema carolinianum Philson, Journ. Elisha Mitchell Sci. Soc. 55(1): 105. 1939.

Tolypothrix delicatula Philson, ibid. 55(1): 106. 1939.

Scytonema subcoactile Jao, Sinensia 10(1–6): 206. 1939.

Scytonema stuposum fa. *nanyohense* Jao, ibid. 10(1–6): 207. 1939.

Scytonema sinense Jao, ibid. 10(1–6): 208. 1939.

Scytonema torulosum Jao, ibid. 10(1–6): 209. 1939.

Tolypothrix distorta var. *breviarticulata* Jao, ibid. 10(1–6): 212. 1939.

Scytonema chiastum fa. *minus* Parukutty (as "minor"), Proc. Indian Acad. Sci. 11(3B): 119. 1940. *S. chiastum* var. *minus* Gupta & Nair (as "minor"), Agra Univ. Journ. Res. 11(3): 232. 1962.

Fremyella Bossei Frémy, Blumea Suppl. 2: 27. 1942. *Fortiea Bossei* Desikachary, Cyanoph., p. 516. 1959.

Fremyella elongata Frémy, ibid. 2: 30. 1942. *Microchaete elongata* Desikachary, ibid., p. 513. 1959.

Scytonema tenuissimum Frémy, ibid. 2: 35. 1942. *S. Fremyi* Desikachary, ibid., p. 474. 1959.

Spelaeopogon Koidzumianus Yoneda (as "Koidzumianum"), Acta Phytotax. & Geobot. 11(4): 331. 1942.

Petalonema alatum var. *indicum* A. R. Rao, Curr. Sci. 13: 260. 1944.

Scytonema Arcangelii var. *longiarticulatum* Jao, Sinensia 15(1–6): 83. 1944.

Scytonema consociatum Jao, loc. cit. 1944.

Scytonema incrassatum Jao, ibid. 15(1–6): 84. 1944.

Scytonema kwangsiense Jao, ibid. 15(1–6): 85. 1944.

Scytonema orientale Jao, loc. cit. 1944.

Scytonema velutinum var. *kwangsiense* Jao, ibid. 15(1–6): 86. 1944.

Tolypothrix lignicola Jao, loc. cit. 1944.

Scytonema julianum fa. *majus* Schwabe (as "maior"), Mitt. Deutsch. Ges. f. Natur- & Völkerk. Ostasiens (Shanghai), Suppl.-Bd. 21: 149. 1944.

Scytonema julianum fa. *minus* Schwabe (as "minor"), loc. cit. 1944.

Tolypothrix Cavanillesiana González Guerrero, Anal. Jard. Bot. Madrid 5: 313. 1945.

Petalonema densum var. *africanum* Cholnoki, Bol. Soc. Portug. Ciênc. Nat., ser. 2, 4(1): 106. 1952.

Tolypothrix distorta var. *garusica* Cholnoki, ibid. 4(1): 109. 1952.

Tolypothrix Hollerbachii Muzafarov, Bot. Mat. Otd. Sporov. Rast. Bot. Inst. Komarova Akad. Nauk SSSR 8: 86. 1952.

Plectonema algeriense Behre, Bull. Soc. d'Hist. Nat. Afr. Nord 44(5–6): 222. 1953.

Fremyella Bossei var. *africana* Serpette, Bull. Inst. Franç. Afr. Noire 17(sér. A) (3): 798. 1955.

Tolypothrix tenuis fa. *cuticularis* Kondratieva, Bot. Mat. Otd. Sporov. Rast. Bot. Inst. Akad. Nauk SSSR 10: 25. 1955.

Tolypothrix Saviczii fa. *paludosa* Kondratieva, ibid. 10: 26. 1955. *Scytonema mirabile* fa. *paludosum* Kondratieva (as "paludosa"), Viznach. Prisnov. Vodor. Ukrainsk. RSR 1(2): 372. 1968.

Seguenzaea Anthonyi Gupta, Trans. & Proc. Bot. Soc. Edinburgh 36(4): 317. 1955.

Aulosira transvaalensis Cholnoky, Hydrobiologia 7(3): 187. 1955.

Scytonema transvaalense Cholnoky (as "transvaalensis"), ibid. 7(3): 193. 1955.

Tolypothrix africana Cholnoky, ibid. 7(3): 196. 1955.

Scytonema Pratii Komárek, Preslia 28: 369. 1956.

Campylonema indicum var. *allhabadii* Gupta, Journ. of Res. Dayanand Anglo-Vedic Coll. (Kanpur) 4(1): 18. 1957.

Tolypothrix Teodorescui Tarnavschi & Mitroiu, Acad. Rep. Pop. Romine Bul. Stiint. Sec. Biol. & Stiinte Agric., ser. Bot. 9(1): 53. 1957.

Tolypothrix tibestiensis Behre in Quézel, Univ. d'Alger Inst. Rech. Sahar., Mission Bot. au Tibesti, p. 18. 1958.

Scytonema simplex fa. *majus* Vasishta (as "major"), Journ. Bombay Nat. Hist. Soc. 58(1): 143. 1961.

Aulosira manipurensis Bharadwaja, Proc. Indian Acad. Sci. 57(4B): 250. 1963.

Campylonema Godwardii Prasad & Srivastava (as "Camptylonema"), Phykos 3: 41. 1964.

Microchaete ghazipurensis Pandey & Mitra, N. Hedwigia 10(1–2): 92. 1965.
Scytonematopsis ghazipurensis Pandey & Mitra, ibid. 10(1–2): 199. 1965.
Scytonema dzschambulicum Obuchova, Bot. Mat. Herb. Inst. Bot. Akad. Nauk Kazakhst. SSR(Alma-Ata) 3: 66. 1965.
Symphyonemopsis katniensis Tiwari & Mitra, Phykos 7: 186. 1969.

Trichomata aeruginea, luteo-viridia, olivacea, fusca, rosea, violacea, vel cinereo-viridia, praecipuius cylindrica atque ad dissepimenta plus minusve constricta, raro partim torulosa, diametro 3–30 μ crassa, partim et passim decrescentia passim increscentia, ambitu recta vel curvata vel spiralia, longitudine indeterminata, per destructionem cellulae intercalaris vel per constrictionem ad dissepimentum frangentia, ad extrema aliquantulum tumescentia aut apices saepe per paucas vel plures cellulas terminales attenuantia. Cellulae diametro trichomatis longiores vel breviores, 3–20μ longae, protoplasmate homogeneo vel granuloso, raro pseudovacuolato, dissepimentis non granulatis; heterocystae intercalares et terminales nonnumquam seriatae, cylindricae vel hemisphaericae vel quasisphaericae vel discoideae, diametro 4–30μ crassae; sporae cylindricae, seriatae, muris lutescentibus vel fuscescentibus; cellulae vegetativae terminales hemisphaericae usque ad fere sphaericas. Materia vaginalis primum hyalina demum lutea vel fusca. Planta trichomata longa vel brevia vulgo in vaginis discretis cylindricis ramosis, saepe laminosis, raro in gelatina distributa comprehens. Figs. 1–27.

Trichomes blue-green, yellow-green, olive, brown, red, violet, or gray-green, chiefly cylindrical and more or less constricted at the cross walls, rarely torulose in part, 3–30μ in diameter, here and there and in part increasing or decreasing in diameter, straight or curved or spiraled in habit, of indeterminate growth in length, breaking by means of the destruction of an intercalary cell or by constriction at a cross wall, often swollen or attenuated through several cells at each tip. Cells longer or shorter than their diameters, 3–20μ long, the protoplasm homogeneous or granulose, the cross walls not granulated; heterocysts intercalary or terminal, sometimes seriate, cylindrical or discoid or hemispherical or almost spherical, 4–30μ in diameter; spores cylindrical, seriate, the walls becoming yellow or brown. Plant consisting of long or short trichomes commonly in cylindrical, discrete, branched, often laminated sheaths, rarely distributed in gelatinous matrices. Figs. 1–27.

Scytonema Hofmannii ranges over the entire earth, from the Arctic regions to Antarctica, from below sea level to altitudes of 4,600 meters or more in the Rocky mountains. It occupies habitats influenced by fresh water and often becomes acclimated to marine waters. Its most widespread ecophene (Fig. 1), with thin, firm, cylindrical sheath material, inhabits soil, rocks, wood, cloth, lichens, mosses, and the bark and leaves of larger plants which receive water, even as dew, intermittently, at least at rare intervals during a year. It does not thrive, however, in desert rain pools quite as ephemeral as some of those in which *Microcoleus vaginatus* (Vauch.) Gom. (Drouet, 1962) is found. It develops as a blue-green or brownish velvety turf, at times variously silted or impregnated with calcium salts or silica. Here, the sheaths become copiously twin-branched (Fig. 1) and single-branched (Figs. 5, 7), the latter especially in wet or submerged locations. In moist places and in well aerated perennial seepage, parts of the sheaths become thicker, more gelatinous, and commonly much laminated, as noted by Royers (1906) and by Jaag (1943), and infested with fungi, as in the Type specimens of *Scytonema figuratum* Ag. (Fig. 20), *S. myochrous* (Dillw.) Ag. (LD), *S. crassum* Näg. (Fig. 15), *S. densum* (A. Br.) Born. & Thur. (Fig. 3), *S. involvens* (A. Br.) Rabenh. (Fig. 9), *S. velutinum* Kütz. (Fig. 24), *S. crustaceum* Ag. (Fig. 17), and *S. alatum* (Carm.) Borzi (Fig. 14). These same kinds of thickened sheaths appear in plants inhabiting seashores and other places where the concentration of dissolved salts becomes very high as the water evaporates. Here we find also diminution of cell division, more deeply constricted trichomes, and excreted colloidal material solidified in the constrictions.

Scytonema Hofmannii is found also as radiately oriented branching threads in permanent bodies of fresh water, where, during growth, each trichome breaks at a heterocyst, or between two vegetative cells, one of which differentiates at once into a heterocyst; and only the segments ending in vegetative cells burst through a side of the sheath. Single branches of a sheath (Figs. 5, 7) are thus produced. Such ecophenes have traditionally been placed in species of the "distinct" genus *Tolypothrix* Kütz. Twin-branching of a sheath, distant from or on both sides of a heterocyst, may also be common in this aquatic ecophene, as Stein (1963) has pointed out. Also, spores are sometimes found in trichomes of aquatic ecophenes, as in the Type specimen of *Aulosira implexa* Born. & Flah. (Fig. 25).

Another ecophene (Fig. 27), described as *Desmonema Wrangelii* (Ag.)

Born. & Flah., consists of one to many parallel trichomes within gelatinous sheath material, attached to rocks, mosses, etc. in well aerated swift streams. Still another, in perennial seepage, contains curved or contorted trichomes within a globular or sac-like matrix of sheath material *(Diplocolon Heppii* Näg., Fig. 13).

In marine habitats of warmer regions, especially in the upper edges of intertidal zones and just above on rocks and on roots and stems of mangroves *(Rhizophora* sp. and *Avicennia* sp.), an ecophene with thick cylindrical sheaths *(Calothrix pilosa* Harv., Fig. 18) is prevalent. A thin-sheathed ecophene *(Scytonema polycystum* Born. & Flah., Fig. 11) is encountered rarely in similar habitats and on corals and on larger algae in deep marine waters.

Trichomes of all the above-mentioned ecophenes are very similar to each other, and the plants which they compose may display the attributes of several or all ecophenes within a distance of one centimeter.

In cultivating these plants in the laboratory, I find that the sheaths can be thickened and made laminated in a short time, as in my collection no. 14992 (PH) from Niagara Falls, Ontario, grown November 18–24, 1969. The same can be achieved by growing freshwater material in sea water, as in my no. 14965 (PH) from Key West, cultivated from October 1, 1969, until May 21, 1970.

In moist tropical and semitropical regions, some plants develop with fungi into species of *Dictyonema* Ag. and other lichens (Bornet, 1873). Fungi commonly infest the sheath material (Fig. 4) of *Scytonema Hofmannii* in natural habitats; they produce conspicuous alterations in color and texture of the sheath materials and are often responsible for the death of cells or trichomes. Bacteria and small algae inhabit gelatinous sheaths, which they may assist in degrading or in further augmenting.

The following specimens, selected from among the some 11,000 studied, are representative of *Scytonema Hofmannii* as described above:

Cultures of Uncertain Origin: Cambridge Botany School 1482/2 (D); Indiana University 424 (PH); Hopkins Marine Station (as *Tolypothrix distorta* var., D); Rutgers University, *D. Rio* (material used for feeding copepods; PH).

Europe: (TYPE of *Lichen velutinus* Ach., LD).

Russia: soil culture, Kirov region, *G. I. Permenova,* ex E. A. Shtina, comm. H. S. Forest SHT017 (PH); Leningrad, *K. Kossinskaja* (D, LD).

Norway: *H. C. Lyngbye* (D, F); Christiania, *Schübeler* (D, F); in Nordre Aurdal (Valders), *N. Wille* (as *Scytonema figuratum* in Wittr. & Nordst., Alg. Exs. 878b, D, F).

Finland: Godby (Alandia), *K. E. Hirn* (D, NY; as *S. crispum* in Wittr. & Nordst., Alg. Exs. 1316b, D, NY).

Sweden: prope Holmiam, Nåttraby (Blekingia), Ramshyttan (Nericia), Lund, and Jäder (Westmannia), *C. A. Agardh* (TYPES of *S. crustaceum* Ag., *S. penicillatum* Ag., *S. myochrous* δ *simplex* Ag., *S. byssoideum* Ag., *S. Hoffman-Bangii* Ag., *Calothrix distorta* β *flaccida* Ag., *Oscillatoria canescens* Ag., *O. crispa* Ag., and *Thorea Wrangelii* Ag., LD); Abisko (Lapland), *H. Skuja* (TYPE of *Fortiea rugulosa* Skuja, Skuja; duplicate: D); Lund, Örtofta, Christianstad, Modbön (Krokstad), Wimla (Sandhem), and Grimstorp, *O. Nordstedt* (as *Scytonema julianum, S. cincinnatum, Tolypothrix aegagropila* β *bicolor, T. distorta,* and *T. Wartmanniana* in Wittr. & Nordst., Alg. Exs. 184a, b, c, 185, 186, 273a, 274, D, F, NY; in Rabenh., Alg. Eur. 2288, D, F; in Aresch., Alg. Scand. Exs., ser. nov. 378, D, F); Bahusia, *S. Åkermark* (as *T. coactilis* in Aresch., Alg. Scand. Exs., ser. nov. 290, D, F); Uppsala, *A. Areschoug* (as *T. coactilis* in Aresch., Alg. Scand. Exs, ser. nov. 291, D,FH); Slite (Gotland) and Uppsala, *P. T. Cleve* (as *T. tenuis* in Wittr. & Nordst., Alg. Exs. 672, D, F); Varberg, *V. Wittrock* (as *T. tenuis* in Wittr. & Nordst., Alg. Exs. 882, D, F); Dalskog (Dalia) (as *Scytonema castaneum* in Aresch., Alg. Scand. Exs., ser. nov. 376, D, F); Stockholm and Välinge (Scania), *G. Lagerheim* (as *S. figuratum, Tolypothrix Wimmeri,* and *T. tenuis* in Wittr. & Nordst., Alg. Exs. 487, 763a, 878c, D, F).

Denmark: *Hornemann* (TYPE of *Conferva distorta* Müll., C); Björnsholm (Jylland), *T. Jensen* (TYPE of *Tolypothrix gracilis* Rabenh., PH); hortus Hauniensis, *O. Nordstedt* (as *Scytonema julianum* and *S. Hofmannii* in Wittr. & Nordst., Alg. Exs. 273b, 876b, D, NY); Hofmansgave (Fyn), *N. Hofman-Bang* (BKL, NY).

Poland: Breslau and Strehlen, *Hilse* (TYPE of *Symphyosiphon Wimmeri* Hilse, in Rabenh., Alg. Eur. 1775, F; TYPE of *Tolypothrix tenuis* fa. *pallescens* Rabenh. in Rabenh., Alg. Eur. 1033, PH; as *T. pulchra* and *Lyngbya Phormidium* in Rabenh., Alg. Sachs./Eur. 930, 1779, D, F); Oswitz bei Breslau, *O. Kirchner* (with *Coelosphaerium Naegelianum* in Rabenh., Alg. Eur. 2423, D, F); Hirschberg and Hitzelberg, *Flotow* (TYPES of *Symploca Flotowiana* Kütz. and *S. Flotowiana* β *tenuior* Kütz., L).

Romania: Grintesulmare river (Neamt), *E. C. Teodorescu 1211* (D, F); Bucharest, *Teodorescu* (as *Scytonema Hofmannii* in Mus. Vindob. Krypt. Exs. Alg. 1341a, D, F).

Germany: Alt-Breisach (Baden), *A. De Bary* (as *S. turicense* in Rabenh., Alg. Sachs. 996b, D, F); Altdöbern (Lausitz), *R. Holla* (as *Spermosira major* in Rabenh., Alg. Sachs. 469, D, F); Arber-See (Baiern), *A. Hansgirg* (D, F, W); Beiersdorf (Franken), *P. Reinsch* (as *Lyngbya cincinnata* in Rabenh., Alg. Eur.

1917, D, FH); Breisgau, *A. Braun* (TYPE of *L. bicolor* A. Br. in Moug. & Nestl., Stirp. Crypt. Vogeso-rhen. 1287, FH); Berlin, *A. Braun* (TYPE of *Symphyosiphon involvens* A. Br., L; TYPE of *Tolypothrix helicophila* Lemm., BREM; as *Drilosiphon julianus* in Rabenh., Alg. Sachs./Eur. 767, 2463, D, F), *De Bary* (TYPE of *Scytonema calothrichoides* fa. *natans* Rabenh., in Rabenh., Alg. Sachs. 248, F), *P. Hennings* (as *S. cincinnatum* in Hauck & Richt., Phyk. Univ. 35B, D, F), *P. Sydow* (D, F); Constanz, *E. Stizenberger* (as *S. helveticum* in Jack, Lein. & Stizenb., Krypt. Badens 103, D, F; as *S. turicense* and *S. helveticum* in Rabenh. Alg. Sachs. 290, 313, D,F), *L. Leiner* (TYPE of *Calothrix Leineri* A. Br., L); Dresden, *C. A. Hantzsch* (as *Drilosiphon julianus* in Rabenh., Alg. Eur. 1151, D, F); Eichstadt (Baiern), *F. Arnold* (TYPE of *Scytonema gracillimum* fa. *crassius* Rabenh. as *S. tenue* in Rabenh., Alg. Sachs. 652, PH); Eller bei Düsseldorf, *H. Royers* (D, F); Flensburg, *R. Häcker* (TYPE of *Tolypothrix distorta* fa. *tenuior* Rabenh. in Rabenh., Alg. Sachs. 824, PH); Frankfort am Main, *De Bary* (as *T. aegagropila β Kneiffii* in Rabenh., Alg. Sachs. 412, D, NY); Freiburg (Baden), *Braun* (LD), *Schnurmann* (as *Lyngbya discolor* in Jack, Lein. & Stizenb., Krypt. Badens 463, D, F); Hainleite (Thüringen), *F. T. Kützing* (TYPE of *Scytonema aerugineo-cinereum* Kütz., L); Halle, *Kützing* (TYPE of *Lyngbya cincinnata* Kütz. in Kütz., Alg. Dec. 5, L); Harz mountains, *Römer* (TYPE of *Symphyosiphon caespitulus β hercynicus* Kütz., L), *Kunze* (TYPE of *Scytonema fasciculatum* Kütz., L); Höllenthal (Baden), *C. Cramer* (TYPE of *S. polymorphum* Näg. & Wartm. in Jack, Lein. & Stizenb., Krypt. Badens 342, W); Jever, *G. H. B. Jürgens* (TYPE of *Oscillatoria Juergensii* Corda, FH); Laacher-See, *Royers* (D, F); Langenoog, *Royers* (D, F); Leipzig, *Auerswald* (as *Tolypothrix aegagropila* in Rabenh., Alg. Sachs. 251, D, NY, and in Wittr. & Nordst., Alg. Exs. 486, D, F), *O. Bulnheim* (TYPE of *T. Bulnheimii* Rabenh. in Rabenh., Alg. Sachs. 393, F; TYPE of *T. pygmaea* fa. *crassior* Rabenh. in Rabenh., Alg. Sachs. 973, F; as *Lyngbya cincinnata* in Rabenh., Alg. Sachs. 557, D, F; as *Tolypothrix bicolor* in Rabenh., Alg. Sachs. 590, D, F), *P. Richter* (TYPE of *Scytonema Hansgirgianum* Richt. in Wittr. & Nordst., Alg. Exs. 674, FH); Merseburg, Kützing (TYPE of *Tolypothrix pygmaea* Kütz., L); Münsterthal (Baden), *Braun* (TYPE of *Scytonema incrustans* fa. *bryophilum* Rabenh., L); Neudamm, *Itzigsohn* (TYPE of *S. tectorum* Itzigs. in Rabenh., Alg. Sachs. 263, PH), *Itzigsohn & Rothe* (as *Tolypothrix muscicola* and *T. Brebissonii* in Rabenh., Alg. Sachs. 297, 312, D, F); Baden, *Kneiff* (TYPE of *Scytonema Kneiffii* Wallr., FI); Neustadt (Baden), *Schnurmann* (as *Hapalosiphon Braunii* and *Tolypothrix pumila* in Jack, Lein. & Stizenb., Krypt. Badens 343, D, F); Nordhausen, *Kützing* (TYPES of *T. coactilis* Kütz., *Symphyosiphon caespitulus* Kütz., *S. crustaceus* Kütz., *S. hirtulus* Kütz., *Scytonema tomentosum* Kütz., and *S. turfosum* Kütz., L); Offenbach am Main, *De Bary* (TYPE of *Tolypothrix flaccida* fa. *tenuior* Rabenh. in Rabenh., Alg. Sachs. 311, F); Plön, *E. Lemmermann* (TYPE of *T. polymorpha* Lemm., BREM); Reichsforst (Franconia), *P. Reinsch* (TYPE of *Calothrix symplocoides* Reinsch, in Rabenh., Alg. Eur. 1923, FH);

Rollsdorf (Mansfeld), *Kützing* (as *C. aegagropila* in Kütz., Alg. Dec. 7, D, F);
Sachswerfen bei Nordhausen, *Kützing* (TYPE of *Scytonema gracillimum* Kütz.,
L); Salem (Baden), *J. Jack* (as *S. Heerianum* in Jack, Lein. & Stizenb., Krypt.
Badens 341, D, F); Schleswig, *C. Jessen* (D, F); Schleusingen, *Kützing* (TYPES
of *Lyngbya cincinnata* β *annosa* Kütz., *Tolypothrix tenuis* Kütz., and *Sclerothrix
Callitrichae* Kütz., L); Sophienstadt (Nieder-Barnim), *Sydow* (as *Aulosira Sydowii*
in Mus. Vindob. Krypt. Exs. Alg. 3136, D, F, US); Spandau, *Steudner* (as
Tolypothrix pulchra in Rabenh., Alg. Sachs. 191, D, F); Tennstädt, *Kützing*
(TYPE of *T. pulchra* Kütz., L); Titisee (Baden), *Braun 15* (TYPE of *Calothrix
cespitosa* Kütz., L); Untersontheim (Schwäbisch-Hall), *Kemmler* (TYPE of
Scytonema incrustans var. *fuscum* Rabenh. in Rabenh., Alg. Sachs. 670, F);
Weissenfels, *Kützing* (TYPE of *Tolypothrix muscicola* Kütz., L; TYPE of *T. tenuis*
fa. *bryophila* Rabenh. in Kütz., Alg. Dec. 6, F); Werratal (Thüringen), *W.
Migula* (as *Scytonema myochrous* in Mig., Crypt. Exs. Alg. 272, D, F, OC); Wesel
(Rheinland), *Royers* (D, F).

Czechoslovakia: Moravia: Mohelno and Třebíč, *F. Nováček* (D); Eisgrub, *G.
de Beck* (as *S. Hofmannii* in Mus. Vindob. Krypt. Exs. Alg. 1341c, D, F).
Slovakia: ad Olassinum Scepusii, *C. Kalchbrenner 23* (D, L); Wallendorf,
Kalchbrenner (as *S. myochrous* in Rabenh., Alg. Sachs. 826, D, F, FH, PH); in der
südlichen Zips, *Kalchbrenner* (as *S. gracile* in Rabenh., Alg. Sachs. 977, D, F).
Bohemia: Eisenbrod, *A. Hansgirg* (TYPE of *Plectonema Tomasinianum* var. *cin-
cinnatum* Hansg., W; as *Scytonema cincinnatum* in Fl. Exs. Austro-Hung. 1596,
MO); Podmorán prope Roztok, *Hansgirg* (as *Scytonema myochrous* in Wittr. &
Nordst., Alg. Exs. 766, D, F, and in Fl. Exs. Austro-Hung, 1595, MO, US);
Praha, *Hansgirg* (as *S. Hofmannii* in Wittr. & Nordst., Alg. Exs. 876a, D, NY,
and in Fl. Exs. Austro-Hung. 1597, MO, FH; TYPES of *S. obscurum* var. *terrestre*
Hansg. and *Coleospermum Goeppertianum* var. *minus* Hansg., W); Smichov,
Hansgirg (TYPE of *Tolypothrix distorta* var. *symplocoides* Hansg., W); Siehdichfür
nächst Neuwelt, *Hansgirg* (Type of *Plectonema phormidioides* Hansg., W); Böh-
misch-Kamnitz, Böhmisch-Leipa, Steinkirchen ad Budweis, Časlav,
Chlumec, Chvatěrub, Doksy, Eisenstein, Eule, Friedland prope Reichenberg,
Hohenfurth, Hořic, Kolín, Königgrätz, Krumlov, Libšic, Lomnice,
Harrachsdorf, Most, Mühlhausen, Neratovic, Opočno, Oužice, Pelhřimov,
Plzné, Přelouč, Liebenau prope Reichenberg, Sázavé, Soběslav, Spin-
delmühle im Riesengebirge, Spitzberg im Böhmerwalde, Tellnitz, Třeboň,
Beroun, Sv. Prokop, Veselí, and Žiželice, *Hansgirg* (D, F, W).

Hungary: Budapest, *Hansgirg* (D, F, W), *S. Mágocsy-Dietz* (as *Scytonema
javanicum* in Mus. Budap. Fl. Hungar. Exs. Alg. 321, D, F).

Austria: Lower Austria: Wien, *Hansgirg* (as *S. Hofmannii* in Mus. Vindob.
Krypt. Exs. Alg. 1341b, D, F; as *S. javanicum* in Wittr. & Nordst., Alg. Exs. 875,
D, F), *C. Rechinger* (D, F); Neuwaldegg, *Rechinger* (as *S. Hofmannii* in Mus.
Vindob. Krypt. Exs. Alg. 1341d, D, F). Upper Austria: Milchdorf, *Schiedermayr*

(as *Calothrix radiosa* in Rabenh., Alg. Eur. 1305, NY); Traunkirchen, *Heufler* (as *Scytonema turicense* in Rabenh., Alg. Sachs. 996c, D, F), *C. de Keissler* (as *S. myochrous* in Mus. Vindob. Krypt. Exs. Alg. 1343, D, F); Windischgarsten, *A. Grunow* (D); Hallstatt, *K. H. Rechinger* (D, W), *de Keissler* (as *Tolypothrix penicillata* in Mus. Vindob. Krypt. Exs. Alg. 1850, D, F). Salzburg: Salzburg, *Sauter* (TYPE of *Scytonema salisburgense* Rabenh. as *S. gracile* in Rabenh., Alg. Sachs. 117a, b, PH; as *S. salisburgense* in Rabenh., Alg. Sachs. 267, D, F), *Zwanziger* (NY). Styria: Aussee, *O. Bulnheim* (as *S. phormidioides* in Rabenh., Alg. Sachs. 532, D, MIN), *F. Ostermeyer & K. Rechinger* (as *S. figuratum* in Mus. Vindob. Krypt. Exs. Alg. 1342, D, F), *C. & L. Rechinger* (as *Tolypothrix penicillata* in Mus. Vindob. Krypt. Exs. Alg. 1850c, D, F), *K. H. Rechinger* (D, W); Graz, *Hansgirg* (D, F, W); Johnsbachtal, *K. & L. Rechinger* (as *Scytonema myochrous* in Mus. Vindob. Krypt. Exs. Alg. 72b, D, F, US); Kainisch, *C. & C. H. Rechinger* (as *S. myochrous* in Mus. Vindob. Krypt. Exs. Alg. 1343, D, F); Leopoldsteiner See, *de Keissler* (as *Tolypothrix penicillata* in Mus. Vindob. Krypt. Exs. Alg. 1850b, D, F); Mariazell, *P. Richter* (D, F); Mürzsteg, *de Beck 660* (D, W); Schladming, *Grunow* (TYPE of *Hydrocoleum calotrichoides* Grun., W). Carinthia: Friesach, Klagenfurt, Pontafel, and Villach, *Hansgirg* (D, F, W); Kleiner Knote, Irschen, Zwickenberg, and Gnoppnitztal, *H. Simmer* (TYPES of *Gloeochlamys Simmeri* Schmidle, *Scytonema Simmeri* Schmidle, *S. bruneum* Schmidle, and *S. figuratum* fa. *minus* Schmidle, W). Tirol: Klobenstein, *de Hausmann* (TYPE of *S. truncicola* b *saxicola* Grun., W); Brixlegg, Hall, Innsbruck, Jenbach, Kufstein, and Patsch, *Hansgirg* (D, F, W).

Yugoslavia: Slovenia: bei Pisino, *A. Hansgirg* (TYPE of *Tolypothrix penicillata* var. *tenuis* Hansg., W); Adelsberg, *F. T. Kützing* (D, L); Abbazia, *G. Beck* (as *Scytonema myochrous* in Mus. Vindob. Krypt. Exs. Alg. 72, D, W), *Wettstein* (as *S. cincinnatum* in Fl. Exs. Austro-Hung. 3598, MO); Pericnik, *F. Hauck* (as *S. myochrous* in Wittr. & Nordst., Alg. Exs. 673b, D, F); Cittanova, Franzdorf, Krainburg, Laibach, Pinguente, Pirano, Podnart, Steinbrück, Veldes, and Zwischenwässern, *Hansgirg* (D, F, W). Croatia: Buccari, *Hansgirg* (D, F, W). Dalmatia: *Stalio* (TYPE of *S. cyaneum* Menegh., FI); Gruda, Ragusa, Scardona, Sebenico, Topla, and Valdinoce, *Hansgirg* (D, F, W).

Albania: Fushes-Dukati (Vloné), *G. de Toni 75* (D, de Toni).

Switzerland: *Schleicher* (TYPE of *S. naideum* Kütz., L); Albula (Engadin), *Hepp* (TYPE of *Ephebella Hegetschweileri* Itzigs., in Rabenh., Alg. Sachs. 598, PH); Fischenthal (Zürich), *H. Schinz* (as *Scytonema alatum* in Mus. Vindob. Krypt. Exs. Alg. 746, D, F); Grimsel (Bern), *J. K. Schundt* (BM, D); Horgen (Zürich), *C. Cramer* (as *Arthrosiphon Grevillei* in Rabenh., Alg. Sachs. 553, D, F); Interlaken, *H. Royers* (D, F); Jolimont bei Bern, *F. T. Kützing* (TYPE of *Scytonema tenue* Kütz., L); Kranckthal (Bern), *R. J. Shuttleworth & Schundt* (BM, D); Staubbach, *Kützing* (TYPE of *S. helveticum* Kütz., L); Liestal (Baselland), *Hepp* (as *S. Kuetzingianum* in Rabenh., Alg. Sachs. 853, D, F); Marin

(Neuchâtel), *A. Braun* (as *S. gracile* in Rabenh., Alg. Eur. 1842, D, F, FH); Neuchâtel, *Hepp* (TYPE of *S. penicillatum* b *muscicola* Hepp as *S. turicense* in Rabenh., Alg. Sachs. 695, F); Pilatus (Luzern), *Royers* (D, F); Ragaz—Pfäffers, *Hepp* (as *Arthrosiphon Grevillei* in Rabenh., Alg. Eur. 1709, D, F); Rheinfall, *K. Nägeli* (TYPES of *Scytonema allochroum* β *rhenanum* Kütz., *S. coalitum* Näg., and *S. turicense* β *rigidum* Kütz., L), *A. Braun* (TYPE of *S. dimorphum* Kütz., L); Rifferschweil (Zürich), *Hepp* (as *S. gracillimum* and *S. turfosum* in Rabenh., Alg. Sachs. 669, 696, D, F); Rigi, *Cramer* (TYPE of *S. pellucidum* Cram., NY); St. Aubin (Neuchâtel), *A. Braun* (TYPES of *Arthrosiphon densus* A. Br. and *A. densus* b *strictus* A. Br., L; TYPES of *A. alatus* d *tinctus* A. Br. and *A. alatus* e *inconspicuus* A. Br. in Rabenh., Alg. Eur. 1843, F); St. Gallen, *Wartmann* (as *Tolypothrix lanata* in Rabenh., Alg. Sachs. 768, D, F, FH; TYPE of *T. Wartmanniana* Rabenh. in Rabenh., Alg. Sachs. 769, F; as *Scytonema chrysochlorum, S. gracillimum*, and *Arthrosiphon Grevillei* in Rabenh., Alg. Eur. 1096, 1097, 1098, D, F); Sierre, *L. W. Riddle* (D, FH); Thuner-See, *Royers* (D, F); Wittingen (Aargau), *Hepp* (TYPE of *Diplocolon Heppii* Näg. in Rabenh., Alg. Sachs. 468, FH); Zug, *Hepp* (as *Scytonema tomentosum* in Rabenh., Alg. Sachs. 595, D, F); Zürich, *Nägeli* (TYPES of *S. crassum* Näg., *S. decumbens* Kütz., *S. gracillimum* β *curvatum* Kütz., *S. Heerianum* Näg., *S. Kuetzingianum* Näg., *S. lignicola* Näg., *S. myochrous* β *tenue* Kütz., *S. Naegelii* Kütz., *S. nigrescens* Kütz., *S. turicense* Näg., *Petalonema alatum* var. *continuum* Näg., *Tolypothrix intricata* Näg., and *T. Naegelii* Kütz., L), *G. Winter* (as *Hydrocoleum lacustre* in Rabenh., Alg. Eur. 2564, D, F; as *Scytonema myochrous* in Wittr. & Nordst., Alg. Exs. 879, D, US), *Hepp* (as *S. clavatum, S. Heerianum*, and *S. turicense* in Rabenh., Alg. Sachs. 594, 597, 996a, F; as *Hassallia byssoidea* a *lignicola* in Wittr. & Nordst., Alg. Exs. 881a, D, F).

Italy: Monte Corno, *G. Meneghini* (TYPE of *Scytonema flexuosum* Menegh., L). Apulia: Otranto, *L. Rabenhorst* (as *Drilosiphon julianus* in Rabenh., Alg. Sachs. 33, D, F). Campania: Ischia, *Nägeli* (TYPE of *Symphyosiphon vaporarius* Näg., L). Emilia: Rio Gamese, Bologna (W). Latium: Filettino, comm. E. Fiorini-Mazzanti (TYPE of *Scytonema asphaltii* Fior.-Mazz., PAD). Liguria: Spotorno, *C. Sbarbaro* (D, F); Genova (TYPE of *S. piligerum* De Not., PAD), *A. Piccone* (as *S. ocellatum* in Erb. Critt. Ital., ser. II, 1044, D, FH). Lombardia: Bormio, *C. Brügger* (TYPE of *S. bormiense* Brügg. in Wartm. & Schenk, Schweiz. Krypt. 241, MIN), *E. Levier* (as *S. bormiense* in Erb. Critt. Ital., ser. II, 712, D, F). Piemonte: Vercelli, *Cesati* (TYPE of *S. truncicola* Rabenh. in Rabenh., Alg. Sachs. 352, FH); Valdieri terme, *E. P. Wright* (D, UC). Toscana: Firenze, *G. Arcangeli* (TYPE of *S. Arcangelii* Born. & Flah. as *S. cinereum* in Erb. Critt. Ital., ser. II, 785, PC; as *S. Arcangelii* in Wittr. & Nordst., Alg. Exs. 874, D, US), *E. Levier* (as *Gonionema pannosum* in Erb. Critt. Ital., ser. II, 1050, D, F); Lucca, *Arcangeli* (as *Sirosiphon ocellatus* in Erb. Critt. Ital., ser. II, 1420, D, F); Pisa, *Meneghini* (TYPE of *Scytonema aureum* Menegh., L), *Arcangeli* (as *S. tomentosum, S. truncicola*, and *Tolypothrix bicolor* in Erb. Critt. Ital., ser. II, 1248, 1249, 1327, D,

F); Terme Juliane, *Meneghini* (TYPE of *Scytonema julianum* Menegh., L).
Venezia Euganea: Abano, *Kützing* (TYPES of *Symphyosiphon spongiosus* Kütz., S.
velutinus Kütz., *Tolypothrix thermalis* Kütz., *Scytonema elegans* Kütz., S. *thermale*
Kütz., and S. *thermale* var. *decumbens* Kütz., L), *Meneghini* (TYPE of S. *intextum*
var. *fasciculatum* Trevis., L), *P. Titius* (as S. *thermale* in Rabenh., Alg. Sachs. 995,
D, F); Belluno, *C. Venturi* (D, F); Colli Euganei, *Naccari 15* (TYPE of S. *intextum*
var. *fibrosum* Trevis., FI), *Meneghini* (TYPE of S. *chrysochlorum* Kütz., L; TYPE
of S. *cinereum* Menegh., PC; TYPES of S. *thermale* var. *hemisphaericum* Menegh.,
Conferva intexta Poll., and *C. muscosa* Begg., FI); Friuli, *Meneghini* (TYPE of
Symphyosiphon Contarenii Kütz., L); Padova, *Meneghini* (TYPES of *Scytonema
crispum* Menegh., S. *maculiforme* Menegh., and S. *parietinum* Menegh., L);
Treviso, *Meneghini* (TYPE of S. *Meneghinianum* Kütz., L); Venezia, *Contarini*
(TYPE of *Tolypothrix subsalsa* Zanard., Mus. Stor. Nat. Venezia). Venezia Giulia:
Barcola, *F. Krasser* (TYPE of *Scytonema Steindachneri* in Mus. Vindob. Krypt.
Exs. Alg. 422, W; duplicates: D, F, NY), *A. Hansgirg* (D, F, W); Contovello,
Görz, Gradisca, Grignano, Miramar, and Trieste, *Hansgirg* (D, F, W); Trieste,
G. de Beck (D, F, W), *Kützing* (TYPES of S. *allochroum* Kütz., S. *incrustans* Kütz.,
S. *pachysiphon* Kütz., and *Drilosiphon muscicola* Kütz., L). Venezia Triden-
tina: Bolzano, *Hausmann* (as *Scytonema tomentosum* and S. *myochrous* in Erb. Critt.
Ital., ser. II, 786, 1045, D, F); Auer, Gossensass, Kardaun, S. Michele,
Neumarkt, S.Margarita, and Trient, *Hansgirg* (D, F, W).

Netherlands: Boeckhorst prope Lochem, *T. Spree* (as *Symphyosiphon Hofman-
nii* in Rabenh., Alg. Eur. 1454, D, F); Botshol near Abcoude, *V. Westoff* (D, L,
PH); Giekerk, *W. F. R. Suringar 90b* (TYPE of *Tolypothrix pulchra* fa. *tenuior* Sur.,
L); Leiden, *J. T. Koster* (D, F, L), *P. W. Leenhouts* (D, F, L); Schiermonnikoog,
C. den Hartog (D, F, L).

Belgium: Malmédy, *M. A. Libert* (TYPE of *Inoconia Michelii* Lib. in Lib., Pl.
Crypt. Arduenna 96, FH).

Scotland: *R. K. Greville* (TYPE of *Oscillatoria rupestris* Grev., LD; TYPE of
Scytonema chlorophaeum Kütz., L); Appin (Argyll), *Carmichael* (TYPES of S.
cirrhosum Carm. and *Oscillatoria alata* Carm., LD; TYPE of *O. lucifuga* Harv., K
in BM; TYPE of *Calothrix interrupta* Carm., BM); Arran, *Greville* (BM); Isle of
Skye (BM); Ben Chiuran (Perth), *G. S. West 38* (D, F, UC).

England: (TYPE of *Byssus aeruginosa* Weis, OXF; TYPE of *Scytonema Sower-
byanum* Ag., LD); near Bristol, *G. H. K. Thwaites* (TYPE of *Petronema fruticulosum*
Thw., BM); Croboro Warren, *E. Jenner* (BM); Henfield, *Jenner* (BM); Kew, *M.
C. Cooke* (TYPE of *Plectonema Kirchneri* Cooke, in Wittr. & Nordst., Alg. Exs.
883, LD); London, *O. Nordstedt* (as *Scytonema Hofmannii* in Wittr. & Nordst.,
Alg. Exs. 765b, D, F, NY); Penzance, *J. Ralfs* (BM); Rydal, *Brady* (BM);
Thornhaugh (Northampton), *M. J. Berkeley* (TYPE of *Conferva cyanea* Sm. &
Sow., LD); Woking (Surrey), *A. Bennett* (D, F).

Wales: (TYPE of *C. mirabilis* Dillw, LD); Dolgelly, *J. Ralfs* (TYPE of

Microcoleus Dillwynii Harv. as *Desmonema Dillwynii* in Ralfs, Brit. Alg. 18, BM);
near Neath, *L. W. Dillwyn* (LINN); Pembroke Castle, *E. M. Holmes* (LD);
Snowdon, *D. Turner* (TYPE of *Conferva myochrous* Dillw., LD), *W. Borrer* (BM);
Swansea, *Ralfs* (TYPE of *Calothrix atroviridis* Harv., BM).

Ireland: Killarney, *W. H. Harvey* (TYPE of *Scytonema contextum* Carm., LD).

France: (TYPE of *Trichophorus lanatus* Desv., PC). Aisne: Berthenicourt-
par-Moy, *J. Mabille* (D, F). Alpes-Maritimes: Antibes, *G. Thuret* (TYPE of
Microchaete tenera Thur., PC), *E. Bornet* (TYPE of *Hassallia byssoidea* fa. *saxicola*
Grun., PC); Cannes, *A. Raphélis* (D, F); Estérel, *Bornet & C. Flahault* (as
Desmonema Dillwynii in Wittr. & Nordst., Alg. Exs. 675, D, F, US); Gorge du
Loup, *Thuret* (D, PC). Aube: Troyes (D, PC). Aveyron: Creissels, *M. Gomont*
(D, PC), *Flahault* (as *Scytonema alatum* in Wittr. & Nordst., Alg. Exs. 880, D, F,
and in Hauck & Richt., Phyk. Univ. 236, D). Bouches-du-Rhône: Marseille,
Lenormand (TYPE of *S. flagelliferum* Kütz., L); St. Chamas, *Lenormand* (TYPE of
Tolypothrix bicolor Kütz., L). Calvados: (as *Calothrix distorta* in Chauvin, Alg.
Normand. 157, D); Falaise, *A. De Brébisson* (TYPES of *C. pulchra* Kütz.,
Scytonema calothrichoides Kütz., *S. castáneum* Kütz., *S. clavatum* Kütz., *S. flexuosum*
β *gallicum* Kütz., *S. gracile* Kütz., *S. gracillimum* γ *obscurum* Kütz., *S.
tolypothrichoides* Kütz., *Arthrosiphon Grevillei* β *nodosus* Kütz., *A. Grevillei* γ *plica-
tulus* Kütz., and *A. Grevillei* var. *cirrhosiphon* Kütz., L; as *A. alatus* var. *plicatulus*
in Rabenh., Alg. Eur. 2183, D, F); Vire, *Lenormand* (TYPE of *Symploca scy-
tonemacea* β *major* Kütz., L). Cantal: Thiézac–St.-Jacques, *Gomont 40* (TYPE of
Tolypothrix fasciculata Gom., PC). Finistère: *A. Tauguy* (D, F); Coatodon, *P. L. &
H. M. Crouan* (TYPE of *Calothrix mirabilis* var. *rupestris* Crouan, Lab. Biol. Mar.
Concarneau). Gironde: Cambes, *C. Roumeguère* (TYPE of *Scytonema burdigalense*
Dur. & Roumeg. in Moug., Dupr. & Roumeg., Alg. France 424, BM).
Hautes-Pyrénées: Bagnères, *de Franqueville* (D, F). Hérault: Agde, *Lespinasse*
(TYPE of *S. aerugineum* Lesp., PC); Montpellier, *C. Flahault* (as *S. thermale, S.
myochrous, S. cincinnatum, S. Hofmannii, S. ocellatum, Tolypothrix penicillata, T.
flaccida, T. tenuis,* and *T. distorta* in Wittr. & Nordst., Alg. Exs. 579, 582, 583,
584, 670, 671, 762, 765, 767, D; TYPE of *T. límbata* Thur., PC), *Grateloup*
(TYPE of *Oscillatoria pannosa* Bory, PC). Isère: Oisans, *Roussel* (D, PC). Lot-
et-Garonne: flumen Lotus, *Lespinasse* (TYPE of *Scytonema vasconicum* Lesp. &
Mont., PC). Maine-et-Loire: Angers, *F. Hy* (NY), *Gomont* (D, PC), *Flahault* (as
S. tolypotrichoides in Wittr. & Nordst., Alg. Exs. 768, D, F, US); Juigné-sur-
Loire, *Hy* (TYPE of *Microchaete striatula* Hy, PC; as *M. striatula* in Wittr. &
Nordst., Alg. Exs. 872, D, F); Montjean, *Hy* (as *Hassallia byssoidea* in Wittr. &
Nordst., Alg. Exs. 861b, D, F). Manche: Saint–Lô, *P. Frémy* (Type of *H.
byssoidea* var. *polyclados* Frémy, NY); Lessay and La Meauffe, *Frémy* (as *Scytonema
tolypotrichoides* and *Tolypothrix tenuis* in Hamel, Alg. France 8, 55, D, F); Mor-
tain, *De Brébisson* (TYPES of *Calothrix radiosa* Kütz. and *C. stuposa* Kütz., L; as
C. stuposa in Rabenh., Alg. Eur. 2185, D, F). Nièvre: Cosne, *Flahault & E. Bornet*
(as *Scytonema cincinnatum* in Wittr. & Nordst., Alg. Exs. 764, D, F). Puy-de-

Dôme: Saint-Nectaire, *Frémy* (as *S. Hofmannii* in Hamel, Alg. France 53, D, MICH). Seine: Paris, *A. N. Desvaux* (TYPE of *Calothrix lanata β fuscescens* Ag., LD). Seine-et-Marne: Melun, *Roussel* (D, PC). Seine-et-Oise: *Bory de Saint-Vincent* (TYPE of *Oscillatoria lanosa* Bory, PC); Magny-en-Vexin, *L. N. Bouteille* (TYPE of *Sirosiphon Bouteillei* Bréb. & Desmaz., PC; as *S. Bouteillei* in Desmaz., Pl. Crypt. France, sér. 2, 140, FH, NY, UC); Meudon, *Gomont* (D, F); Trappes, *E. Jeanpert* (D, F). Vaucluse: Vaucluse, *Requien* (PC).

Portugal: Coimbra and Serra de Cabaca, *F. Welwitsch 30, 227* (BM, D).

Ethiopia: Abita, Sciotel (Bogos), *O. Beccari* (W); Museri island (Dahlak islands), *Y. Lipkin 2305* (PH).

Kenya: westen van Nairobi, *R. A. Maas Geesteranus 4371* (L); Thika, *E. A. Mearns 1137* (D, F, MO, US); near Kitale, *I. B. Talling* (D); Mt. Elgon, *Talling* (D); Tsavo national park, *F. M. Isaac 4526* (PH).

Uganda: Murchison Falls park and Fort Portal—Lake George, *Talling* (D).

Tanzania: Dar Es Salaam, *G. W. Lawson A3031* (GC, PH); Usambara, *C. Holst* (TYPES of *Scytonema Holstii* Hieron. and *Hassallia usambarensis* Hieron., BM).

Rhodesia: Lundi river, *A. D. Harrison* (TYPE of *Scytonema Masonianum* Welsh, BM); Lake Kariba, *A. J. McLachlan 1, 10* (PH).

Swaziland: Usutu river south of Mbabane, *J. D. Agnew* (TYPE of *Coleodesmium swazilandicum* Welsh, BM).

South Africa: Cape: Grahamstown, *E. S. Twyman* (TYPE of *Scytonema Twymanianum* Welsh, BM); mouth of Riet river, *S. C. Seagrief* (TYPE of *S. Seagriefianum* Welsh, BM); Vryheid, *J. W. Bews 18* (TYPE of *S. Bewsii* Fritsch & Rich); Drakensberg, *Bews 109, 113* (TYPES of *S. splendens* Fritsch & Rich and *Calothrix parietina* var. *africana* Fritsch, BM); Maritzburg, *Bews 21* (TYPE of *Homoeothrix aequalis* Fritsch & Rich, BM).

Tchad: Lac Tchad, delta du Chari, *J. Léonard 3806* (TYPE of *Fortiea africana* Comp., BR; duplicate, D).

Cameroon: Bibamdi and Cap Debrunscha, *R. Jüngner 38a, 61* (LD); ins. Mandoleh ad Victoriam, *P. Dusén 5* (LD; as *Scytonema guyanense* in Wittr., Nordst. & Lagerh., Alg. Exs. 1509, L); Victoria, *Dusén 6* (as *S. javanicum* in Wittr., Nordst. & Lagerh., Alg. Exs. 1510, NY).

São Tomé: *A. Moller* (LD).

The Congo: Ruashi (Katanga), *Duvigneaud 2851* (BRLU, PH).

Angola: Benguella, *F. Welwitsch* (TYPE of *S. Welwitschii* Rabenh., L); Golungo Alto, *Welwitsch* (TYPE of *S. insigne* W. & G. S. West, UC; TYPE of *Tolypothrix phyllophila* W. & G. S. West, BM); Huilla, *Welwitsch 180* (D, F, UC); Pungo Andongo, *Welwitsch* (TYPES of *T. arenophila* W. & G. S. West, *T. crassa* W. & G. S. West, *Calothrix breviarticulata* W. & G. S. West, and *Scytonema myochrous* var. *chorographicum* W. & G. S. West, UC; TYPE of *S. cincinnatum* var. *aethiopicum* W. & G. S. West, BM); Moçâmedes, *A. Roseira* (PH, PO).

South West Africa: confluence of Omatako Omuramba and Okavango rivers, *B. J. Cholnoky 187* (TYPE of *S. bivaginatum* Welsh, BM); Tsheye, *Cholnoky 174* (TYPE of *Coleodesmium Scottianum* Welsh, BM).

Tunisia: Gorges de la Seldja, *M. Serpette T11* (D, Serpette).

Nigeria: Benin, *M. Fox 53* (D); Bonny River, *A. Grant* (BM, D); Ibadan, *Fox III* (D); Ile-Ife, *T. F. Allen 8, 12* (Allen, PH); Ikogosi warm springs, *Allen E* (Allen, PH); Shaki, *D. J. Hambler 1103, 1104* (BM, D).

Togo: Lake Togo, *M. Fox Nielsen 192* (D).

Ghana: Achimota, *G. W. Lawson 159* (D, GC), *A1462* (GC, PH); Akotokyir, *J. B. Hall A29, A56* (Hall, PH); Brimsu, *Hall A84* (Hall, PH); Cape Coast, *Hall A9* (Hall, PH); Komenda, *Lawson A1272* (D, GC); Kumasi, *Lawson A1688* (GC, PH); Legon *Lawson 1874* (GC, PH); Prince Town, *Lawson A1644* (GC, PH); Sampa, *Lawson A1697* (GC, PH); Amedikaw near Tefle, *Hall A38* (Hall, PH).

Ivory Coast: Forêt de Divo, *H. C. D. de Wit 5860* (L).

Liberia: Gbanga, *D. H. Linder 462, 590* (D, FH).

Morocco: Istaf east of Tinerhir, *R. E. Cameron, E. E. Staffeldt & J. Martes 809* (Cameron, PH).

Iceland: Hallormstadir, *J. B. Petersen 70* (TYPE of *Tolypothrix tenuis* fa. *terrestris* B. Peters., C); Reikum, *Steenstrup* (TYPE of *Scytonema chthonoplastes* Liebm., C).

Greenland: Igdlorbaid, herb. Wensk. (W. R. Taylor); Nunatarssuaq, *W. S. Benninghoff & H. C. Robbins 53-13* (D).

Madeira: *Holl* (TYPE of *S. lichenicola* Kunze, PC), *Masson* (BM).

Bermuda: Bailey's bay, *T.A. & A. Stephenson* BRMI1 (D, F); Castle Island, *W. R. Taylor & A. J. Bernatowicz 49-55* (D, MICH); near Causeway, *F. S. Collins 7121* (D, UC); Church Cave, *E. G. Britton 1083* (D, NY); Fairyland, *W. G. Farlow* (TYPE of *Scytonema junipericola* Farl., FH; in Coll., Hold. & Setch., Phyc. Bor.-Amer. 756, D, F, NY, Taylor); Gravelly Bay, *Bernatowicz 49-1585* (D, F, MICH); Harrington Sound, *A. B. Hervey* (as *S. myochrous* in Coll., Hold. & Setch., Phyc. Bor.-Amer. 1902, D, NY); Nonsuch island, *Taylor* (W. R. Taylor); Paget, *Farlow* (as *S. ocellatum* in Coll., Hold. & Setch., Phyc. Bor.-Amer. 711, D, NY); Pear Rock, *Bernatowicz 49-786* (D, F, MICH); Walsingham, *H. H. Whetzel* (D, NY); Whalebone Bay, *Taylor 49-96* (D, F, MICH).

Nova Scotia: Upper Brookside, Colchester county, *L. E. Wehmeyer 8057* (D, F).

New Brunswick: salt springs, Sussex, *H. Habeeb 13477* (D, F, Habeeb); Murray lake south of Dalhousie, *M. Le Mesurier 6* (D, F); Lake Edward, Grand Falls, Ennishone, California, and Blue Bell (Victoria county), *Habeeb 10522, 10956, 11163, 11418, 11611* (D, Habeeb).

Maine: Piscataquis river, *F. S. Collins 3422* (D, UC); Eagle island, *Collins 2683* (D, F); Mount Desert, *W. R. Taylor* (Taylor).

New Hampshire: Chocorua and Shelburne, *W. G. Farlow* (D, FH); Enfield,

L. H. Flint (D, F, PH); culture from Hanover, *J. Donaldson* (D); Ossipee, *A. Driscoll* (PH).

Vermont: Mount Mansfield, *C. G. Pringle* (D, F, UC).

Massachusetts: Brewster, *L. Walp* (D, F); Cambridge, *Farlow* (D, FH); Concord, *F. C. Seymour* (D, F); Cuttyhunk, *F. Drouet 2139, 2163* (D, FH); Eastham, *F. S. Collins* (as *Tolypothrix tenuis* in Coll., Hold. & Setch., Phyc. Bor.-Amer. 1715, D, F); Falmouth, *H. Croasdale* (D), *Drouet 1493, 1911, 2170* (D); Lynn, *Collins 1933* (D, UC); Mashpee, *H. I. Brown & F. C. Seymour* (D, F); Medford, *Collins* (as *T. lanata* in Coll., Hold. & Setch., Phyc. Bor.-Amer. 209a, D, F, MICH); Melrose, *Collins* (as *Scytonema Hofmannii* in Coll., Hold. & Setch., Phyc. Bor.-Amer. 404, D, F); Nantucket, *Drouet 2164, 2220* (D), *W. A. Setchell 669* (D, F, UC), *J. Schramm* (as *S. crispum* in Coll., Hold. & Setch., Phyc. Bor.-Amer. 1613, D, NY); Naushon island, *Drouet 2127* (BM, D, FH, L, NY, S), *H. Croasdale* (D); Newton, *Farlow* (D, FH); Nonamesset island, *Croasdale* (D); North Falmouth, *Croasdale* (D); Pasque island, *Drouet 1121* (D, W), *C.-C. Jao 72* (D); Sharon, *W. A. Setchell* (as *S. ocellatum* in Coll., Hold. & Setch., Phyc. Bor.-Amer. 210, D, F); Sippewisset, *W. R. Taylor* (D, Taylor); Stoneham, *Collins 1899* (D, UC); Truro, *Walp 41616* (D, F); Waverley, *A. B. Seymour* (D, F); Woods Hole, *Drouet 3540* (D, F), *Farlow* (D, FH).

Rhode Island: Coventry Centre, *W. J. V. Osterhout 483* (D, F, UC); Cranston, *Collins* (as *Tolypothrix distorta* in Coll., Hold. & Setch., Phyc. Bor.-Amer. 1761, D, F); Knightsville, *Osterhout 436* (D, UC); Watch Hill, *Setchell 6424a* (D, F, UC).

Connecticut: Bantam lake, *R. D. Wood 588149* (D, Wood); Bridgeport, *I. Holden* (as *T. lanata* in Coll., Hold. & Setch., Phyc. Bor.-Amer. 209b, D, F); Cheshire, *Setchell* (as *Desmonema Wrangelii* in Coll., Hold. & Setch., Phyc. Bor.-Amer. 108, D, NY); Gaylordsville, *Holden* (as *Scytonema figuratum* in Coll., Hold. & Setch., Phyc. Bor.-Amer. 857, D, F); Lake Saltonstall, *G. E. Nichols* (D, YU); Lisbon, *Setchell 705a* (D, UC), *unnumbered* (as *Hassallia byssoidea* in Coll., Hold. & Setch., Phyc. Bor.-Amer. 258a, D, F); Mount Carmel, *Setchell 400* (D, F, UC); Mudge lake (Litchfield county), *H. K. Phinney 1127* (D, F, Phinney); New Haven, *F. Drouet 1965* (D), cultures, *R. A. Lewin 1-2b, 2-1b* (D, F); North Branford, *Drouet 2190* (BM, D, FH, L, NY, S); North Bridgeport, *L. N. Johnson 110* (D, F); North Haven, *Setchell* (as *Scytonema crispum* in Coll., Hold. & Setch., Phyc. Bor.-Amer. 655, D, NY); Plainfield, *F. C. Seymour* (D, F); Putney, *Johnson 115* (D, F); Salisbury, *Holden* (as *S. myochrous* in Coll., Hold. & Setch., Phyc. Bor.-Amer. 109, D, NY, US), *Setchell & Holden* (TYPE of *Tolypothrix Setchellii* Coll. in Coll., Hold. & Setch., Phyc. Bor.–Amer. 310, NY), *O. P. Phelps* (as *Scytonema crustaceum* in Coll., Hold. & Setch., Phyc. Bor.-Amer. 1358, D, NY); Lake Wononskopomuc (Litchfield county), *Phinney 1115a* (D, F, Phinney); Woodbridge, *Setchell 380* (D, F, UC).

Franklin District: Cape Baird, Ellesmere Land, *Feilden, Moss & Hart* (TYPE of *Tolypothrix glacialis* Dick., BM); Cornwallis island, *J. Michéa 1* (D, F); Truelove river, Devon island, *D. R. Oliver 17* (D).

Quebec: Baie Kayak, Baie Kopaluk, Ile Ikarasakiktut, rivière Kogaluk, and lac Payne (Ungava), *J. Rousseau 63, 209, 517, 834, 1134, 1513, 1535* (D, F, MTJB); Aqueduc and Beattie lakes (Montcalm county), *P. Giraudon* (D, F); Mingan islands, Mont-Royal, Rawdon, and Richelieu river, *J. Brunel 276, 355, 452, 623* (D, F, MT); Montreal, *A. F. Kemp* (FH); Koksoak river near Fort Chimo, *H. A. Senn 3458a* (D, DAO); Grand River (Gaspé), *J. F. Collins 3621A* (D); Marymac lake (57° N, 68° 33′ W), *J. F. Grayson 6028* (D, MICH); Mount Albert (Gaspé), *H. Habeeb 1752* (D, F, Habeeb); Trois-Pistoles, *G. Préfontaine 378* (D, F).

New York: Allegany state park (Cattaraugus county), Depew, East Aurora, Protection, and Springville (Erie county), Mendon Ponds park (Monroe county), and Eagle and North Java (Wyoming count), *J. Blum 164, 121, 189, 284, 292, 320, 404, 402* (D, F, Blum); Buffalo, *B. M. Davis* (D, MICH); Jamesville, *O. F. Cooke* (D, F, US); Keene valley (Adirondack mountains), *W. G. Farlow* (D, FH); Lake Mohansic (Westchester county), *H. C. Bold B122* (D, F, Bold); Niagara Falls, *F. Drouet & H. B. Louderback 14983* (PH), *J. W. Bailey* (TYPE of *Arthrosiphon Grevillei* fa. *americanus* Lesp., L), *H. C. Wood* (TYPE of *Scytonema cataracta* Wood, PH), *F. Wolle* (as *S. cataracta* in Rabenh., Alg. Eur. 2492, D, F); Rensselaerville, *M. S. Markle 16* (D, EAR, F); Suffern, *Mary Ursula* (D, F); Warrensburg, *J. Bader 177* (D, F); West Point, *S. Ashmead 16* (PH); cultures from Watkins Glen, *M. P. Kraus* (PH).

New Jersey: Andover, Brookside, Lake Hopatcong, Huntsburg, Springdale, and Millburn, *H. Habeeb 3629, 3729, 4081, 4096, 4098, 4175* (D, Habeeb); Hopatcong lake, *F. Wolle* (TYPE of *Tolypothrix bombycina* Wolle, D); Morris pond, *Wolle* (TYPE of *Scytonema gracile* var. *tolypotrichoides* Wolle in Wittr. & Nordst., Alg. Exs. 389, FH); Browns Mills, *Wolle* (D); Little Falls, *C. F. Austin* (TYPE of *S. Austinii* Wood, D); East Millstone, *R. Renlund & E. T. Moul 6658* (D, RUT); Lebanon state forest (Burlington county), *Moul 7510* (D, RUT); New Brunswick, *B. D. Halstead* (as *Hassallia byssoidea* in Coll., Hold. & Setch., Phyc. Bor.-Amer. 258, D, NY); Saddle River, *H. C. Bold B107* (Bold, D, F); Hammonton, *K. Y. Lee* (PH).

Pennsylvania: *F. Wolle* (TYPE of *Scytonema simplice* Wood, PC); Bethlehem, *Wolle* (D, F; TYPE of *Symphyosiphon Wollei* Born., PC); Daylesford, *W. R. Taylor* (Taylor); Delaware Water Gap, *C. F. Austin* (FH), *Wolle* (TYPE of *Tolypothrix rupestris* Wolle, D; duplicates in Rabenh., Alg. Eur. 2573, D, F); Philadelphia, *W. R. Taylor* (D, F, Taylor), *F. Drouet et al. 14892, 14975, 15022* (PH); Quakertown, *Wolle* (D); Stroudsburg, *Drouet & H. B. Louderback 14910* (PH).

Maryland: Baltimore, *J. E. Humphrey* (as *T. tenuis* in Coll., Hold. & Setch., Phyc. Bor.-Amer. 257, D, F); Cabin John, *Drouet et al. 4894a* (D, F, US); Great

Falls island, *A. J. Pieters* (D, F); Soldiers Delight (Baltimore county), *A Rutledge 75* (D, F); Thurmont state park, *K. Y. Lee* (PH).

Virginia: Cobham's Wharf and Claremont (Surry county), Mountain Lake, New Bohemia, White Top mountain (Grayson county), and Williamsburg, *J. C. Strickland 400, 541, 567a, 1209, 1406, 1426* (D, F, Strickland); Damascus, Dot, Farmville, Galax, Hilton, Jonesville, Petersburg, Sebrell, Yale, and Zuni, *Strickland & E. S. Luttrell 770, 864, 878, 884, 1357, 1360, 1372, 1377, 1389, 1448* (D, F, Strickland); Bull Run mountains, *H. A. Allard 10507C* (D, US); Charlottesville, *B. F. D. Runk & Strickland 552* (D, F, Strickland); Conway and Shenandoah Caverns, *C. F. Reed 57496, 75640* (PH, Reed); west of Staunton, *Luttrell 3436* (D, F, Strickland); Great Bridge, *C. M. Wilson* (D, F, Strickland); Hot Springs, *R. B. Patterson 2801* (D, F, Strickland); Natural Bridge, *R. Thaxter* (D, FH); Richmond, *E. I. Loving* (D, F, Strickland).

West Virginia: Oakvale, *R. F. Smart 128* (D, F, Strickland); Athens, *E. M. McNeill 34* (D).

North Carolina: Beaufort, *F. Drouet 14743* (PH), *H. J. Humm* (D, Humm), *K. C. Fan 10723* (D, Fan); Cashiers, *E. S. Luttrell* (D, F, Strickland); Cullowhee, *R. Thaxter* (D, FH); Franklin, *C. Nielsen & W. Culberson 1895* (D, FSU); Mount Whitesides, Highlands, Cullasaja Gorge, and Tryon, *H. C. Bold H96a, H103, H163, H403* (D, F, Bold); Raleigh, *R. C. Phillips* (D); Winston-Salem, *L. D. de Schweinitz* (PH).

South Carolina: *H. W. Ravenel 183* (TYPE of *Sirosiphon scytonematoides* Wood, FH), *218* (TYPE of *Symphyosiphon Bornetianus* Wolle, FH), *47* (TYPE of *Scytonema cortex* Wood, FH), *65* (TYPE of *S. cortex* var. *bruneum* Wolle, D); Aiken, *Ravenel 115, 211* (BM, D); Myrtle Beach, *P. J. Philson SC40* (D, F); St. Johns near Charleston, *Ravenel 412* (FH); Society Hill, *Ravenel* (TYPE of *S. Ravenellii* Wood, NY); Walhalla, *H. C. Bold H132* (D, F, Bold).

Georgia: Butler, *P. C. Standley 92566* (D, F); Cumberland island, *T. L. Mead* (D, F); Darien, *Ravenel 81* (BM), *3159* (BM, D); Little Stone Mountain, *J. K. Small 5023a* (D, F); Tallulah Gorge and Tiger, *H. C. Bold H302, H305* (D, F, Bold).

Florida: Hollywood, Cedar Key, and Tallahassee, *F. Drouet 10286, 12253, 12272* (D, F); Palm Beach, Wakulla Springs, Key West, and Marathon, *Drouet & H. B. Louderback 10215, 12293, 14965, 15029* (D, PH); Florida Caverns state park, Quincy, Gulf Beach, Pensacola Beach, West Bay, Apalachicola, and Chiefland, *Drouet, C. S. Nielsen, et al. 10389, 10422, 10539, 10592, 10871a, 10978, 11215* (D, F, FSU); Silver Springs, Leesburg, Hesperides, and Orange Lake, *Drouet & M. A. Brannon 11022, 11041, 11088, 11099* (D, F); St. Marks lighthouse, *Drouet & G. C. Madsen 11768* (D, F, FSU); Key West, *W. H. Harvey* (TYPE of *Calothrix pilosa* Harv., BM); Gainesville, *H. W. Ravenel* (TYPE of *Tolypothrix Ravenelii* Wolle, BM; TYPE of *Scytonema mirabile* Wolle, D); "Florida," *J. D. Smith* (TYPE of *S. cortex* var. *corrugatum* Wolle, D); Cottondale,

Nielsen & Madsen 853 (D, F, FSU); Pineola and Rock Hill, *R. C. Phillips 370, 721* (D, FSU); Bonita Beach, Highland Hammock state park, Kissimmee, Lamont, Venice, and Marco Island, *P. C. Standley 73224, 92684, 92728, 92757, 92780, 92839* (D, F); La Belle, Hernando, Olustee, Gainesville, and Orlando, *M. A. Brannon 20, 89, 227, 263, 342* (D, F, PC); Cocoanut Grove, Daytona Beach, and Ocala, *R. Thaxter* (D, FH); Aspalaga (Gadsden county), *Phillips & L. R. Almodóvar 126* (D, FSU); Port St. Joe, *A. H. Johnston* (D, FSU); Lake Tsala Apopka, *S. A. Earle & R. D. Suttkus 367* (D); Perry, *Madsen, A. L. Pates et al. 1078* (D, F, FSU); Bradenton, *C. B. Stifler* (D, F); Hillsborough state park, *P. O. Schallert 1151* (D, F); Camp Jackson, *N. L. Britton* (D, NY); Lakeland, *L. D. Ober 8* (D, FSU); Key West, *M. A. Howe 1635* (D, NY); Cudjoes key, *C. L. Pollard et al. 92* (D, NY); Big Pine key, *E. P. Killip 42002* (D, F, US); Sanibel island, *Killip 44259* (D, F, US); Goulds, *C. Jackson 3* (D, FSU); Cox Hammock (Dade county), *J. K. Small & C. A. Mosier 5835* (D, F); Miami, *H. J. Humm* (D, F, Humm); Eustis, *W. G. Farlow* (D, FH); Homestead, *A. M. Scott 85* (D).

Ontario: Niagara Falls, *F. Drouet & H. B. Louderback 15008* (PH), *G. T. Velasquez 77* (D, F); Kingston, *J. H. Wallace 26* (D, PH); Superior Shoal in Lake Superior, *E. F. Stoermer* (PH).

Ohio: Buzzard's Rock (Adams county) and McArthur, *W. A. Daily 125, 146* (D, Daily, F); Paint, Catawba island, Cantwell Cliffs state park, and Constitution, *F. K. & W. A. Daily 318, 588, 686, 905* (D, Daily, F); Cincinnati, *J. B. Lackey* (D, F); Fort Hill, *A. T. Cross & J. Lambert,* (D, Daily, F); Foster, *J. H. Hoskins & Daily 228* (D, Daily, F); Kellys Island, *L. H. Tiffany* (D); North Appalachian experimental watershed (Coshocton county), *E. Clark 677* (D, EAR, F); Oberlin, *P. Smith 34* (D, OC); Seven Caves (Highland county), *W. B. Cooke* (D, Daily, FH, NY); Shade, *A. H. Blickle* (D, Daily, F).

Kentucky: Torrent, *R. Kosanke & F. K. & W. A. Daily 496* (D, Daily, F); Biggs, *C. F. Reed 57494* (PH, Reed); Natural Bridge state park and Henderson, *B. B. McInteer* (D, Daily, F).

Tennessee: Ramsey falls (Great Smokies national park), Knoxville, Falls Creek falls (Van Buren county), Copperhill, Scottsboro, Ashland City, and Memphis, *H. Silva 135, 513, 677, 1759, 1777, 1832, 2131* (D, F, TENN); Radnor lake (Davidson county), Harrogate, Kingston Springs, Nashville, and Snail Shell cave (Rutherford county), *H. C. Bold 10A, 21, B133, B142, 3942* (D, F, Bold); Burbank, *R. Thaxter* (D, FH); Shady Grove, *A. E. Clebsch 1911* (D, F, TENN).

Alabama: Mobile, Point Clear, and Fairhope, *F. Drouet & H. B. Louderback 10106, 10136, 10151* (D, F); Montgomery, *R. P. Burke* (MO); Moulton, *J. M. Peters* (FH); Scottsboro, *H. Silva 1772* (D, F, TENN).

Michigan: Beaver island, *Drouet & Louderback 12474, 12914* (D); Germanfask, *A. B. Ackley 153C* (TYPE of *Microchaete spiralis* Ackl., Ackley); Douglas lake (Cheboygan county), Wycamp lake (Emmet county), Au Train, Munising, Marquette, Bois Blanc island, Tahquamenon Falls, and Albion, *H. K. Phinney 3M40/3, 19M41/1, 21M40/17, 21M40/21, 21M40/27, 36, 154, 396* (D, F, Phinney); South lake (Washtenaw county), *F. M. Begres* (PH); Round lake (Emmet county), *G. W. Prescott 3M5* (D, EMC, F); Lake Lansing (Ingham county), *D. Jackson* (EMC, PH); Bloomfield Hills, *S. A. Cain C4* (D, F); Grand Traverse bay, *J. H. Hoskins* (D, Daily, F); Hammond bay (Presque Isle county), *A. H. Gustafson* (D, F); Lake Odessa, *W. E. Wade 1* (D, F).

Indiana: Miller and Beverly Shores, *F. Drouet 5835a, 5863* (D, F); Dunes state park, *Drouet & H. B. Louderback 5224* (D, F); Turkey Run state park, *Drouet & D. Richards 2489, 5872* (D, F); Mount Vernon and Vernon, *W. A. Daily 36A, 60* (D, Daily, F); Clifty Falls state park, Monticello, St. Paul, Indianapolis, Pine Hills (Montgomery county), McCormicks Creek state park, Brown County state park, and Modoc, *F. K. & W. A. Daily 674, 1018, 1140, 1177, 1447, 1785, 2830, 2959* (D, Daily); Centerville, Williamsburg, Richmond, and Blowing Cave (Harrison county), *L. J. King 4, 157, 190, 278* (D, EAR, F); Bass lake (Steuben county), *C. M. Palmer* (BUT, D); Lake Maxinkuckee, *H. W. Clark & B. W. Evermann 76* (D, F, US).

Wisconsin: Pensaukee–Brookside, *F. Drouet & H. B. Louderback 12551* (D); Prairie du Chien, *Drouet 5069* (D, F); Arbor Vitae lake (Vilas county), Sweeney lake (Oneida county), Eagle, and Rhinelander, *G. W. Prescott 3W7, 3W69, 3W97, 3W128* (D, EMC, F); Drells pond and Clear lake near Bradley, *F. C. Seymour* (D, F); Madison, *L. H. Pammel* (MO), *R. H. True* (D, F); Richland Center, *R. A. Harper* (BKL, D); Minocqua, *J. Rubinstein* (D, F).

Illinois: Pontiac, Geneva, St. Charles, Chicago Heights, Savanna, and Galena, *Drouet & Louderback 5259, 5400, 5445, 5461, 5683, 5694, 13300* (D); Olympia Fields, *Drouet & H. Rubinstein 5814* (D); Lemont, *G. T. Velasquez, D. Richards & Drouet 2499* (D, F); Robbsville (Pope county) and Evanston, *H, K. Phinney* (D, F, Phinney); Glenview, *M. E. Britton 461* (D, Daily); culture from Urbana, *L. R. Hoffman* (PH); Chicago, *K. C. Fan 10587* (D, Fan), *L. J. King* (D, F); Athens, *E. Hall* (D, F); Payson, *E. T. & S. A. Harper* (D, F).

Mississippi: Hendersons Point and Biloxi, *Drouet 9870, 9995* (D, F); Sandy Hook, *Drouet & L. H. Flint 9677* (D, F); Magnolia state park, *Drouet & R. L. Caylor 9925* (D, F); Goodman and Ocean Springs, *Caylor 45-11*, 9B53 (D, F); Biloxi, *Flint* (D); Hattiesburg, *G. J. Hollenberg* (D, Hollenberg).

Minnesota: Spring Park, North St. Paul, Oak Terrace, Itasca state park, Lake Itasca, and highway 113 at highway 71 (Becker county), *F. Drouet 5026, 5058a, 5599, 12073, 12160, 14855* (D); Minneapolis and Stillwater, *J. E. Tilden* (as *Tolypothrix distorta, T. tenuis,* and *Scytonema crispum* in Tild., Amer. Alg. 82, 628, 632, D); Minneapolis, *C. M. Crosby* (as *S. figuratum* in Tild., Amer. Alg. 396, D); Itasca state park, *K. C. Fan 10503* (D, Fan); Big Thunder lake south of Remer, *D. Richards 1044* (D, F); Elk lake (Clearwater county), *S. Eddy* (D, Daily, F).

Iowa: Marquette, *Drouet & H. B. Louderback 5159* (D, F); Fayette, *B. Fink* (as *S. myochrous* in Tild., Amer. Alg. 290, D, F); Fairport, *G. W. Prescott* (D); West Lake Okoboji, *C. W. Reimer* (PH); Clear lake (Cerro Gordo county), *F. M. Begres* (PH).

Missouri: Ha Ha Tonka, Bigsprings, Waynesville, Gravois Mills, Wayland, Wilton, Ashland, Columbia, and Alley Spring state park, *F. Drouet et al. 726, 911, 1235, 1244, 1763, 1766, 2417, 13485* (D); Portageville, St. Anthony, Capps, Zora, Calhoun, Montevallo, and New Frankfort, *C. Shoop 30, 61, 76, 139, 187, 233, 391* (D, F); Big spring on Current river and Warsaw, *W. Trelease* (MO); Birch Tree, *W. Sharp 122* (D, MO); Knob lake (Madison county), *C. Russell 24* (D, F, UC); St. Louis and Thayer, *N. L. Gardner 1394, 1416* (D, F, UC); Tom Sauk mountain, Rockbridge (Ozark county), and Montier, *J. A. Steyermark 26533, 27096, 28851* (D, F); Basswood lakes (Platte county), *L. J. Gier 424* (D, WJC); Bat Cove (Franklin county), *H. H. Iltis & R. E. Woodson 1984* (D, MO); Moberly, *E. Adams* (D, UMO); Callaway county, *W. B. Drew 1040A* (D, F).

Arkansas: Lake Hamilton (Garland county) and Lake Catherine (Hot Springs county), *N. E. Gray 140, 285* (D, F); Arkansas Post and Mount Ida, *D. Demaree 25467, 45284* (D); Berryville, *C. Wilton* (D, F); Eureka Springs, *S. C. Dellinger et al. 15010* (D, UARK); Marianna–Helena, *B. C. Marshall* (D); Hot Springs, *H. H. Iltis* (D, F, UARK), *D. Richards & C. F. Massey 16* (D).

Louisiana: Cameron, New Iberia–Morbihan, and Theriot, *F. Drouet 8897, 8929, 9293* (D, F); Emma Station, Avery Island, Jefferson Island, Abbeville, Weeks Island, and Chauvin–Montegut, *Drouet & R. P. Ehrhardt 8964, 8975, 9047, 9118, 9198, 9247* (D, F); Valentine–Larose and Grand Isle, *Drouet & P. Viosca Jr. 9417, 9525* (D, F); Mandeville, Lacombe, Abita Springs, Varnado, Angia, and Covington, *Drouet & L. H. Flint 9552, 9557, 9654, 9748, 9761, 9776* (D, F); New Orleans, *Drouet & H. B. Louderback 15045* (D); Holden, Greensburg, Reeves, Baton Rouge, Spanish lake (Ascension parish), and Bogalusa, *L. H. Flint* (D, LSU); "St. Ubert," *A. B. Langlois* (US); Lafayette, *Bro. Neon 1724* (D, FH); Crowley, *G. W. Prescott La89* (D, EMC).

Manitoba: Fort Churchill, *J. M. Gillett 2092* (D, F).

South Dakota: Lake Alice, *DA. Saunders* (MO); Big Stone lake, *Saunders* (as *Tolypothrix tenuis* in Tild., Amer. Alg. 397, D, F).

Nebraska: Ansley, Arapahoe, Arthur, Belmont, Benkelman, Chadron state park, Champion, Cody, Columbus, Culbertson, David City, Enders, Falls City, Franklin, Fremont, Grand Island, Gretna, Haigler, Halsey, Homer, Imperial, Johnson reservoir (Gosper county), Kearney, Lincoln, McCook, Ogallala, Omaha, Oxford, Parks, Peru, Phillips, Red Cloud, Riverton, Rulo, Schuyler, Seneca, South Bend, Spencer, Stratton, Valentine, Wahoo, Weeping Water, Wellfleet, and Whitney, *W. Kiener 10493, 10515, 10632, 10845, 10925, 11053, 11432, 11457, 11538, 11648, 11900, 11944, 12019, 12580, 12791,*

12797, 13050, 13053, 13593, 13690, 13833, 14005, 14901, 15125, 15325, 15413, 15961, 16555, 16744, 17049, 17992, 18014, 19262, 20021, 20425, 20718, 21327, 21579, 21588, 21783, 21829, 21942, 22183c, 24255 (D, F, NEB); Long Pine, *B. W. Evermann* (D, F); Rock Bluff, *G. E. Condra 12622* (D, F, NEB).

Kansas: Lawrence, Baxter Springs, and Yates Center, *K. C. Fan 10672, 10694, 10698* (D, Fan).

Oklahoma: Stroud, *W. E. Booth 4* (D, F); Cheyenne, *B. O. Osborn RR6, RR24* (D, F); Broken Bow and Bethel, *C. E. Taft 301, 313* (D); Elk mountain (Comanche county), *E. O. Hughes 606B* (D); Turner falls (Murray county), *A. J. Sharp S415* (D, TENN), *W. Kiener 12370* (D, NEB), *J. F. Macbride et al. 8155* (D, F).

Texas: Lewisville, Boot Springs (Chisos mountains), Hot Springs (Brewster county), Long mountain (Llano county), Dallas, and Juno, *E. Whitehouse 22716, 24634, 24644, 24879, 24918, 25118* (D, SMU); Enchanted Rock (Llano county), Bastrop state park, and laboratory cultures from Austin and Taylor, *H. C. Bold* (Bold, D); Alvin and Galveston, *H. K. Phinney* (D, F, Phinney); San Antonio, *W. G. Farlow* (D, FH); Jacksboro, *B. O. Osborn 153* (D); Terry ranch (Mitchell county), *R. W. Pohl 4699* (D, F); Rabb palm Grove (Cameron county), *R. Runyon 3849* (D, F, Runyon); San Jacinto river in Harris county, *E. S. Deevey* (D, F).

Saskatchewan: laboratory culture of soil, *J. McEown 13* (D, F).

Montana: Missoula, *F. A. Barkley* (D, F, MONTU); Whitehall and Frenchtown, *F. H. Rose 4020, 4262* (D, F); Rattlesnake valley (Missoula county), *S. R. Ames 2* (D, F); Lost lake (Glacier national park), *S. Wright* (D, F); Columbia Falls, *R. S. Williams 290b* (D, F); Kalispell, *H. F. Buell 465* (Buell, D, F).

Wyoming: Towner lake (Albany county), *W. G. Solheim* (D, Daily, F); West Thumb Station (Yellowstone national park), *W. A. Setchell 1919* (D, F, UC); Park county, *F. H. Rose & F. A. Barkley 4023* (D, F); Fremont county, *A. Nelson* (D, FH).

Colorado: *T. S. Brandegee* (TYPE of *Scytonema Brandegei* Wolle, D); Manitou Springs, *F. Drouet et al. 4067* (D, F); Cañon City, *Brandegee* (PC); Dolores river (Montezuma county), *G. Piranian* (D, F); Longs Peak, at 13,800 feet alt., *W. Kiener 1288* (D, F); South Colony creek (Custer county), *Kiener 10275* (D, F, NEB).

New Mexico: Las Vegas, Montezuma, and Pecos, *F. Drouet & D. Richards 2575, 2640, 2692* (D, F); Gila hot springs north of Silver City and San Francisco hot springs and Whitewater creek near Glenwood, *H. Habeeb* (Habeeb, PH); Carrizozo and White Sands national monument, *L. M. Shields* (D); Black River Village, *J. F. Macbride et al. 8170* (D, F); Santa Fe, *J. B. Routien* (D, F).

Alberta: Banff, *E. Butler & J. M. Polley* (as *S. mirabile* var. *Leprieurii* in Coll.,

Hold. & Setch., Phyc. Bor.-Amer. 1014, D, W. R. Taylor), *H. M. Richards 3* (D, F, UC), *O. & I. Degener 26807* (D, Degener).

Idaho: Ponces (Kootenay county), *J. H. Sandberg* (D, F); Lolo Divide (Clearwater county), *F. H. Rose & F. A. Barkley 4044* (D, F, MONTU).

Utah: White Pine lake (Cache county), *G. Piranian* (D, F); Big Cottonwood, *C. T. D. Brown F199* (D, F); Rainbow Bridge national monument, *W. S. Phillips & C. T. Mason* (D).

Arizona: Sabino canyon near Tucson, Ruby, Saguaro national monument, Quijotoa, Madera canyon (Santa Cruz county), and Sycamore canyon (Santa Cruz county), *F. Drouet et al. 2733a, 14399, 14428, 14529, 14601, 14702* (D); Tucson, *J. D. Wien* (D), culture of atmospheric dust, *E. Luty & R. W. Hoshaw 16* (PH); Santa Rita mountains, *C. G. Pringle* (FH); Marble canyon (Coconino county), *E. U. Clover 2241* (D, MICH).

Nevada: Yucca Flat (Nye county), *F. Drouet 13806, 13914* (D); Walti hot springs (Lander county), *I. La Rivers 2625* (D, RENO); Railroad valley and Toquima range (Nye county), *T. C. Frantz 2863, 2982* (D, RENO).

Yukon: Dawson, *R. S. Williams* (D, F).

Alaska: St. Matthew island, Harding lake near Anchorage, and near Richardson highway 248 miles south of Fairbanks, *D. Hilliard* (D); Unalaska, *W. A. Setchell & A. A. Lawson* (as *Tolypothrix lanata* in Coll., Hold. & Setch., Phyc. Bor.-Amer. 956, D, MICH); Amakurak island (Unalaska), *Setchell & Lawson 4005* (D, F, UC); St. Michael, *Setchell* (D, F, UC); Donnelly Moraine, *W. S. Benninghoff 9033* (D).

British Columbia: Baird point (near Victoria), *J. E. Tilden* (TYPE of *Hassallia byssoidea* fa. *cylindrica* Tild. in Tild., Amer. Alg. 398, MIN); Departure Bay (near Nanaimo), *J. Macoun 141* (D, FH, UC); Beavermouth, *W. R. Taylor 23, 55* (Taylor).

Washington: Seattle, *T. C. D. Kincaid 768* (TYPE of *Microchaete robusta* Setch. & Gardn., UC; duplicates: D, F); Lake Washington, *Tilden* (as *Calothrix Braunii* in Tild., Amer. Alg. 286a, D, F); Falls lake (Lower Grand Coulee), *R. W. Castenholz 11, 21* (D, F); Preston creek (Chelan county), *R. G. Genoway 32* (D, F); Priest rapids of Columbia river (Yakima county) and Richland, *C. C. Palmiter* (D, F).

California: Bells Station (Santa Clara county), Pacheco pass (Merced county), Los Banos, Coarsegold, South Pass west of Needles, Vidal, Shavers Summit (Riverside county), Desert Center, Hopkins Well, Brawley, Jacumba, and Torrey Pines state park, *F. Drouet & J. F. Macbride 4340, 4349, 4392, 4403, 4604, 4646, 4701, 4710, 4718, 4775, 4787, 4852* (D, F); Dunsmuir, Weaverville, Palo Alto, and Stanford University, *Drouet & D. Richards 4212, 4274, 4300, 4335* (D, F); Porterville and Springville, *Drouet & M. J. Groesbeck 4418, 4470* (D, F); Arrowhead Hot Springs, San Francisco, Folsom, and Tassajara, and many laboratory cultures from California, *N. L. Gardner 51,*

6765, 6924, 7030, 7948 (D, F, UC); Fort Ross, *Gardner* (as *Scytonema ocellatum* in Coll., Hold. & Setch., Phyc. Bor.-Amer. 2210, D, NY); Berkeley, *Gardner* (as *S. guyanense* in Coll., Hold. & Setch., Phyc. Bor.-Amer. 1716a, D, NY); Mount Tamalpais, *Gardner* (as *S. guyanense* and *S. Hofmannii* in Coll., Hold. & Setch., Phyc. Bor.-Amer. 1716b, 1717a, D, F); Marin county, *W. J. V. Osterhout & Gardner* (as *S. Hofmannii* in Coll., Hold. & Setch., Phyc. Bor.-Amer. 1717b, D, F); Fresno, *Osterhout 1387* (D, F, UC); Bolinas Ridge, *W. A. Setchell* (as *S. Hofmannii* in Coll., Hold. & Setch., Phyc. Bor.-Amer. 803, D, F); Yosemite national park, Arrowhead and Waterman hot springs, and El Portal, *Setchell 1387, 1537, 1549, 6511* (D, UC); St. Helena, *Setchell 1095* (TYPE of *S. occidentale* Setch., UC; duplicate: PC); Waterman hot springs, *S. B. Parish* (TYPE of *S. caldarium* Setch., UC; duplicates in Coll., Hold. & Setch., Phyc. Bor.-Amer. 559, D, F, W. R. Taylor); La Verne, Manker Flats near Mount Baldy, and Snow creek (Imperial county), *G. J. Hollenberg 1643, 1657, 1658* (D, Hollenberg); Georgetown (Eldorado county) and Camptonville (Nevada county), *G. H. Giles 25, 48* (D); Palm Springs and Aguanga, *Macbride 7701, 8552* (D, F); La Jolla, *W. E. Allen 7360* (D, F, UC), *E. Y. Dawson 195a* (D, US); San Diego, *M. Reed 1* (D, F, UC); Nevares spring (Death Valley), *P. A. Munz 2280* (D, Hollenberg); Yosemite national park, *A. Carter 1690* (D, F, UC); Sequoia national park, *R. Prettyman* (D, BUT).

Bahama Islands: Nassau, Silver cay, Rose island, Great Stirrups cay (Berry islands), Little Galliot cay (Exuma chain), Cat island, and Watling island, *M. A. Howe 3009A, 3032, 3396, 3597, 4040, 5077, 5149* (D, NY, UC); Garden cay (Great Bahama), Andros, and New Providence, *L. J. K. Brace 3664, 5168, 9700* (D, NY, UC); Great Abaco island and Gorda cay, *C. W. Reimer 1, 5* (PH); Bimini islands, *H. J. Humm* (D, Humm); New Providence, *E. G. Britton 3256* (D, NY), *P. Wagenaar Hummelinck 547* (D, L); Andros, *P. Bartsch* (D, US), *Humm* (D, Humm), *R. N. Ginsburg 8077.5* (PH), *C. Monty FCII2* (PH); Watling island, *N. L. Britton & C. F. Millspaugh 6124* (D, NY).

Puerto Rico: Manati, *W. C. Earle* (TYPE of *Tolypothrix papyracea* Gardn., NY); Las Marias, *F. L. Stevens 1175* (TYPE of *Scytonema guyanense* var. *epiphyllum* Gardn., NY); Rio Piedras, *N. Wille 108, 118, 209a, 1932* (TYPES of *S. subgelatinosum* Gardn., *Tolypothrix penicillata* var. *brevis* Gardn., *T. Willei* Gardn., and *T. amoena* Gardn., NY); Coamo, *Wille 292c, 1870, 1895, 1913* (TYPES of *Scytonema ocellatum* var. *purpureum* Gardn., *Hassallia discoidea* Gardn., *H. granulata* Gardn., and *Calothrix conica* Gardn., NY); Caguas, *Wille 439b* (TYPE of *Synechococcus intermedius* Gardn., NY); San Lorenzo, *Wille 517, 534* (TYPES of *Scytonema spirulinoides* Gardn. and *Hassallia rugulosa* Gardn., NY); Fajardo, *Wille 659* (TYPE of *H. brevis* Gardn., NY); Palmer, *Wille 747c* (TYPE of *H. fragilis* Gardn., NY); Laguna Tortuguero, *Wille 827, 867* (TYPES of *Tolypothrix robusta* Gardn. and *Scytonema ocellatum* var. *majus* Gardn., NY); Mayaguez, *Wille 899a, 1000, 1087a, 1191* (TYPES of *S. javanicum* var. *pallidum* Gardn., *S. javanicum* var. *distortum* Gardn., *Hassallia scytonematoides* Gardn., and *H. heterogenea* Gardn., NY); Sabana Grande, *Wille 936a* (TYPE of *Scytonema tenellum* Gardn., NY); Maricao, *Wille 1149a, 1166, 1229a, 1253* (TYPES of *S. variabile* Gardn., *S. Millei* var. *majus* Gardn., *S. longiarticulatum* Gardn., and *S. ocellatum* var. *constrictum* Gardn., NY); Hatillo–Arecibo, *Wille 1392b* (TYPE of *Lyngbya Martensiana* var. *minor* Gardn., NY); Arecibo–U-tuado, *Wille 1455, 1482* (TYPES of *Scytonema guyanense* var. *minus* Gardn. and *S. evanescens* Gardn., NY); Utuado, *Wille 1527, 1537a, 1556, 1574, 1607, 1609, 1617, 1658* (TYPES of *S. multiramosum* Gardn., *S. capitatum* Gardn., *S. catenulum* Gardn., *S. pulchellum* Gardn., *S. lyngbyoides* Gardn., *S. magnum* Gardn., *S. mirabile* var. *majus* Gardn., and *S. punctatum* Gardn., NY); Santurce, Cayey, Humacao, La Juanita, Adjuntas, Ponce, Jayuya, Guanica, Descalabrado river, Aibonito, and San Juan, *Wille 34, 431, 630, 907, 1666, 1680, 1766, 1842, 1850, 1987, 2015* (D, NY); Maricao, Minillas, San Germán, Sabana Grande, Mayaguez, Coamo, Adjuntas, Ponce, Cayey, and El Yunque, *L. R. Almodóvar 173, 229, 253, 291, 397, 522, 608, 620, 681, 779* (D, Almodóvar); "Puerto Rico," *M. A. Howe* (as *Calothrix pilosa* in Coll., Hold. & Setch., Phyc. Bor.-Amer. 1167, D, F); Sierra de Naguabo, *J. A. Shafer 3350, 3780* (NY); Culebra island, *N. L. Britton* (D, UC); Desecheo island, Cayo Muertos, and Barros, *Britton et al. 1627, 5083, 9667a* (D, NY).

Virgin Islands: St. John, *F. Børgesen 131* (TYPE of *Scytonema Boergesenii* Gardn., NY); St. Thomas (as *S. Leprieurii* in Hohenacker, Meeresalg. 458a, FH), *E. G. Britton & D. W. Marble 448* (MO); Anegada, *W. G. D'Arcy 5147* (MO, PH); Tortola, *P. T. Cleve* (TYPE of *S. Clevei* Grun., PC).

Anguilla: *W. R. Elliott 70* (D, F, UC).

St. Kitts: *Breutel* (TYPE of *Tolypothrix Selaginellae* Rabenh., L).

Guadeloupe: *Perrottet* (TYPE of *Scytonema coactile* Mont., L); Dolé, *H. Mazé & A. Schramm 182* (TYPE of *S. cyanescens* Crouan, PC); Gourbeyre, *Mazé & Schramm 960* (TYPE of *S. parietinum* Crouan, PC); Matouba, *Mazé & Schramm 587* (TYPE of *Calothrix conferta* Crouan, PC); Morne de la GrandeDécouverte, *Mazé & Schramm 150* (TYPE of *Scytonema elegans* var. *antillarum* Crouan, PC); Moule, *Mazé & Schramm 256, 1160* (TYPES of *S. coactile* var. *radians* Crouan and *Tolypothrix guadelupensis* Crouan, PC); Vieux-Fort, *Mazé & Schramm 144* (TYPE of *Lyngbya decipiens* Crouan, PC); *l'Herminier* (TYPES of *Dictyonema membranaceum* var. *guadelupense* Rabenh. and *Calothrix calibaea* Rabenh. in Rabenh., Alg. Eur. 2361, 2362, PH); Marie–Galante, *P. Allorge 40A* (PC), *A. Questel 1927* (D, F, MO); Basse Terre, *Allorge 25B* (TYPE of *Tolypothrix Rober-ti-Lamii* Bourr., PC); Massif Central, Soufrière, and Baino Jaunes, *Questel 1871, 1876, 1908* (D, F, MO).

Dominica: *W. R. Elliott 903* (TYPE of *Scytonema amplum* W. & G. S. West, with that of *Symploca cuspidata* W. & G. S. West, UC), *1240, 2171* (BM, D); St. Georges parish, *C. F. Rhyne & F. R. Fosberg 288* (MICH).

Martinique: pitons Didier (BM, D).

Barbados: Fryer Well point and Paynes bay, *W. R. Taylor, M. Goldstein & D. Patriquin 66-17, 66-30* (MICH).

St. Vincent: Lomond bay, *Elliott 132* (BM, D).

Grenada: Grand Étang, *R. Thaxter* (D, FH).

Trinidad and Tobago: Port of Spain and Roxborough Bay, *Thaxter* (D, FH); Trinidad, *W. E. Broadway* (D, F, MO); Walter Field, *W. L. White & C. C. Yeager 2* (D, F); Maracas Falls, Caura Royal Road, and Aripo savanna (Trinidad), *M. R. Crosby 2443, 2445, 2446* (DUKE, PH).

Dominican Republic: San Pedro de Macorís, *J. G. Scarff* (D).

Haiti: Baie des Flamonds, Aux Cayes—Torbeck, Port Salut, Carpentier, Tiburon, Cap-à-Foux, Rivière Won Sardina, and Abricots, *C. R. Orcutt 9094a, 9601, 9661, 9814, 10029, 10120, 10136, 10142* (D, US); Cap Haitien, *A. W. Evans* (D, YU); Coffe station west of S. Nichol, *C. H. Arndt 343* (W. R. Taylor).

Cuba: *Ramon de la Sagra* (TYPE of *Scytonema rubrum* Mont., PC); Monte Toro, *C. Wright* (FH). Camaguey: San Miguel de los Baños, Sabana Cromo and Camino Antón near Camaguey, Minas, and Trinidad Finca, *J. F. Macbride & B. E. Dahlgren 2, 47, 69, 150, 189* (D, F). Habana: Guines, *Hno. León 2742a* (D, F, LS); Bahía de Siguanea, *E. P. Killip 42997* (D, US); Cañada mountains (Isle of Pines), *L. Ross* (D, F). Matanzas: Bacunayagua, *M. Díaz-Piferrer 5073* (FPDB, PH). Oriente: Pico Turquino, *Hno. León 11270* (D, F, LS); Santiago, *Fr. Clément 2111* (D, F, LS); Loma San Juan, El Guaso, and Loma del Gato, *Hno. Hioram 11804, 12855, 13597* (D, F, LS); Baracoa, *C. L. Pollard et al. 84* (D, F); Gibara, San Antonio, and Punta Peregrina, *Díaz-Piferrer 4692, 5198, 5815* (FPDB, PH). Pinar del Río: Mariél and Arroyos de Mantua, *Díaz-Piferrer 4251, 6147* (FPDB, PH); Vinales, *Killip 13598* (D, F, US). Villas: Cienfuegos and Trinidad mountains, *A. O. Dahl* (D); Venegas, *Hno. León 16212* (D, F, LS).

Jamaica: Navy island, *J. E. Humphrey* (TYPE of *S. conchophilum* Humphr. in Coll., Hold. & Setch., Phyc. Bor.-Amer. LII, NY); Castleton and Bath, *Humphrey* (as *S. crispum* and *S. Hofmannii* in Coll., Hold. & Setch., Phyc. Bor.-Amer. 60, 1258, D, NY); Fern Hill, Westphalia, Gray's Inn (St. Marys), Chepstowe, Troja, Robins Bay, Buff Bay, Arntully, Morces Gap, Farm Hill Works, Moy Hall, Stone Valley River, Darliston, Weireka, Ferry River, Kingston, and Glenburnie Mountain, *C. R. Orcutt 3148, 3757, 4088a, 4516, 4638, 4682, 4769, 5108, 5564, 5597, 5843, 5853, 6217, 6267, 6276, 6465, 7726* (D, US); Blue Mountain peak, *W. R. Maxon & E. P. Killip 1169* (US); "Jamaica," Mavis Bank Road, and Cinchona, *I. F. Lewis* (as *S. Millei, S. mirabilis* var. *Leprieurii*, and *S. ocellatum* in Coll., Hold. & Setch., Phyc. Bor.-Amer. 1405, 1557a, b, 1558, 1559a, b, D, NY); Constant Spring, *G. Lagerheim* (as *S. ocellatum* in Wittr., Nordst. & Lagerh., Alg. Exs. 1322b, D, F); Hope Gardens and Palisadoes, *M. A. Howe* (D, NY); Port Antonio, *C. E. Pease & E. Butler* (D, NY); Falmouth, *F. A. Barkley 22J033* (D, F); Hardware Gap and

86 REVISION OF THE NOSTOCACEAE

Bath, *W. R. Taylor 56-552b, 56-689* (D, MICH); Jim Crow peak, *L. M. Underwood* (D, UC); Catherines Peak, *Eggers 3557* (D, C); Mandeville, *S. Brown 293* (PH); Geddy Hall, *J. Maxwell 10* (BM, D).

Netherlands Antilles: Playa Boca Canoa, Playa Abau, and Manantial de San Pedro (Curaçao), *M. Díaz-Piferrer 1132, 1243, 2000* (D, FPDB, L).

Mexico: Baja California: "Puerto Escondido," *J. Hempel* (AHFH, D, G, W); Bahía de los Angeles, *M. D. Brown 329* (PH). Colima: Clarion island (Revilla Gigedo islands), *W. R. Taylor 57* (AHFH, D, MICH). Jalisco: Guadalajara and Lake Chapala, *W. Kiener 18208, 18225, 18251, 18309* (D, F, NEB). Michoacán: Uruapan, *Kiener 18392* (D, F, NEB). Nuevo León: El Nagalar and Santa Catarina, *F. A. Barkley 1462, 14611* (D, F, TEX); Monterrey, *M. M. Lacas 483* (D, F). Orizaba: Río Blanco, *F. Müller* (TYPE of *Tolypothrix mexicana* Kütz., L). Sonora: Hermosillo, Villa de Seris, Ures, Baviácora, Pilares de Nacozari, Nacozari, Jécori, Cumpas, Magdalena, Guaymas, Navajoa, and Alamos, *F. Drouet, D. Richards, et al. 2800, 2860, 2942, 2954, 2988, 2991, 3011, 3012, 3072, 3091, 3168, 3255* (D, F); Elegante crater (Sierra del Pinacate), *W. F. Faust 13* (PH). Vera Cruz: Vera Cruz, *F. Müller* (TYPE of *T. flavo-viridis* Kütz., L). Yucatán: Lake Zotz, *C. L. Lundell 3290* (D, W. R. Taylor); Campeche Banks, *S. Springer* (D); Chichankanab, *G. F. Gaumer* (D, F). Zacatecas: Fresnillo, *M. D. Brown 373, 377* (PH).

Guatemala: Antigua, Río Guacalate (Escuintla), San Martín Jilotepeque, El Rancho (Progreso), Cobán, Quiriguá, Zacapa, Chiquimula, Jutiapa, Jalapa, Guazacapán, Fiscal, Huehuetenango, Cumbre del Aire (Totonicapán), Zunil, and Retalhuleu, *P. C. Standley 58539, 60191, 64405, 69037, 71567, 72498, 74213, 74290, 75357, 76763, 79715, 80354a, 82749, 83115, 84832, 88543* (D, F); Agua Blanca—Amatillo, Volcán Zunil, Volcán Tajumulco, Lanquín, Cerro Ceibal (Petén), and San Ildefonso Ixtahuacán—Cuilco, *J. A. Steyermark 30393, 35356, 37314, 44089, 46141, 50710C* (D, F); Lakes Atitlán and Amatitlán, *W. A. Kellerman* (D, F); Tikal (Petén), *O. & I. Degener 26825* (D, Degener).

Honduras: Las Casitas, Yuscarán, El Zamorano, Comayagua, Siguatepeque, La Lima, San Alejo, Suyapa, and Tela, *Standley et al. 581, 1191, 1328, 5826, 6157, 7127, 7965, 26256, 55344* (D, F); Guinope (El Paraíso), *J. Valerio R. 1810a* (D, EAP); Cortez, *F. A. Barkley 40447* (PH).

El Salvador: Ahuachapán, *Standley & E. Padilla V. 2520, 2575, 2813* (D, F).

Nicaragua: El Crucero, La Libertad, Juigalpa, Jinotega, Las Mercedes, and Chichigalpa, *P. C. Standley 8237, 9000, 9352a, 9784, 10802, 11461* (D, F).

Costa Rica: Cerro de la Muerte, Los Chiles (Río Frío), and Cascajal, *R. W. Holm & H. H. Iltis 1005b, 1023, 1033* (D, MO); Río Sándoval—Río Tigre and Río Sándalo, *C. W. Dodge 4537A, 10429* (D, MO); Finca Guayabillos (San José), *Dodge & V.F. Georger* (D, MO); La Palma de San Ramón (Alajuela), *A. B. Brenes 176, 177* (CR, D); Bahía Ballena, *E. Y. Dawson 16737* (D, US); Puerto

Limón, Playa Manuel García, and Punta Quepos, *Dawson & W. Rudersdorf 24006, 24108, 24289, 24308a* (PH, US).

Panama: Colón, *G. Lagerheim* (as *Scytonema Hofmannii* and *S. ocellatum* in Wittr., Nordst. & Lagerh., Alg. Exs. 1318, 1322a, D, NY); Barro Colorado island and Gigante bay (Gatun lake), *C. W. Dodge* (D, MO); Salamanca Hydrographic Station, *Dodge, J. A. Steyermark & P. H. Allen 16973* (D, MO); Río Indio, *Dodge & Allen 8853* (D, MO); Summit, *Dodge & A. A. Hunter 8823* (D, MO); Almirante, *Dodge & J. I. Permar 4157* (D, MO); Río Chagres and Río Boquerón, *Steyermark & Allen 16811, 16816, 17266* (D, MO); Madden Dam road, Río Las Lajas, Río Pilón, and Gatun Locks, *G. W. Prescott CZ23, CZ122, CZ150, CZ157* (D, EMC); El Valle, *Prescott & R. L. Caylor CZ106* (D, EMC); Balboa, *G. W. Martin 2032* (D).

Venezuela: Cristóbal Colón, *W. E. Broadway 547* (D, F); El Valle, *A. F. Blakeslee* (D, FH); Juan Griego (Margarita), *P. Wagenaar Hummelinck 18* (D, L); Cubagua island, *W. R. Taylor 39-475c* (AHFH, D, MICH); Puerto Ayacucho (Amazonas), *L. Williams 16072* (D, F); Chimantá Massif, Canaguá (Mérida), Bolero (Mérida), Upata (Bolívar), Esmeralda (Amazonas), Sanariapo (Amazonas), Santa Teresita de Kavanayén (Bolívar), Ijigua (Anzoátegui), Cerro San José (Anzoátegui), Cerro de la Cueva de Doña Anita (Monagas), Guácharo (Monagas), Cuchivano (Sucre), Cerro Venamo (Bolívar), El Dorado (Bolívar), Borburata (Carabobo), Cerro de Río Arriba (Sucre), Peña Blanca (Lara), and El Cantón (Barinas), *J. A. Steyermark et al. 613, 56428a, 56746, 57522, 57759, 58509, 60495, 61268, 61550, 61928, 62293, 62808, 92412, 92985, 95490, 96259, 97452, 102214* (D or PH, VEN); Isla Harapo (Anzoátegui), Golfo de Santa Fé (Sucre), and Isla Cubagua, *M. Díaz-Piferrer 20283, 20319, 20728* (FPDB, PH).

French Guyana: Cayenne, *Leprieur* (TYPES of *Symphyosiphon guyanensis* Mont., PC; *Scytonema Leprieurii* Kütz., L; and *S. vinosum* Mont., PC), *Mille* (TYPE of *S. Millei* Born. & Thur., PC).

Brazil: *A. Löfgren* (as *S. Hofmannii* in Wittr. & Nordst., Alg. Exs. 765c, D, F). Amazonas: São Paulo (Rio Solimões), *J. W. H. Trail 133* (BM, D). Baía: *Gaudichaud* (TYPE of *S. Panicii* Mont., L); Formosa (Jeremoabo), *G. & L. T. Eiten 4980* (PH, SP). Ceará: Urubú (Fortaleza), Aroeiras (Quixadá), and Euzebio (Aquiraz), *F. Drouet 1383, 1401, 1486* (D); Açudes Bom Successo (Maranguape) and Forquilha (Sobral), *S. Wright* (D, FH, NY, RB); Guaramiranga (Pacotí), *H. C. Cutler 8357a* (D, F). Goias: east of Formoso, *E. Y. Dawson 15202* (D, LAM). Guanabara: Rio de Janeiro, *Riedel* (CN). Maranhão: Sacavém (São Luiz), *Drouet 1320* (D, FH, NY, RB). Mato Grosso: Serra de São Jerônimo and São José, *C. A. M. Lindman* (D, F, S). Minas Gerais: Ouro Fino, *H. Kleerekoper 253* (D, F). Pará: Museu Paraense and Fonte Maguary (Belém), *Drouet 1288, 1536* (D, FH, L, NY). Paraíba: Campina Grande, *Wright 2035* (D, W). Rio Grande do Norte: Natal, *Drouet 1527* (D, FH,

L, NY, S, W). Rio Grande do Sul: Pelotas, *G. A. Malme 41* (D, F, S). São Paulo:
São Paulo, *A. Löfgren 1506* (NY); Apiahy, *Puiggari 1280, 1355* (W); Piras-
sununga, *Löfgren 162* (TYPE of *S. coactile* var. *brasiliense* Nordst., LD); Cam-
pinas, *Löfgren 205* (as *Tolypothrix aegagropila β bicolor* in Wittr. & Nordst., Alg.
Exs. 580, D, F); Emas (Pirassununga), *Kleerekoper 23* (D, F); São Paulo,
Ubatuba, Tabatinga (Caraguatatuba), Ilha Santo Amaro, Peruíbe, and
Itanhaen, *A. B. Joly 7, 11, 13 (1953), 324, 412, 441* (D, SPF); Praia de Tenorio
(Ubatuba), *M. Díaz-Piferrer 16779* (FPDB, PH).

Paraguay: Paraguarí, *B. Balansa* (as *Scytonema* sp. in Moug., Dupr. &
Roumeg., Alg. d. Eaux Douc. 1124, BM).

Uruguay: Montevideo, *J. Arechavaleta* (TYPE of *Aulosira implexa* Born. &
Flah., PC; duplicates in Wittr. & Nordst., Alg. Exs. 787, D, NY; as *Tolypothrix
tenuis* and *Scytonema cincinnatum* in Wittr. & Nordst., Alg. Exs. 763b and 873, D,
F).

Argentina: laboratory cultures from Buenos Aires, *D. Rabinovich de Halperín
CP160, 35, 82* (D, de Halperín); Lago Guillermo (Río Negro) and Buenos
Aires, *S. A. Guarrera 2061, 3343* (D, F); Loreto (Missiones), *P. Moreau 1993* (D,
F).

Colombia: Santa Marta, *C. Acleto 1369, 1370* (PH, USM).

Ecuador: Baños (Tungurahua), *G. Lagerheim* (as *S. alatum* and *S. crispum* in
Wittr., Nordst. & Lagerh., Alg. Exs. 1315, 1316a, D, F); Puente de Chimbo
(Chimborazo), *Lagerheim* (as *S. guyanense* and *S. javanicum* in Wittr., Nordst. &
Lagerh., Alg. Exs. 1317, 1319a, D, F); San Nicolas (Pichincha), *Lagerheim* (as *S.
javanicum* and *S. mirabile* in Wittr., Nordst. & Lagerh., Alg. Exs. 1319b, 1320, D,
F); Chorrera de Agoyan (Tungurahua), *Lagerheim* (as *S. myochrous* in Wittr.,
Nordst. & Lagerh., Alg. Exs. 1321, D, F); Narborough island (Colón), *J. T.
Howell 819* (TYPE of *S. guyanense* var. *marinum* Setch. & Gardn., CAS), *W. R.
Taylor 154* (AHFH, D, MICH); Isabela island, *Taylor 94A, 96* (AHFH, D,
MICH); Paccha—Puente Grande and Sambotambo (El Oro), *J. A. Steyermark
54170, 54183* (D, F).

Peru: Ancaşh: La Puna de Recuay and Lagunas Paron and Conococha, *A.
Aldave P. 1037, 1076, 1098* (HUT, PH). Ayacucho: Aina, *E. P. Killip & A. C.
Smith 22601* (D, F). Huánuco: Muña, *G. S. Bryan 508* (D, F); Tingo María, *C.
Acleto 289* (PH, USM). La Libertad: Pataz and Santiago de Chuco, *Aldave 11,
237* (HUT, PH); Laguna del Toro and Huamachuco, *M. Fernández H. 196, 198*
(HUT, PH). Lima: Lima, *A. Maldonado 16* (D, F); Cascadas de Barranco,
Pachacamac, and Laguna de Villa, *Acleto A47, A178, A187* (PH, USM).
Pasco: Laguna de Cerro Pasco, *Fernández 102* (HUT, PH).

Bolivia: Sorata prope Milipaya (Larecaja), *G. Mandon 1836* (S);
Chuquiaguillo (La Paz), *Weddell* (TYPE of *Tolypothrix andina* Mont., PC).

Chile: Port Corral (Valdivia), *R. Thaxter* (D, FH); Puerto Montt, *G. H.
Schwabe 2* (TYPE of *T. fusca* Schwabe, D); laboratory cultures from Llanos de

Diabolo y de Cardona (Arica), *E. A. Flint 1a, 2b, 2c* (PH); Puerto Bueno (Magellanes), *J. J. Engel 5623E* (MSC, PH).

Clipperton Island: *M.-H. Sachet 463, 471b, 472, 473, 474* (D, US).

Hawaii: Sandwich islands, *R. Hitchcock* (D, F). Hawaii: Hamakua, *J. E. Tilden* (as *Nostoc commune* in Tild., Amer. Alg. 486, D, F); Waialuka river, *Tilden* (as *Scytonema mirabile* in Tild., Amer. Alg. 631, D); Volcanoes national park, *O. & I. Degener 30595* (Degener, PH); Cocoanut island near Hilo, *W. A. Setchell 5217* (D, F, UC); Hilo, *M. Doty 13580* (D, Doty); Kau desert, *G. A. Smathers 1a* (PH); Kealamoku, *Degener 30681* (Degener, PH); Kulani Prison, *Degener 27675* (Degener, PH); Pahala Plantation beach, *Tilden* (TYPE of *S. hawaiianum* Tild. in Tild., Amer. Alg. 573, MIN); Puna, *Tilden* (TYPE of *S. azureum* Tild. in Tild., Amer. Alg. 630, MIN); Iilewa and Kamaili, *Doty 13212, 13565* (D, Doty); Kumakahi point, *Degener 30629* (Degener, PH); Keei, *M. Reed 20* (D, F, UC); Napau crater, *Degener 30674* (Degener, PH); Hookena (South Kona), *Doty & A. J. Bernatowicz 13421* (D, Doty); Waiohinu, *Degener 23697* (D, Degener). Kauai: Hanapepe river, *A. A. Heller* (MO); Manaa, *Reed 379* (D, UC); Barking Sands, *Setchell 10684* (D, UC); Waimea drainage, *C. N. Forbes 1407K* (D, F, MO). Lanai: Kaumalapau harbor and Luahiwa petroglyphs, *Degener 24319* (D, Degener), *28780* (Degener, PH). Molokai: Waikolu gulch, *F. R. Fosberg 8981aa* (D, F). Oahu: *S. Berggren 30* (LD); Honolulu, *Berggren* (TYPE of *S. pulvinatum* Nordst., LD); Kaipapau valley, Manoa Falls, Blowhole, and Kaena, *W. J. Newhouse 53-2, 53-31, 53-98* (D, F, Newhouse), *2000* (Newhouse, PH); Camp Erdman (Mokuleia), *A. Kruckeberg* (D, F); Koolauloa, *Tilden* (as *Tolypothrix distorta, Scytonema rivulare,* and *S. cincinnatum* in Tild., Amer. Alg. 478, 479, 480, D, F; TYPE of *S. hawaiianum* var. *terrestre* Tild. in Tild., Amer. Alg. 574, MIN); Kawela, Hanauma bay, Kawailoa, Palolo, Punaluu, and Koko Head, *Doty 8040, 8648, 8911, 9810, 10381, 10640* (D, Doty); Nuuanu stream, *K. Yanagihara 19851* (Doty, PH); Opaeula (Waialua), *Degener 24902* (BISH, D, Degener, NY).

Line Islands: Washington island, *Elschner* (D, UC); Paradise island (Palmyra atoll), *C. R. Long 1806* (PH); Menge island (Palmyra atoll), *E. Y. Dawson 19855, 19859* (D, US).

Marquesas Islands: Valée d'Ikoei, Noukahiva, *E. Jardin 168* (LD).

Tuamotu Islands: Oromea, Opaneke, Teuriamote, Oneroa, Otetou, and Ngarumaoa (Raroia atoll), *M. S. Doty & J. Newhouse 11233, 11379, 11434, 11665, 11884, 11981* (D, Doty).

Society Islands: Tahiti, *C. Crossland 7189, 7190a* (D, UC); between Papenu and Huau (Tahiti), *Setchell 5103* (D, UC).

Rapa Island: Kaimaru, *H. St. John* (D, F, UC).

Phoenix Islands: Canton island, *O. Degener 21338, 21342, 21347* (D, Degener), *O. & I. Degener 24572, 24581a, 24585* (D, Degener).

Samoa: *T. Powell 51* (D, F); Fagatoga (Tutuila), *W. A. Setchell 1005* (FH);

Saneva (Savaii), *K. Rechinger 2969* (TYPE of *Tolypothrix distorta* var. *samoensis* Wille (O); Upolu, *Rechinger* (TYPES of *Scytonema samoense* Wille, *S. coactile* var. *minus* Wille, *Hassallia Rechingeri* Wille, and *H. Rechingeri* fa. *saxicola* Wille, O); Utumapu (Upolu), *L. & K. Rechinger* (as *Scytonema ocellatum* in Mus. Vindob. Krypt. Exs. Alg. 2142, D, F).

Antarctica: Arrival Heights (Ross island), *G. A. Llano 2170* (PH).

Marshall Islands: Arno: Jilang and Ine islands, *L. Horwitz 9101, 9665* (D, F). Bikini: Amen island, *R. F. Palumbo Y28* (D, F); Eniirikku, Enyu, Namu, and Airukiraru islands, *W. R. Taylor 46-41, 46-121, 46-128, 46-244* (D, F, MICH). Eniwetok: Bogombogo and Rujoru islands, *Taylor 46-341, 46-342* (D, F, MICH); Japtan island, *M. S. Doty 12489* (D, Doty); Aniyaanii island, *E. Y. Dawson 13955* (D, US); Aniyaanii-Giriinien lagoon, *M. Gilmartin 519* (D). Jaluit: Jaluit island, *Dawson 13151* (D, US). Jemo Island: *F. R. Fosberg 33904* (D, F). Kwajelein: Kwajelein island, *Doty 18147* (D, Doty). Likiep: Lado islet, *Fosberg 33816* (D, F); Likiep island, *D. P. Rogers 1773* (D, F). Lae: Lae islet, *Fosberg 34104* (D, F). Majuro: Uliga, Dalap, and Enierippu islands, *Dawson 12707, 12754, 12763* (D). Rongelap: Eniaetok, Rongelap, and Romuraru islands, *Taylor 46-475, 46-511, 46-621* (D, F, MICH). Rongerik: Bock island, *Taylor 46-527* (D, F, MICH). Taka: Taka islet, *Fosberg 33733* (D, F). Ujae: Rua islet, *Fosberg 34379* (D, F). Ujelang: Daisu island, *Doty 18145* (D, Doty); Bokan and Morina islets, *Fosberg 34167, 34170* (D, F). Wotho: Wotho, Enejelto, and Kabben islets, *Fosberg 34258, 34425a, 34432* (D, F).

Gilbert Islands: *R. Catala 48* (D, F, NSW); North and South islands and Taburarorae (Onotoa), *E. T. Moul 8182, 8215, 8241B* (D, RUT).

Fiji: Cacatau road, Tamavua, and Virea, *H. E. Parks 20958, 20960, 20968* (D, F, UC).

Kermadec Islands: Raoul island, *V. J. Chapman 38* (D).

Marianas Islands: Guam, *W. L. White & C. C. Yeager 7, 10* (D, F).

Caroline Islands: Ponape, Kalap (Mokil), and Nikalap (Ant), *S. F. Glassman 2558, 2588, 2822* (CHI, D, F); Mogmog island (Ulithi), *B. A. Young Jr.* (D, F).

Kapingamarangi: Touhou, Parakahi, Warua, and Kakuniu islets, *W. J. Newhouse 1635, 1655, 1656, 1657* (D, Newhouse); Nunakita and Wehua islets, *W. Niering 1131, 1571* (D, Newhouse).

Admiralty Islands: Los Negros island off Manus, *W. L. White 4a* (D, F).

Solomon Islands: Rawak (Bougainville), *Gaudichaud* (TYPE of *Scytonema figuratum* Ag., LD); Munda (New Georgia), *White 148* (D, F); Wickham island off New Georgia, *H. B. S. Womersley & A. Bailey 504* (AD-U, PH).

New Hebrides: Point Uanouro (Efate), *J. Y. Bagot* (PH).

New Caledonia: Nouméa, *A. Grunow* (TYPE of *S. polycystum* Born. & Flah., PC); Wagap, *Vieillard 2184* (D, L); Isle Ouin, *J. T. Buchholz 1667a* (D, F); Ile aux Canards, *V. May 310* (D, NSW).

New Zealand: laboratory cultures from Cass, Alexandra, and near Lake

Taupo, *E. A. Flint* (D or PH); Wairakei, Taupo, and Waimangu, *W. A. Setchell* (D, F, UC); Keri Keri, *V. W. Lindauer* (as *S. figuratum* in Lindauer, Alg. Nov.-Zel. Exs. 151, D, F); Paterson inlet (Stewart island), *I. B. Warnock 798* (D, F); Otapai swamp, *H. Oswald 5* (D, H. Skuja); Three Kings islands north of North Cape, *G. Edwards* (D, F); Auckland and Hen island, *K. W. Loach* (D); Waitakere and Whatipu (Auckland), *J. Trevarthen 10, 29* (D).

Japan: Kato (Fukui), Shirahama (Wakayama), Amami-oshima, O-shima (Amami islands), and Kikaijima (Amami islands), *I. Umezaki 55, 389, 1837, 2345, 2378* (D, Umezaki); Ryu Kyu islands, *C. Wright* (TYPE of *Lyngbya effusa* Harv., US); Osaka, *R. Hitchcock* (D, F); Sukiram (Okinawa), *W. D. Field & O. G. Loew 130g* (D, F); Nakano-shima (Tokano islands), *E. Ogata 1141* (D, Umezaki); Hatakejima island (Wakayama), *H. Uchinomi 2118* (D, Umezaki).

Philippines: Agusan: Mount Hilonghilong, *C. M. Weber 1334* (D, F). Albay: Coal Harbor (Cacraray island), *C. B. Robinson 199* (D, F, UC); Libog, Tabaco, and Legaspi, *G. T. Velasquez 4327, 4342, 4348* (D, PUH). Antique: San José and Bugasong, *J. D. Soriano 1532, 1827* (D, F, PUH). Bataan: Lomao river, *R. S. Williams 849a* (D, F); Balanga and Pilar, *Velasquez 2405, 2416* (D, F, PUH). Batangas: Lipa, *Velasquez 602* (D, F, PUH); Mabini and Maricaban, *P. Makabenta 3843, 3850* (D, PUH). Bohol: Tagbilaran, *Makabenta 75* (D, PUH). Bulacan: Calumpit, *Velasquez 428* (D, F, PUH). Camarines Sur: Bato, *Velasquez 4395* (D, PUH). Capiz: Roxas City, Ivisan, and Sigma, *Soriano 1501, 1516, 1551* (D, F, PUH). Cebu: Mandaue and Cebu City, *Makabenta 35, 47* (D, PUH). Iloilo: Iloilo City, Oton, Passi, Dueños, Cabatuan, Janiuay, and Jordan, *Soriano 1058, 1059, 1076, 1078, 1105, 1844, 1857* (D, F, PUH). Laguna: San Pedro, *Velasquez 654* (D, F, PUH); Los Baños, *E. Quisumbing 77303* (D, F). Leyte: San Joaquin, *M. E. Britton 142* (D, F); Biliran, *R. C. McGregor 4768* (D, F, UC); Capoocan, Abuyog, and Hinugayan, *Makabenta 10, 23, 73* (D, PUH). Manila: Vito-Cruz, Blumentritt, and Sampaloc, *Velasquez 553, 2433, 2438* (D, F, PUH). Mindoro: Medio island, Tabinay Maliit, Puerto Galera, and Naujan lake, *Velasquez 972, 1116a, 1928, 3929* (D, PUH). Mountain: Benguet, *Williams 1857a* (D, F); Bangad–Lubuagan, *A. C. Herre* (D, F). Negros Occidental: Santa Fé and Murcia, *Soriano 1872, 2013* (D, F, PUH). Negros Oriental: San Carlos, Palinpinon, and Siaton, *Soriano 1882, 1895, 2006* (D, F, PUH). Palawan: Cuyo, Iwahig penal colony, Tapun, Babuyan, and Araceli, *Velasquez 2823, 2891, 2892b, 2912b, 2920* (D, F, PUH). Pampanga: Arayat, *Velasquez 520* (D, F, PUH). Pangasinan: Labrador, Quezon island, and de la Cruz island, *Makabenta 4408, 4420, 4438* (D, PUH). Rizal: Montalban, *H. H. Bartlett 14679* (D, MICH); Quezon City, *M. B. Valero 73* (D); Mandaluyong, Tanay, Pandacan, Navotas, and Antipolo, *Velasquez 167, 436, 562, 618, 667* (D, F, PUH). Samar: Basey, *Makabenta 56* (D, PUH). Sorsogon: Irosin, *A. D. E. Elmer 17136b* (D, F); Sorsogon, *Velasquez 4379* (D, PUH). Sulu: Bungao, Sanga-Sanga, Jolo, and Siasi, *Velasquez 3212, 3219, 3245, 3233* (D, F,

PUH). Tayabas: Baler, *J. V. Santos 334* (D, PUH). Zambales: Castillejos, *L. E. Ebalo 24* (D, YU).

Territory of New Guinea: Lae, *W. L. White & C. C. Yeager 1* (D, F); Finschhafen, *L. B. Martin* (D, F).

Papua: Boridi, *C. E. Carr 13557* (D, NY); Mount Albert Edward, *L. J. Brass 4491* (D, F).

China: Fukien: University of Amoy, Amoy island, and Kuliang, Kushan, and Huangchun near Foochow, *H. H. Chung A71, A75, A295, A327a, A371, A387, A425b* (TYPES of *Tolypothrix Chungii* Gardn., *T. tenella* Gardn., *T. consociata* Gardn., *T. curta* Gardn., *Scytonema crassum* var. *majus* Gardn., and *S. hyalinum* Gardn., FH). Hopeh: *M. S. Clemens 4016b, 4028* (D, UC); Tang Ho, *Clemens 6005* (D, UC); Peiping, *Y. C. H. Wang 225* (D, F, UC). Hunan: Tschangscha, *H. Handel-Mazzetti 11510, 11602* (D, W). Kiangsu: Nanking, *Y. C. H. Wang 98* (TYPE of *S. crispum* var. *minus* Li, D); Tsishiasin and Wuhsien, *C. C. Wang 350, 366* (D, F, UC). Kwangtung: Nodoa, Ha Kung Leng (Hainan), *F. A. McClure 1693* (D, UC); Fong Ngau Po, Kan-en (Hainan), *S. K. Lau 5493* (D, FH). Szechwan: Tjiaodjio and Lemoka (TYPE of *S. praegnans* Skuja, W) east of Ningyüen, and Datioku, *Handel-Mazzetti 1553, 1620, 2238, 5308* (D, W); Nan-chwan and Omei, *H.-J. Chu 849, 1336* (D, F, Chu). Yunnan: Yünnanfu, Mudidjin south of Yungning (TYPE of *Tolypothrix metamorpha* Skuja, W), Yaotou, and Dsutong near Likiang, *Handel-Mazzetti 1970, 3196, 5980, 7833* (D, W).

Hong Kong: *C. Wright 6.4.4.* (D, US).

North Viet Nam: Laogai, *Handel-Mazzetti 1* (TYPE of *Handeliella Stockmayeri* Skuja, W; duplicates: D, FH, PC); Dong-Dang, *Lucas* (PC).

South Viet Nam: Cauda, *E. Y. Dawson 11065* (D, F, US).

Cambodia: Lem Dan (Koh-Chang), *J. Schmidt X* (TYPE of *Scytonema Schmidtii* Gom., PC; duplicate: C).

Thailand: Bangkok, *O. & I. Degener 29010* (Degener, PH); Li (northern Thailand), *M. S. Doty 16522* (D, Doty).

Malaysia: Penang (Malacca), *H. Möller 92* (LD); Marop (Sarawak), *E. Beccari 6* (TYPE of *Tolypothrix flexuosa* Zanard., FI), *21* (FI); Monte Mattang, *Beccari 33* (TYPE of *Calothrix maculiformis* Zanard., FI); Mount Kinabalo (North Borneo), *J. & M. S. Clemens 32395* (NY).

Indonesia: Amboina: *C. B. Robinson 2403* (NY). Banda Islands: Banda, *W. A. Setchell* (D, UC). Borneo: Kuala Kwajan, *A. H. G. Alston 13232* (BM, D). Celebes: Maros, *A. Weber-van Bosse 789* (D, F, L). Java: comm. Montagne (TYPE of *Scytonema varium* Kütz., L), *H. Zollinger 902* (BM), *1355* (TYPE of *Symphyosiphon javanicus* Kütz., L); Bogor (Buitenzorg), *J. Massart 993* (TYPE of *Scytonema dubium* De Wildem., L), *H. Möller 13* (PC), *S. Kurz* (D, F, L), *Alston 13487* (BM, D); *G. Wilis, J. A. Derks* (L); Tjiapoes, *Massart 817* (TYPE of *S. foliicola* De Wildem., L); Zuiderbergite south of Toeren, *P. Groenhart* (D, F, L); Salak, *E. Nyman 85a* (D, F); Pelabuan Ratu, *Alston 12979* (BM, D). Saleyer

Island: *Weber-van Bosse* (TYPE of *S. saleyerense* Web.-v. B., L). Schouten Islands: Biak, *W. L. White 3, 5* (D, F). Sumatra: Singkarah and Ajer Tegenang, *Weber-van Bosse 602, 649* (D, L).

Australia: New South Wales: Sydney, *E. Gordon* (NSW, PH); Cape Banks (Botany bay), *V. May* (NSW, PH); National Park, *A. B. Cribb 141-3* (BRIU, D). Queensland: Brampton island off Mackay, *May 2701* (D, NSW); Brisbane, Southport, Ipswich, Mount Cordeaux, Enoggera reservoir (Brisbane), and Cunninghams Gap, *Cribb H, J, K, U, 104, 145-2* (D, F, BRIU); Lady Musgrave island (Great Barrier Reef), *Cribb 651.54* (BRIU, PH); Clump point east of Tully, *E. Wollaston 674* (BRIU, PH). South Australia: near Cantara (Coorong), *J. G. Wood 12* (AD-U, D). Tasmania: Port Arthur, *Cribb 115-9* (BRIU, D).

Burma: Rangoon, Palay kweng, and Yomah (Pegu), *S. Kurz* (TYPES of *S. Kurzianum* Zell. and *S. murale* Zell. in Rabenh., Alg. Eur. 2343, 2344, NY; as *S. gracile* and *S. myochrous* in Rabenh., Alg. Sachs./Eur. 673a, 2339, D, FH); Pegu, *Kurz 3146, 3156, 3168, 3235, 3352* (TYPES of *S. fulvum* Zell., *S. parvulum* Zell., *S. subclavatum* Zell., *S. olivaceum* Zell., and *S. Zellerianum* Br. & Bisw., L); Akyab, *Kurz 3287* (BM); University Estate (Rangoon) and Kemmendine, *L. P. Khanna 359, 433* (TYPES of *S. pseudopunctatum* Skuja and *S. burmanicum* Skuja, D); 10th Mile of Prome Road, Hmawbi (Pegu), Mandalay, Rangoon, and Maymyo, *Khanna 404, 421, 482, 599, 639* (D, F).

Kerguelen Island: *640* (TYPE of *Calothrix olivacea* Hook. & Harv., TCD; duplicates: K in BM, PC); slide no. 20 (TYPE of *Aulosira implexa* fa. *kerguelensis* Wille, O).

India: Assam: comm. M. J. Berkeley (TYPE of *Calothrix indica* Mont., PC). Bihar: Ranchi, *J. P. Sinha D1* (PH). Bombay: Matheran, *A. Hansgirg* (D, F, W); Mahim, *Hansgirg* (TYPE of *Stigonema indicum* Schmidle with *Trentepohlia Monilia* fa. *hyalina* in Mus. Vindob. Krypt. Exs. Alg. 858, W). Bengal: Calcutta, *S. Kurz 2675* (TYPE of *Scytonema Vieillardii* Mart., L). Punjab: Himaehal Pradesh, *R. A. Maas Geesteranus 14136* (L); Simla, *S. L. Ghose* (TYPE of *Tolypothrix conglutinata* var. *colorata* Ghose, BM). United Provinces: Agra, *Lorenz* (W); Allahabad, *P. Maheshwari 95* (FH).

Ceylon: Wellematta, *A. Grunow* (W); Veyangoda, *W. Ferguson 264* (W); "Ceylon," *Ferguson 374* (L), *E. Beccari* (TYPE of *Symphyosiphon multistratus* Zanard., L).

Maldive Islands: island south of Fedu (Addu), Maro—Mafilefuri (Fadiffolu), and Filadu (Tilladummati), *H. E. Hackett* (PH).

Chagos Islands: near East Point village (Diego Garcia), *C. Rhyne 516, 623, 697* (PH).

Pakistan: Jamshoro (Hyderabad), *I. I. Chaudhri* (D).

Kazakhstan: laboratory cultures of soil from Bet-tak-dala, *N. V. Sdobnikova*, comm. H. S. Forest Sk1, Sk14, Sk15 (PH).

Seychelles: Anse Bateau (Praslin), *W. E. Isaac 4468* (PH).

Aldabra Atoll: La Passe Femme (West island), *C. Rhyne 997, 1278* (PH); Takamaka well (South island), *F. R. Fosberg 49267* (PH); back of settlement (West island), *Fosberg 49503* (PH).

Mauritius: *Bory de Saint-Vincent* (TYPE of *Scytonema torridum* Ag., LD), *M. de Robillard* (L), *J. McGregor* (BM, D); Flacq, *N. Pike* (D, NY, YU).

Iraq: Durbendikan (Kirkuk), Gurna (Basrah), and Howear canal (Amara), *F. A. Barkley et al. 7434, 8920, 8921* (BUA, PH).

Arabia: laboratory culture, comm. H. C. Bold (D).

Socotra: *B. Balfour* (TYPE of *Lyngbya scabrosa* Dick., BM), *1369* (BM, D).

Calothrix Agardh

[*Radiella* Nees von Esenbeck, Alg. d. Süss. Wassers, p. 47. 1814.] *Gloeotrichia* J. Agardh [Alg. Mar. Medit. & Adriat., p. 8. 1842] ex Bornet & Flahault, Ann. Sci. Nat. VII. Bot. 4: 365. 1887. *Rivularia* Sect. *Gloeotrichia* Kirchner in Engler & Prantl, Natürl. Pflanzenfam. 1(1A): 90. 1900. —Type species: *Rivularia angulosa* Roth.

Calothrix Agardh [Syst. Algar., p. xxiv. 1824] ex Bornet & Flahault, ibid. 3: 345. 1886. [*Calothrix* Trib. *Confervicolae* Agardh, ibid., p. 70. 1824.] *Leibleinia* Endlicher [(as "Leiblinia"), Gen. Pl. Sec. Ord. Nat. Disp., p. 5. 1836] ex Bornet & Flahault, ibid. 3: 351. 1886. *Lyngbya* Subgen. *Leibleinia* Gomont [Journ. de Bot. 4(20): 354. 1890] ex Gomont, Ann. Sci. Nat. VII. Bot. 16: 120. 1892. —TYPE species: *Conferva Mucor* Roth.

[*Calothrix* Trib. *Scopulorum* Agardh, loc. cit. 1824.] *Conishymene* Schousboe *pro synon.* ex Bornet, Mém. Soc. Nat. Sci. Nat. & Math. Cherbourg 28: 187. 1892. —TYPE species: *Conferva scopulorum* Web. & Mohr.

[*Stylobasis* Schwabe *pro synon.* in Sprengel, Linn. Syst. Veget., ed. 16, 4(1): 372. 1827.] —Type species: *Linckia amblyonema* Spreng.

[*Rivularia* Sect. *Scytochloria* Harvey in Hooker, Brit. Fl. 2(Engl. Fl. 5): 393. 1833.] [*Scytochloria* Harvey, loc. cit. 1833.] —Type species: *Rivularia nitida* Ag.

Aphanizomenon Morren [Bull. Acad. R. Sci. Bruxelles 3: 430. 1836] ex Bornet & Flahault, ibid. 7: 241. 1888. —Type species: *A. incurvum* Morr.

Mastigonema Schwabe [Linnaea 1837: 112. 1837] ex Bornet & Flahault, Ann. Sci. Nat. VII. Bot. 3: 368. 1886. —Type species: *Oscillatoria subulata* Corda.

Zonotrichia J. Agardh [Alg. Mar. Medit. & Adriat., p. 9. 1842] ex Bornet & Flahault, ibid. 4: 345. 1887. —Type species: *Z. hemispherica* J. Ag.

Diplotrichia J. Agardh [ibid., p. 10. 1842] ex Bornet & Flahault, loc. cit. 1887.
—Type species: *D. polyotis* J. Ag.

Trichormus Allman [Ann. & Mag. of Nat. Hist. 11: 163. 1843] ex Bornet & Flahault, ibid. 7: 224. 1888. —Type species: *T. incurvus* Allm.

Mastichothrix Kützing [Phyc. Gener., p. 232. 1843] ex Bornet & Flahault (as "Mastigothrix"), ibid. 3: 345. 1886. *Calothrix* Sect. *Calothrix* (as "Eucalothrix") Subsect. *Mastichothrix* Hansgirg, Prodr. Algenfl. Böhm. 2: 51. 1892. —Type species: *Mastichothrix fusca* Kütz.

Schizosiphon Kützing [ibid., p. 233. 1843] ex Bornet & Flahault, loc. cit. 1886. [*Schizosiphon* Sect. *Gymnopodii* Kützing, loc. cit. 1843.] *Calothrix* Sect. *Calothrix* (as "Eucalothrix") Subsect. *Schizosiphon* Hansgirg, ibid. 2: 48. 1892. —Type species: *Scytonema salinum* Kütz.

[*Schizosiphon* Sect. *Chaetopodii* Kützing, ibid., p. 234. 1843.] —Type species: *S. chaetopus* Kütz.

[*Schizosiphon* Sect. *Fastigiatim-ramosi* Kützing, loc. cit. 1843.] —Type species: *S. gypsophilus* Kütz.

[*Schizosiphon* Sect. *Fasciculati* Kützing, loc. cit. 1843.] —Type species: *S. flagelliformis* Kütz.

Geocyclus Kützing [Phyc. Gener., p. 235. 1843] ex Bornet & Flahault, Ann. Sci. Nat. VII. Bot. 4: 345. 1887. —Type species: *G. oscillarinus* Kütz.

Physactis Kützing [loc. cit. 1843] ex Bornet & Flahault, loc. cit. 1887. —Type species: *P. lobata* Kütz.

Heteractis Kützing [ibid., p. 236. 1843] ex Bornet & Flahault, loc. cit. 1887. —Type species: *H. mesenterica* Kütz.

Chalaractis Kützing [ibid., p. 236. 1843] ex Bornet & Flahault, ibid. 4: 365. 1887. —Type species: *C. villosa* Kütz.

Limnactis Kützing [ibid., p. 237. 1843] ex Bornet & Flahault, loc. cit. 1887. —Type species: *Tremella globulosa* Hedw.

Ainactis Kützing [loc. cit. 1843] ex Bornet & Flahault, ibid. 4: 345. 1887. —Type species: *A. alpina* Kütz.

Dasyactis Kützing [ibid., p. 239. 1843] ex Bornet & Flahault, loc. cit. 1887. —Type species: *D. salina* Kütz.

Euactis Kützing [ibid., p. 240. 1843] ex Bornet & Flahault, loc. cit. 1887. —Type species: *E. marina* Kütz.

[*Rivularia* Sect. *Lithonema* Hassall, Hist. Brit. Freshw. Alg. 1: 265. 1845.] *Lithonema* Hassall [ibid. 1: 455. 1845] ex Forti, Syll. Myxoph., p. 670. 1907. —Type species: *Rivularia calcarea* Sm. & Sow.

Homoeoactis Zanardini [Notizie Intorno alle Cellulari Marine delle Lagune e de'Litorali di Venezia, p. 77. 1847] ex Forti, ibid., p. 610. 1907. —Type species: *Rivularia Contarenii* Zanard.

Dichothrix Zanardini [Mem. Ist. Veneto d. Sci., Lett. & Arti 7(2): 297. 1858] ex Bornet & Flahault, Ann. Sci. Nat. VII. Bot. 3: 373. 1886. *Calothrix* Sect. *Dichothrix* Hansgirg, Prodr. Algenfl. Böhm. 2: 52. 1892. —Type species: *Dichothrix penicillata* Zanard.

Thrichocladia Zanardini [Phyc. Indic. Pugill., p. 25. 1872] ex Forti (as "Trichocladia"), ibid., p. 661. 1907. —Type species: *T. nostocoides* Zanard.

Polythrix Zanardini [ibid., p. 32. 1872] ex Bornet & Flahault, ibid. 3: 380. 1886. *Gardnerula* J. de Toni, Noter. di Nomencl. Algol. 8: [5]. 1936. —Type species: *Polythrix spongiosa* Zanard.

Microchaete Thuret [Ann. Sci. Nat. VI. Bot. 1: 375. 1875] ex Bornet & Flahault, ibid. 5: 83. 1887. *Fremyella* J. de Toni, ibid. 8: [3]. 1936. —Type species: *Microchaete grisea* Thur.

Aulosira Kirchner [Krypt.-Fl. Schles. 2(1): 238. 1878] ex Bornet & Flahault, ibid. 7: 256. 1888. —Type species: *A. laxa* Kirchn.

Leptochaete Borzi [N. Giorn. Bot. Ital. 14(4): 298. 1882] ex Bornet & Flahault, Ann. Sci. Nat. VII. Bot. 3: 341. 1886. *Leptochaete* Sect. *Diaphanochaete* Forti, Syll. Myxoph., p. 598. 1907. —Type species: *L. crustacea* Borzi.

Calothrix Sect. *Homoeothrix* Thuret ex Bornet & Flahault, ibid. 3: 345. 1886. *Homoeothrix* Thuret ex Kirchner in Engler & Prantl, Natürl. Pflanzenfam. 1(1A): 87. 1900. —Type species: *Lyngbya juliana* Menegh.

Scytomene Schousboe *pro synon.* ex Bornet, Mém. Soc. Nat. Sci. Nat. & Math. Cherbourg 28: 189. 1892. —Type species: *Rivularia atra* Roth.

Gloeotrichia Sect. *Sclerothrichia* Hansgirg, Prodr. Algenfl. Böhm. 2: 44. 1892. —Type species: *Rivularia Pisum* Ag.

Ammatoidea W. & G. S. West, Journ. R. Microsc. Soc. 1897: 506. 1897. [Variant spelling "Hammatoidea" by Lemmermann, Krypt.-Fl. Mark Brandenb. 3(1): 256. 1907.] —Type species: *A. Normanii* W. & G. S. West.

Diplothrix Bornet & Flahault *pro synon.* ex Kirchner in Engler & Prantl, loc. cit. 1900. —Type species: *Dichothrix Nordstedtii* Born. & Flah.

Richelia Schmidt in Ostenfeld & Schmidt, Vidensk. Medd. Naturh. Fören. Kjöbenhavn 1901: 146. 1902. —Type species: *R. intracellularis* Schm.

Tildenia Kossinskaia, Bot. Mat. Inst. Sporov. Rast. Glavn. Bot. Sada SSSR 4(5–6): 76. 1926. *Setchelliella* J. de Toni, Noter. di Nomencl. Algol. 8: [6]. 1936. —Type species: *Scytonema fuliginosum* Tild.

Kyrtuthrix Ercegović, Arch. f. Protistenk. 66(1): 170. 1929. —Type species: *K. dalmatica* Erceg.

Camptylonemopsis Desikachary, Proc. Indian Acad. Sci 28(2B): 35. 1948. —Type species: *Campylonema lahorense* Ghose.

Ghosea Cholnoki, Bol. Soc. Portug. Ciênc. Nat., ser. 2, 4(1): 92. 1952. —Type species: *Aulosira fertilissima* Ghose.

* * *

The original material of the type species of the following genera and subgeneric taxa has not been available to me for study:

Potarcus Rafinesque [Journ. de Phys. 89: 107. 1819] ex Kuntze (as "Portacus"), Rev. Gen. Pl. 2: 911. 1891. —Type species: *P. bicolor* Raf.

Gaillardotella Bory de Saint-Vincent [in Mougeot & Nestler, Stirp. Crypt. Vogeso-rhen. 8: 796. 1823] ex Bornet & Flahault, Ann. Sci. Nat. VII. Bot. 4: 369. 1887. [*Rivularia* Sect. *Raphidia* Carmichael in Harvey in Hooker, Brit. Fl. 2(Engl. Fl. 5): 394. 1833.] *Raphidia* Carmichael [loc. cit. 1833] ex Bornet & Flahault, ibid. 4: 370. 1887. *Gloeotrichia* Sect. *Malacothrichia* Hansgirg, Prodr. Algenfl. Böhm. 2: 45. 1892. —Type species: *Tremella natans* Hedw.

Listia Meyen [Verh. K. Leop.-Carol. Akad. Naturf. 14(2): 469. 1829] ex Forti, Syll. Myxoph., p. 632. 1907. —Type species: *L. crustacea* Meyen.

[*Siradium* Liebmann, Naturh. Tidskr. 2: 489. 1838–39.] —Type species: *S. intricatum* Liebm.

Agonium Oersted [De Region. Marin., p. 44. 1844] ex Kirchner in Engler & Prantl, Natürl. Pflanzenfam. 1(1A): 92. 1900. —Type species: *A. centrale* Oerst.

Arthrotilum Rabenhorst [Fl. Eur. Algar. 2: 230. 1865] ex Forti, ibid., p. 603. 1907.—Type species: *Amphithrix papillosa* Rabenh.

Calothrix Sect. *Rivulariopsis* Kirchner in Engler & Prantl, ibid. 1(1A): 87. 1900. *Rivulariopsis* Voronikhin, Bot. Mat. Inst. Sporov. Rast. Glavn. Bot. Sada Petrograd 2(8): 115. 1923. —Type species: *Calothrix wembaerensis* Hieron. & Schmidle.

Scytonematopsis Kisseleva, Zhurn. Russk. Bot. Obshch. 15: 169. 1930. —Type species: *S. Woronichinii* Kiss.

Montanoa González Guerrero, Anal. Jard. Bot. Madrid 8: 267. 1948. —Type species: *M. castellana* Gonz. G.

Vindobona Claus, Verh. Zool.-bot. Ges. Wien 97: 95. 1957. —Type species: *Gloeotrichia Andreanszkyana* Claus.

Fortiella Claus, N. Hedwigia 4(1–2): 63. 1962. —Type species: *F. Subaiana* Claus.

Trichomata cylindrica vel plus minusve torulosa, unum extremum vel utrinque extrema saepe per aliquot cellulas attenuantia atque decolorantia vel tumefacientia, septata, ambitu recta vel curvantia vel spiralia, cellulis terminalibus vegetativis primum hemisphaericis demum obtuse conicis aut cylindricis atque ad extrema rotundis; heterocystis plus minusve sphaericis vel cylindraceis terminalibus vel intercalaribus; sporis plus minusve cylindricis vel ovoideis. Materia vaginalis dispersa vel mucosa vel discreta. Planta trichomata nuda vel in muco vel in vaginis cylindraceis discretis saepe ramosis comprehens.

Trichomes cylindrical or more or less torulose, attenuating and often becoming hyaline through several cells, or becoming swollen, at one or both ends, septate, straight or curving or spiraled, the terminal vegetative cells at first hemispherical becoming obtuse-conical or cylindrical and at the tips rotund; heterocysts terminal or intercalary, more or less spherical or cylindrical; spores more or less spherical or ovoid. Sheath material dispersed or mucous or discrete. Plant composed of naked trichomes or trichomes in mucous or discrete, more or less cylindrical, often branched sheaths.

Fan (1956), using many of the specimens studied here, attempted to delineate the species of this genus on the bases of presence or absence of heterocysts, size of cells, and type of habitat. Of his seven species, one is actually an ecophene of *Scytonema Hofmannii*. Three others are from fresh water, and three are from marine habitats; each of these groups corresponds to one species of Calothrix recognized in this revision. Umezaki (1858) made a perceptive study of large numbers of specimens of an ecophene of *C. crustacea* [*Kyrtuthrix maculans* (Gom.) Umez.] from all parts of its range, including the Type specimens and the synonymy involved.

Key to Species of Calothrix

Trichomes attenuated gradually to the undifferentiated tips; plants of fresh-water habitats..*C. parietina*, p. 99.
Trichomes abruptly attenuated near the undifferentiated tips; plants of marine habitats..*C. crustacea*, p. 163.

Calothrix parietina (Nägeli) Thuret

Tremella globulosa Hedwig [Theoria Generat. & Fructif. Pl. Crypt. Linn., Retract. & Aucta, p. 217. 1798] ex Agardh, Syst. Algar., p. 29. 1824. *Rivularia dura* Roth, N. Beytr. z. Bot. 1: 273. 1802. *R. dura β utriculata* Roth, ibid. 1: 277. 1802. [*Linckia dura* Lyngbye, Tent. Hydroph. Dan., p. 197. 1819.] *Rivularia Pisum β dura* Agardh, ibid., p. 25. 1824. [*Linckiella dura* Gaillon, Mém. Soc. d'Emulation d'Abbeville, p. 473. 1883.] *Limnactis Lyngbyana* Kützing [Phyc. Gener., p. 237. 1843] ex Bornet & Flahault, Ann. Sci. Nat. VII. Bot. 4: 369. 1887. *Rivularia radians* d *Lyngbyana* Kirchner, Krypt.-Fl. Schles. 2(1): 223. 1878. *R. Pisum* var. *dura* Agardh ex Bornet, Bull. Soc. Bot. France 36: 149. 1889. *R. radians* var. *Lyngbyana* Kirchner ex Forti, Syll. Myxoph., p. 650. 1907.—TYPE specimen, here designated in the absence of Hedwig's material, presumably from Berlin, Germany: "Rivularia dura teste Roth. In Seen, ex Willdenow," in herb. Grunow (W).

Rivularia Linckia Roth (as "Linkia," corrected to "Linckia" by Roth in Catal. Bot. 3: 336. 1806), N. Beytr. z. Bot. 1: 265. 1802. *R. gigantea* Trentepohl *pro synon.* in Roth, loc. cit. 1802. *Nostoc confusum* Agardh [Syst. Algar., p. 22. 1824] ex Bornet & Flahault, Ann. Sci. Nat. VII. Bot. 7: 192. 1888. [*Undina confusa* Fries, Syst. Orb. Veget. 1(Pl. Homon.): 348. 1825.] *Nostoc Linckia* Bornet & Thuret [Notes Algol. 2: 86. 1880] ex Bornet & Flahault, loc. cit. 1888. *Statonostoc Linckia* Elenkin, Monogr. Alg. Cyanoph., Pars Spec. 1: 595. 1938. —TYPE specimen, from Oldenburg, Germany: "Rivularia Linkia, dedit Roth ipse," in herb. Agardh (LD).

Rivularia angulosa Roth, N. Beytr. z. Bot. 1: 283. 1802. *Gloeotrichia angulosa* J. Agardh [Alg. Mar. Medit. & Adriat., p. 8. 1842] ex Bornet & Flahault, Ann. Sci. Nat. VII. Bot. 4: 370. 1887. *Raphidia angulosa* Hassall [Hist. Brit. Freshw. Alg. 1: 264. 1845] ex Bornet & Flahault, loc. cit. 1887. *Gloeotrichia Sprengelii* Kützing [ex Reinsch, Algenfl. mittl. Th. Franken, p. 50. 1867] ex Bornet & Flahault, loc. cit. 1887. [*Gloeotrichia natans* c *angulosa* Kirchner, Krypt.-Fl. Schles. 2(1): 222. 1878.] *G. natans* var. *angulosa* Kirchner ex Hansgirg, Prodr. Algenfl. Böhm. 2:45. 1892. —TYPE specimen , from Oldenburg, Germany: "Rivularia angulosa Roth, ded. Mertens 1808," in herb. Bory (PC).

Conferva echinulata Smith & Sowerby, Engl. Bot. 36: 1378. 1804. *Rivularia echinulata* Sowerby in Smith & Sowerby, ibid., Gen. Indexes, Alphab. Index. 1814. *Echinella articulata* Agardh, Syst. Algar., p. 16. 1824. *Rivularia echinata* Smith & Sowerby *pro synon.* ex Agardh, loc. cit. 1824. *Conferva echinata* Smith & Sowerby ex Harvey in Hooker, Brit. Fl. 2 (Engl. Fl. 5): 398. 1833. *Gloeotrichia echinulata* Richter, Forschungsber. biol. Sta. Plön 21: 31. 1894. —TYPE specimen from Wales: lake in Anglesey, *Davies*. This material is attached to no. 1378 of the original plates of *English Flora* drawn by Sowerby (BM).

Rivularia Linckia β simplex Roth, Catal. Bot. 3: 336. 1806. —TYPE specimen with that of *R. Linckia* above, in herb. Agardh (LD).

Batrachospermum haematites Lamarck & De Candolle, Synops. Pl. in Fl. Gallica Descr., p. 11. 1806. *Conferva haematites* Ramond *pro synon.* in Lamarck & De Candolle, loc. cit. 1806. *Chaetophora haematites* Bory de Saint-Vincent (as "hematites"), Dict. Class. d'Hist. Nat. 3: 431. 1823. *Rivularia haematites* Agardh, Syst. Algar., p. 26. 1824. *Zonotrichia haematites* Endlicher [Mant. Bot. Altera Gen. Pl. Suppl. 3: 13. 1843] ex Bornet & Flahault, Ann. Sci. Nat. VII. Bot. 4: 350. 1887. *Ainactis haematites* Trevisan [Alghe d. Tenere Udinese, p. 19. 1844] ex Forti, Syll. Myxoph., p. 669. 1907. *Rivularia rivularis* Wittrock & Nordstedt *pro synon.* ex Bornet & Flahault, ibid. 4: 351. 1887. *Euactis amnigena* Stizenberger *pro synon.* ex Bornet & Flahault, ibid 4: 350. 1887.—TYPE specimen from the Pyrenees, France: "Batrachospermum haematites B. G." in "Herbier de la Flore Française (Bot. Gall.) donné au Muséum par. A. F. De Candolle, 1822" (PC). Fig. 28.

Rivularia calcarea Smith & Sowerby, Engl. Bot., pl. 1799. 1807. [*Linckia dura β calcarea* Greville, Fl. Edin., p. 322. 1824.] *Zonotrichia calcarea* Endlicher [Mant. Bot. Altera Gen. Pl. Suppl. 3: 13. 1843] ex Bornet & Flahault, loc. cit. 1887.

Lithonema calcareum Hassall [Hist. Brit. Freshw. Alg. 1: 265. 1845] ex Forti, ibid., p. 670. 1907. *Ainactis calcarea* Kützing [Bot. Zeit. 5(11): 178. 1847] ex Bornet & Flahault, loc. cit. 1887. —TYPE specimen, from Queens County, Ireland: "a Sowerby, misit Borrer," in herb. Agardh (LD).

Scytonema compactum Agardh [Disp. Alg. Suec., p. 39. 1812] ex Bornet & Flahault, Ann. Sci. Nat. VII. Bot. 3: 378. 1886. *Conferva compacta* Sommerfelt, Suppl. Fl. Lappon., p. 191. 1826. *Sirosiphon compactus* Kützing (as "compactum") [Phyc. Germ., p. 178. 1845] ex Bornet & Flahault, ibid. 5: 68. 1887. *Hassallia compacta* Hassall [ibid. 1: 232. 1845] ex Bornet & Flahault, ibid. 5: 117. 1887. *Stigonema compactum* Kirchner [Krypt.-fl. Schles. 2(1): 230. 1878] ex Bornet & Flahault, ibid. 5: 68. 1887. *Dichothrix compacta* Bornet & Flahault [Mém. Soc. Nat. Sci. Nat. & Math. Cherbourg 25: 204. 1885] ex Bornet & Flahault, Ann. Sci. Nat. VII. Bot. 3: 378. 1886. *Calothrix compacta* Hansgirg, Sitzungsber. K. Böhm. Ges. Wiss., Math.-Nat. Cl. 1890(2): 126. 1891. *Stigonema panniforme* var. *compactum* Hansgirg, Prodr. Algenfl. Böhm. 2: 23. 1892. —TYPE specimen from mountains in Sweden: "saxicola" in herb. Agardh (LD). Fig. 43.

Linckia hypnicola Lyngbye [Tent. Hydroph. Dan., p. 197. 1819] ex Bornet & Flahault, ibid. 4: 366. 1887. *Rivularia hypnicola* Lyngbye ex Kützing, Sp. Algar., p. 337. 1849. —TYPE specimen from Denmark: in paludosis ad lacum Lyngbye, *Lyngbye*, 8 Jul. 1815 (C).

[*Linckia dura* var. *lutescens* Lyngbye, Tent. Hydroph. Dan., p. 197. 1819.] —TYPE specimen from Denmark: ad littus Fuursöe Sellandiae, in herb. Lyngbye (C).

Rivularia Pisum Agardh, Syst. Algar., p. 25. 1824. *R. dura* β *Pisum* Duby, Bot. Gallic., ed. 2, 2: 961. 1829. *Physactis Pisum* Kützing [Sp. Algar., p. 333. 1849] ex Bornet & Flahault, Ann. Sci. Nat. VII. Bot. 4: 367. 1887. *Gloeotrichia Pisum* Thuret [Ann. Sci. Nat. VI. Bot. 1: 382. 1875] ex Bornet & Flahault, ibid. 4: 366. 1887. *G. Pisum* fa. *parvula* Wittrock & Nordstedt *pro synon.* ex Bornet & Flahault, ibid. 4: 367. 1887. *Potarcus Pisum* Kuntze (as "Portacus"), Rev. Gen. Pl. 2: 911. 1891. —TYPE specimen, here designated in the absence of the original material from Lake Mälaren, from Sweden: prope Lundam, in herb. Agardh (LD).

Scytonema pulverulentum Agardh [ibid., p. 40. 1824] ex Bornet & Flahault, ibid. 3: 378. 1886. *S. cinereum* d *pulverulentum* Rabenhorst [Fl. Eur. Algar. 2: 248. 1865] ex Bornet & Flahault, ibid. 5: 113. 1887. —TYPE specimen from Sweden: Stockholm ad saxa, in herb. Agardh (LD).

Oscillatoria Mougeotiana Agardh [ibid., p. 61. 1824] ex Gomont, Ann. Sci. Nat. VII. Bot. 16: 243. 1892. *Linckia vivax* Mougeot *pro synon.* [in Agardh, loc. cit. 1824] ex Forti, Syll. Myxoph., p. 634. 1907. *Schizosiphon Mougeotianus* Kützing [Tab. Phyc. 2: 18. 1850–52] ex Forti, loc. cit. 1907. *Rivularia Mougeotiana* Gaillardot in A. Mougeot & Roumeguère in Louis, Le Départ. des

Vosges 2: 590. 1887. —TYPE specimen from France: ad saxa lacus Longemer Vogesorum, in herb. Agardh (LD).

[*Linckia amblyonema* Sprengel, Linn. Syst. Veget., ed. 16, 4(1): 371. 1827.] [*Stylobasis stylocarpa* Schwabe *pro synon.* in Sprengel, ibid. 4(1): 372. 1827.] *Rivularia Schwabeana* Wallroth, Fl. Crypt. Germ. 2: 9. 1833. —TYPE, presumably from Dessau, Germany: "Sphaerobasis stylocarpa," in herb. Sprengel (F).

Calothrix Berkeleyana Carmichael [in Harvey in Hooker, Engl. Fl. 5(Brit. Fl. 2): 367. 1833] ex Bornet & Flahault, Ann. Sci. Nat. VII. Bot. 3: 370. 1886. [*Tolypothrix Berkeleyana* Carmichael in Hassall, Hist. Brit. Freshw. Alg. 1: 241. 1845.]—In the absence of Berkeley's original collection, a TYPE, labeled thus by Carmichael, from Appin, Argyll, Scotland, is here designated (K in BM).

Rivularia botryoides Carmichael in Harvey, ibid., p. 392. 1833. *Potarcus botryoides* Kuntze (as "Portacus"), Rev. Gen. Pl. 2:911.1891.—TYPE specimen here designated in the absence of the original material, from Wales: Dolgelley, *Ralfs,* in herb. Hooker (K in BM).

Rivularia granulifera Carmichael in Harvey, ibid., p. 393. 1833. —TYPE specimen from Argyll, Scotland: Appin, *Carmichael,* in herb. Hooker (K in BM).

Rivularia crustacea Carmichael in Harvey, loc. cit. 1833. *Lithonema crustaceum* Hassall [ibid. 1: 266. 1845] ex Forti, Syll. Myxoph., p. 676. 1907. —TYPE specimen, presumably from Argyll, Scotland: "Rivularia *crustacea* Car. mss.," in herb. Jenner (BM).

Oscillatoria subulata Corda [in de Carro, Almanach de Carlsbad 5: 183. 1835] ex Gomont, Ann. Sci. Nat. VII. Bot. 16: 240. 1892. *Mastigonema thermale* Schwabe [Linnaea 1837: 112. 1837] ex Bornet & Flahault, Ann. Sci. Nat. VII. Bot. 3: 368. 1886. *Calothrix thermalis* Hansgirg [Österr. Bot. Zeitschr. 34(10): 357. 1884] ex Bornet & Flahault, loc. cit. 1886. —In the absence of the original material by Corda (see Hansgirg, Prodr. Algenfl. Böhm. 1: 6. 1886), a TYPE is here designated from the original locality, Karlovary, Czechoslovakia: am Sprudelkorb, *Schwabe,* Oct. 1836, ex Grunow in herb. Bornet-Thuret (PC). Fig. 30.

Scytonema salinum Kützing [Alg. Aq. Dulc. Germ. Dec. 14: 136. 1836] ex Bornet & Flahault, ibid. 3: 366. 1886. *Schizosiphon salinus* Kützing [Phyc. Gener., p. 233. 1843] ex Bornet & Flahault, loc. cit. 1886. *Calothrix salina* Hansgirg [Österr. Bot. Zeitschr. 34: 353. 1884] ex Hansgirg, Prodr. Algenfl. Böhm. 2: 49. 1892. *C. parietina* var. *salina* Hansgirg, loc. cit. 1892. *C. parietina* fa. *salina* Hansgirg ex Forti, Syll. Myxoph., p. 623. 1907. —TYPE specimen from Thuringia, Germany: ad salinas prope Artern, Sept. 1833, with *Chthonoblastus salinus* Kütz. [= *Schizothrix arenaria* (Berk.) Gom.] in Kützing, Algae Aquae Dulcis Decas 14: 136 (L).

Aphanizomenon incurvum Morren [Bull. Acad. R. Sci. Bruxelles 3: 430. 1836] ex Bornet & Flahault, Ann. Sci. Nat. VII. Bot. 7: 242. 1888.—TYPE specimen, here designated, from France: Étang de St.-Quentin près Versailles, *G. Thuret 61,* 25 Sept. 1851, in herb. Bornet—Thuret (PC).

Calothrix brevis Liebmann [Naturh. Tidskr. 2(5): 491. 1838–39] ex Bornet & Flahault, ibid. 3: 370. 1886. [*Leibleinia brevis* Fries (as "Leiblinia"), Summa Veget. Scand., p. 134. 1846.] —TYPE specimen from Denmark: in *Conferva fracta* in amne ad Soeborghuus, *Liebmann* (C).

Rivularia Biasolettiana Meneghini in Zanardini, Bibl. Ital. 99: 197. 1840. —TYPE specimen from Italy: Tergesti, *Biasoletto,* in herb. Meneghini (FI). Fig. 45.

Lyngbya juliana Meneghini [Giorn. Tosc. Sci. Med., Fis. e Nat. 1: 187. 1840] ex Gomont, Ann. Sci. Nat. VII. Bot. 16: 154. 1892. *Leibleinia juliana* Kützing [Bot. Zeit. 5: 194. 1847] ex Bornet & Flahault, ibid. 3: 348. 1886. *Phormidium julianum* Rabenhorst [Fl. Eur. Algar. 2: 118. 1865] ex Gomont, ibid. 16: 141. 1892. *Calothrix juliana* Bornet & Flahault [Mém. Soc. Nat. Sci. Nat. & Math. Cherbourg 25: 200. 1885] ex Bornet & Flahault, Ann. Sci. Nat. VII. Bot. 3: 348. 1886. *Homoeothrix juliana* Kirchner in Engler & Prantl, Natürl. Pflanzen-fam. 1(1a): 87. 1900. —TYPE specimen from Toscana, Italy: Therm. Julian., *Meneghini,* in herb. Kützing (L). Fig. 49, 52.

Scytonema furcatum Meneghini [in Trevisan, Prosp. Fl. Eugan., p. 55. 1842] ex Bornet & Flahault, ibid. 5: 111. 1887. *Schizothrix furcata* Rabenhorst (as "fur-catum") [ibid. 2: 269. 1865] ex Gomont, ibid. 15: 327. 1892. *Microcoleus furcatus* Hansgirg, Prodr. Algenfl. Böhm. 2: 79. 1892. —TYPE specimen from Venezia Euganea, Italy: Therm. Eugan., *Meneghini,* in herb. Bornet—Thuret (PC).

Mastichothrix aeruginea Kützing [Phyc. Gener., p. 232. 1843] ex Bornet & Flahault, ibid. 3: 364. 1886. [*M. aeruginosa* Kützing ex Rabenhorst, Krypt.-Fl. Sachs. 1: 104. 1863.] *Mastigonema aerugineum* Kirchner [Krypt.-Fl. Schles. 2(1): 220. 1878] ex Forti, Syll. Myxoph., p. 617. 1907. —TYPE specimen from Sachsen, Germany: unter *Chaetophora elegans,* Weissenfels, in herb. Kützing (L).

Mastichothrix fusca Kützing [Phyc. Gener., p. 232. 1843] ex Bornet & Flahault (as "Mastigothrix"), ibid. 3: 345. 1886. [*Mastigonema aerugineum* b *fuscum* Kirchner, loc. cit. 1878.] *Calothrix solitaria* Kirchner [Mikrosk. Pflan-zenw., p. 37. 1885] ex Hansgirg, Prodr. Algenfl. Böhm. 2: 51. 1892. *C. fusca* Bornet & Flahault [Mém. Soc. Nat. Sci. Nat. & Math. Cherbourg 25: 202. 1885] ex Bornet & Flahault, Ann. Sci. Nat. VII. Bot. 3: 364. 1886. —TYPE specimen from Sachsen, Germany: an *Limnaea stagnalis,* Nordhausen, Aug. 1841, in herb. Kützing (L).

Mastigonema paradoxum Kützing (as "Mastichonema") [Phyc. Gener., p. 233. 1843] ex Forti, Syll. Myxoph., p. 632. 1907. —TYPE specimen from Sachsen, Germany: Nordhausen, in aqua dulce, in herb. Kützing (L).

Schizosiphon gypsophilus Kützing [ibid., p. 234. 1843] ex Bornet & Flahault, ibid. 3: 377. 1886. *Calothrix gypsophila* Thuret [Ann. Sci. Nat. VI. Bot. 1: 381. 1875] ex Bornet & Flahault, loc. cit. 1886. *Dichothrix gypsophila* Bornet & Flahault [Mém. Soc. Nat. Sci. Nat. & Math. Cherbourg 25: 204. 1885] ex Bornet & Flahault, Ann. Sci. Nat. VII. Bot. 3: 377. 1886. —TYPE specimen from Sachsen, Germany: an nassen Gypsfelsen bei Sachswerfen (Nordhausen), in herb. Kützing (L). Fig. 42.

Geocyclus oscillarinus Kützing [ibid., p. 235. 1843] ex Bornet & Flahault, ibid. 4: 352. 1887. *Rivularia Biasolettiana* fa. *terrestris* Hauck in Wittrock & Nordstedt, Alg. Aq. Dulc. Exs. 12: 577. 1883. —TYPE specimen from Venezia Giulia, Italy: Montfalcone, in herb. Kützing (L).

Physactis saccata Kützing [loc. cit. 1843] ex Bornet & Flahault, ibid. 4: 370. 1887. *Physactis Pisum β saccata* Kützing [Sp. Algar., p. 333. 1849.] ex Bornet & Flahault, loc. cit. 1887. *Rivularia Pisum* var. *saccata* Kirchner ex Forti, Syll. Myxoph., p. 655. 1907. —TYPE specimen, here designated, the only material labeled thus in the Kützing herbarium, from Germany: bei Berlin in d. Spree, *A. Braun,* Jul., Aug. 1851 (L).

Limnactis minutula Kützing [Phyc. Gener., p. 237. 1843] ex Bornet & Flahault, ibid. 4: 348. 1887. *Rivularia radians* b *minutula* Kirchner, Krypt.-Fl. Schles. 2(1): 223. 1878. *R. radians* var. *minutula* Kirchner ex Wolle, Bull. Torrey Bot. Club 8: 38. 1881. *R. minutula* Bornet & Flahault, loc. cit. 1887. —TYPE specimen from Thuringia, Germany: Heringen, *Wallroth,* in herb. Kützing (L). Fig. 46.

Chalaractis villosa Kützing [loc. cit. 1843] ex Bornet & Flahault, Ann. Sci. Nat. VII. Bot. 4: 366. 1887. *Physactis villosa* Kützing [Sp. Algar., p. 333. 1849] ex Bornet & Flahault, loc. cit. 1887. *Rivularia villosa* Rabenhorst, Krypt.-Fl. Sachs. 1: 102. 1863. [*Gloeotrichia Pisum* d *villosa* Kirchner, Krypt.-Fl. Schles. 2(1): 222. 1878.] *G. Pisum* var. *villosa* Kirchner ex Hansgirg, Prodr. Algenfl. Böhm. 2: 44. 1892. *Rivularia Pisum* var. *villosa* Kirchner ex Forti, Syll. Myxoph., p. 655. 1907. —TYPE specimen, in which the trichomes are infested with *Schizothrix calcicola* (Ag.) Gom., from Germany: Merseburg, in herb. Kützing (L).

[*Chalaractis mutila* Kützing, Phyc. Gener., p. 237. 1843.] *Physactis mutila* Kützing [Sp. Algar., p. 333. 1849] ex Bornet & Flahault, loc. cit. 1887. *Rivularia mutila* Kützing ex Ainé, Pl. Crypt.-cell. d. Saône-et-Loire, p. 254. 1863. *Potarcus mutilus* Kuntze (as "Portacus"), Rev. Gen. Pl. 2: 912. 1891. —TYPE specimen, with that of *Chalaractis villosa,* from Germany: Merseburg, in herb. Kützing (L).

Ainactis alpina Kützing [Phyc. Gener., p. 237. 1843] ex Bornet & Flahault, Ann. Sci. Nat. VII. Bot. 4: 350. 1887. *A. baldensis* Meneghini *pro synon.* [in Kützing, Sp. Algar., p. 335. 1849] ex Forti, ibid., p. 670. 1907. *Zonotrichia alpina* Wittrock [in Rabenhorst, Alg. Eur. 236–237: 2360. 1873] ex Bornet & Flahault, ibid. 4: 351. 1887. —TYPE specimen from Italy: in rivulis rapide fluent. in Forojulii (Friuli), *Meneghini,* in herb. Kützing (L).

Rivularia salina Kützing, Phyc. Gener., p. 238. 1843. *Gloeotrichia salina* Rabenhorst [Krypt.-Fl. Sachs. 1: 101. 1863] ex Bornet & Flahault, ibid. 4: 368. 1887. *Potarcus salinus* Kuntze (as "Portacus"), loc. cit. 1891. *Gloeotrichia natans* fa. *salina* Poliansky ex Elenkin, Monogr. Alg. Cyanoph., Pars Spec. 2: 1175. 1949. —TYPE specimen from Sachsen, Germany: Mansfelder See, in herb. Kützing (L). Fig. 33.

Rivularia Brauniana Kützing, loc. cit. 1843. *R. gigantea* b. *Brauniana* Rabenhorst, Deutschl. Krypt.-Fl. 2(2): 93. 1847. *Gloeotrichia Brauniana* Rabenhorst [Krypt.-Fl. Sachs. 1: 101. 1863] ex Bornet & Flahault, Ann. Sci. Nat. VII. Bot. 4: 370. 1887. *G. natans* d *Brauniana* Kirchner [Krypt.-Fl. Schlesien 2(1): 222. 1878] ex Forti, Syll. Myxoph., p. 651. 1907. *G. natans* var. *Brauniana* Kirchner ex Hansgirg, Prodr. Algenfl. Böhm. 2: 45. 1892. —TYPE specimen from Germany: Fl. Badens., in herb. Kützing (L).

Rivularia gigantea Trentepohl ex Kützing, loc. cit. 1843. *Gloeotrichia gigantea* Rabenhorst [Fl. Eur. Algar. 2: 201. 1865] ex Bornet & Flahault, loc. cit. 1887. *G. natans* b *gigantea* Kirchner [Krypt.-Fl. Schles. 2(1): 222. 1878] ex Forti, loc. cit. 1907. *G. natans* var. *gigantea* Kirchner ex Hansgirg, loc. cit. 1892. *G. natans* fa. *gigantea* Kirchner ex Poliansky in Elenkin, Monogr. Alg. Cyanoph., Pars Spec. 2: 1174. 1949. —TYPE specimen from Germany: Carlsruhe, *A. Braun,* in herb. Kützing (L).

Rivularia minuta Kützing, Phyc. Gener., p. 239. 1843. —TYPE specimen from Sachsen, Germany: Eilenburg, in herb. Kützing (L).

Rivularia Lens Meneghini, Mem. R. Accad. Sci. Torino, ser. 2, 5: 134. 1843; in Kützing, loc. cit. 1843. *Gloeotrichia Lens* Endlicher [Mant. Bot. Altera Gen. Pl. Suppl. 3: 12. 1843] ex Bornet & Flahault, Ann. Sci. Nat. VII. Bot. 4: 366. 1887. *G. Pisum* var. *Lens* Hansgirg, ibid. 2: 44. 1892. *Rivularia Pisum* var. *Lens* Hansgirg ex Forti, ibid., p. 655. 1907. —TYPE specimen from Italy: foglie della Nymphaea e della *Trapa natans,* Orto Botanico, Padova, Aug. 1838, in herb. Meneghini (FI).

Rivularia lenticula Kützing, loc. cit. 1843. *Gloeotrichia lenticula* Rabenhorst [Fl. Eur. Algar. 2: 204. 1865] ex Bornet & Flahault, ibid. 4: 367. 1887. [*G. Pisum* e *lenticula* Kirchner, Jahresh. Ver. Vaterl. Naturk. Württemb. 36: 193. 1880.] *Potarcus lenticula* Kuntze (as "Portacus"), Rev. Gen. Pl. 2: 912. 1891. —TYPE specimen from Sachsen, Germany: an Nymphaea, bei Sachswerfen (Nordhausen), in herb. Kützing (L).

Rivularia Sprengeliana Kützing, loc. cit. 1843. *R. gigantea* c *Sprengeliana* Rabenhorst, Deutschl. Krypt.-Fl. 2(2): 93. 1847. *Gloeotrichia Sprengeliana* Rabenhorst [Krypt.-Fl. Sachs. 1: 101. 1863] ex Bornet & Flahault, Ann. Sci. Nat. VII. Bot. 4: 370. 1887. *Potarcus Sprengelianus* Kuntze (as "Portacus"), Rev. Gen. Pl. 2: 912. 1891. —TYPE specimen from Germany: Brückdorf bei Halle, in herb. Kützing (L).

Dasyactis salina Kützing [Phyc. Gener., p. 239. 1843] ex Bornet & Flahault, ibid. 4: 352. 1887. *Limnactis salina* Rabenhorst [Fl. Eur. Algar. 2: 212. 1865] ex

Bornet & Flahault, loc. cit. 1887. —TYPE specimen from Germany: Rollsdorf bei Halle, in herb. Kützing (L).

Dasyactis Kunzeana Kützing [ibid., p. 240. 1843] ex Forti, Syll. Myxoph., p. 668. 1907. *Zonotrichia Kunzeana* Rabenhorst [ibid. 2: 218. 1865] ex Forti, loc. cit. 1907. *Rivularia Kunzeana* Forti, loc. cit. 1907. —TYPE specimen from Sachsen, Germany: Dömecken bei Wanzleben, *Kunze*, in herb. Kützing (L).

Euactis chrysocoma Kützing [ibid., p. 242. 1843] ex Forti, ibid., p. 669. 1907. *Rivularia dura* γ *cespitosa* Lyngbye *pro synon.* ex Kützing, loc. cit. 1843. *Zonotrichia chrysocoma* Rabenhorst [Alg. Sachs. 15: 145. 1852] ex Bornet & Flahault, ibid. 4: 350. 1887. *Rivularia chrysocoma* Kirchner, Jahresh. Ver. Vaterl. Naturk. Württemb. 36: 193. 1880. —TYPE specimen from Austria: Salzburg, *Meneghini*, in herb. Kützing (L).

Rivularia rudis Meneghini, Mem. R. Accad. Sci. Torino, ser. 2, 5(Sci. Fis. e Mat.): 140. 1843. *Zonotrichia rudis* Endlicher [Mant. Bot. Altera Gen. Pl. Suppl. 3: 13. 1843] ex Bornet & Flahault, Ann. Sci. Nat. VII. Bot. 4: 350. 1887. *Amphithrix rudis* Meneghini [in Kützing, Sp. Algar., p. 275. 1849] ex Forti, Syll. Myxoph., p. 602. 1907. *Arthrotilum rude* Rabenhorst (as "rudis") [Fl. Eur. Algar. 2: 230. 1865] ex Forti, loc. cit. 1907.—TYPE specimen from Venezia Euganea, Italy: Euganeis, *Meneghini*, in herb. Kützing (L).

Trichormus incurvus Allman [Ann. & Mag. of Nat. Hist. 11: 163. 1843] ex Bornet & Flahault, ibid. 7: 229. 1888. [*Anabaena incurva* Trevisan, Nomencl. Algar. 1: 38. 1845.] *A. cupressaphila* Wolle [Fresh-w. Alg. U. S., p. 288. 1887] ex Stokes, Analyt. Keys to Gen. & Sp. Fresh-w. Alg. U. S., p. 67. 1893. —TYPE specimen from Ireland: in the Grand Canal Dock, Dublin, *Allman* (K in BM).

Rivularia pygmaea Kützing, Phyc. Germ., p. 188. 1845. *Gloeotrichia pygmaea* Rabenhorst [ibid. 2: 206. 1865] ex Bornet & Flahault, ibid. 4: 366. 1887. *Potarcus pygmaeus* Kuntze (as "Portacus"), Rev. Gen. Pl. 2: 912. 1891. —TYPE specimen from Germany: Mecklenburg, *Flotow*, in herb. Kützing (L).

Euactis calcivora Kützing [ibid., p. 190. 1845] ex Bornet & Flahault, Ann. Sci. Nat. VII. Bot. 4: 350. 1887. *Rivularia calcivora* Rabenhorst, Deutschl. Krypt.-Fl. 2(2): 92. 1847. *Zonotrichia calcivora* Rabenhorst [Fl. Eur. Algar. 2: 214. 1865] ex Bornet & Flahault, ibid. 4: 351. 1887. —TYPE specimen from Switzerland: Neuchateler See bei Concise und Onnens, *A. Braun 142*, Sept., in herb. Kützing (L).

Euactis scardonitana Kützing [loc. cit. 1845] ex Bornet & Flahault, ibid. 4: 350. 1887. *Zonotrichia scardonitana* Rabenhorst [ibid. 2: 215. 1865] ex Bornet & Flahault, ibid. 4: 351. 1887. —TYPE specimen from Dalmatia, Yugoslavia: in flumine Scardona, *Meneghini*, in herb. Kützing (L).

Mastigonema caespitosum Kützing [ibid., p. 184. 1845; in Roemer, Alg. Deutschl., p. 32. 1845] ex Bornet & Flahault, ibid. 3: 366. 1886. *Calothrix caespitosa* Hansgirg [Österr. Bot. Zeitschr. 34: 390. 1884] ex Dalla Torre & Sarnthein, Alg. Tirol. Vorarlb. & Liechtenst., p. 132. 1901. *Isactis caespitosa* Wolle, Fresh-w. Alg. U.S., p. 245. 1887. *Calothrix parietina* var. *caespitosa*

Hansgirg, Prodr. Algenfl. Böhm. 2: 49. 1892. *C. parietina* fa. *caespitosa* Hansgirg ex Forti, Syll. Myxoph., p. 623. 1907.—TYPE specimen from the Harz mountains, Germany: Clausthal, *Römer 108*, 20 Aug. 1844, in herb. Kützing (L).

[*Schizosiphon curvulus* Kützing ex Roemer, loc. cit. 1845.] —TYPE specimen from Germany: Soole Salzboden bei Salzdettfurth bei Hildesheim, *Roemer*, in herb. Vigener (NY).

Physactis chalybea Kützing [Bot. Zeit. 5(11): 178. 1847] ex Bornet & Flahault, Ann. Sci. Nat. VII. Bot. 4: 370. 1887. *Rivularia Pisum* c *chalybea* Rabenhorst, Fl. Eur. Algar. 2: 206. 1865. —TYPE from Schleswig, Germany: Husbyer Moor, an Grashalmen, *Hansen*, in herb. Kützing (L).

Schizosiphon Meneghinianus Kützing [loc. cit. 1847] ex Forti, ibid., p. 641. 1907. *Calothrix Meneghiniana* Kirchner [Krypt.-Fl. Schles. 2(1): 220. 1878] ex Bornet & Flahault, ibid. 3: 370. 1886. *Dichothrix Meneghiniana* Forti, loc. cit. 1907. —TYPE specimen from Venezia Euganea, Italy: in flumine Tartaro (insula Cola) ad plantas, *Meneghini*, in herb. Kützing (L).

Schizosiphon julianus Kützing [ibid., p. 179. 1847] ex Forti, Syll. Myxoph., p. 635. 1907. —TYPE specimen (corrected by Kützing in Sp. Algar., p. 329. 1849) from Toscana, Italy: Therm. Julian., *Meneghini*, in herb. Kützing (L).

Mastigonema Orsinianum Kützing (as "Mastichonema") [Bot. Zeit. 5(11): 179. 1847] ex Bornet & Flahault, ibid. 3: 376. 1886. *Limnactis Orsiniana* Meneghini *pro synon.* [in Kützing, loc. cit. 1847] ex Forti, ibid., p. 673. 1907. *Calothrix Orsiniana* Thuret [Ann. Sci. Nat. VI. Bot. 1: 381. 1875] ex Bornet & Flahault, ibid. 3: 377. 1886. *Rivularia radians* e *Orsiniana* Kirchner, Jahresh. Ver. Vaterl. Naturk. Württemb. 36: 193. 1880. *Dichothrix Orsiniana* Bornet & Flahault [Mém. Soc. Nat. Sci. Nat. & Math. Cherbourg 25: 204. 1885] ex Bornet & Flahault, Ann. Sci. Nat. VII. Bot. 3: 376. 1886. *Rivularia minutula* var. *Orsiniana* Meneghini ex Forti, loc. cit. 1907. *Calothrix gypsophila* fa. *Orsiniana* Poliansky in Elenkin, Monogr. Alg. Cyanoph., Pars Spec. 2: 1091. 1949. —TYPE specimen from Italy: Fonti del Coppo, *Meneghini*, in herb. Kützing (L). Fig. 51.

Mastigonema fasciculatum Kützing (as "Mastichonema") [Bot. Zeit. 5(11): 179. 1847] ex Forti, Syll. Myxoph., p. 630. 1907. —TYPE specimen from Germany: Kohnstein, Nordhausen, in herb. Kützing (L).

Amphithrix incrustata Kützing [ibid. 5: 195. 1847] ex Forti, ibid., p. 602. 1907. *Lophopodium incrustatum* Kützing [ex Rabenhorst, Fl. Eur. Algar. 2: 232. 1865] ex Forti, loc. cit. 1907. —TYPE specimen from Venezia Euganea, Italy: Therm. Eugan., *Meneghini*, in herb. Kützing (L).

Amphithrix crustacea Kützing [Bot. Zeit. 5: 195. 1847] ex Forti, ibid., p. 603. 1907. *Lophopodium crustaceum* Kützing [ex Rabenhorst, loc. cit. 1865] ex Forti, loc. cit. 1907. [*Dasyactis crustacea* Crouan, Fl. Finistère, p. 116. 1867.] —TYPE specimen, here designated, the only material labeled thus in the Kützing herbarium, from Germany: ad lapides in piscina, Nordhausen (L).

Hypheothrix turicensis Nägeli [in Kützing, Sp. Algar., p. 303. 1849] ex

Gomont, Ann. Sci. Nat. VII. Bot. 15: 329. 1892. [*Lyngbya turicensis* Hansgirg, Bot. Centralbl. 22: 291. 1885.] *Schizothrix turicensis* Geitler, Rabenh. Krypt.-Fl., ed. 2, 14: 1083. 1932. —TYPE specimen from Switzerland: Zürich, an feuchten Felsen, *Nägeli 284,* in herb. Kützing (L).

Limnochlide hercynica Kützing [ibid., p. 286. 1849] ex Bornet & Flahault, Ann. Sci. Nat. VII. Bot. 7: 242. 1888. *Aphanizomenon cyaneum* Ralfs [Ann. & Mag. of Nat. Hist., ser. 2, 5: 341. 1850] ex Bornet & Flahault, loc. cit. 1888. [*Limnochlide flos-aquae* β *hercynica* Kützing ex Ralfs *pro synon.,* loc. cit. 1850.] *Aphanizomenon flos-aquae* var. *hercynicum* Kützing (as "hercynica") ex Forti, Syll. Myxoph., p. 470. 1907. —TYPE specimen from Harz mountains, Germany: Clausthal, *Römer,* in herb. Kützing (L).

Scytonema melanopleurum Meneghini [in Kützing, Sp. Algar., p. 303. 1849] ex Bornet & Flahault (as "melanopleuron"), ibid. 5: 113. 1887. —TYPE specimen from Venezia Euganea, Italy: ad rupes Eugan., *Meneghini,* in herb. Kützing (L).

[*Merizomyria flagelliformis* Kützing, ibid., p. 325. 1849.] [*Schizosiphon Brebissonii* Crouan, Fl. Finistère, p. 115. 1867.] —TYPE specimen from Calvados, France: Falaise, *De Brébisson 468,* in herb. Kützing (L).

Schizosiphon apiculatus β *rupestris* Kützing [ibid., p. 327. 1849] ex Forti, ibid., p. 633. 1907. *S. rupestris* Kützing [ex Rabenhorst, Krypt.-Fl. Sachs. 1: 105. 1863] ex Forti, loc. cit. 1907. —TYPE specimen from Calvados, France: Falaise, *De Brébisson,* in herb. Kützing (L).

Schizosiphon decoloratus Nägeli [in Kützing, Sp. Algar., p. 327. 1849] ex Bornet & Flahault, Ann. Sci. Nat. VII. Bot. 3: 366. 1886. *Calothrix parietina* var. *decolorata* Hansgirg, Prodr. Algenfl. Böhm. 2: 49. 1892. *C. parietina* fa. *decolorata* Hansgirg ex Forti, Syll. Myxoph., p. 623. 1907. —TYPE specimen from Switzerland: Zürich, feuchte Felsen, *Nägeli 378,* in herb. Kützing (L).

Schizosiphon parietinus Nägeli [in Kützing, loc. cit. 1849] ex Bornet & Flahault, loc. cit. 1886. *C. parietina* Thuret [Ann. Sci. Nat. VI. Bot. 1: 381. 1875] ex Bornet & Flahault, loc. cit. 1886. —TYPE specimen from Switzerland: Zürich, *Nägeli 107,* in herb. Kützing (L). Fig. 40.

Schizosiphon aponinus Meneghini [in Kützing, ibid., p. 328. 1849] ex Forti, ibid., p. 635. 1907. —TYPE specimen, mixed with *Schizothrix arenaria* (Berk.) Gom., from Venezia Euganea, Italy: in thermis Euganeorum ad terram, *Meneghini,* in herb. Kützing (L).

Schizosiphon crustiformis Nägeli [in Kützing, loc. cit. 1849] ex Bornet & Flahault, loc. cit. 1886. —TYPE specimen from Switzerland: Ct. Schwytz, an einem überrieselten Felsen, Schindellegi, *Nägeli 100,* in herb. Kützing (L).

Schizosiphon Kuetzingianus Nägeli (as "Kützingianus") [in Kützing, Sp. Algar., p. 329. 1849] ex Bornet & Flahault, Ann. Sci. Nat. VII. Bot. 3: 377. 1886. *Calothrix Kuetzingiana* Hansgirg, Physiol. & Algol. Stud., p. 168. 1887. —TYPE specimen from Switzerland: Zürich, an feuchten Felsen, *Nägeli 256,* in herb. Kützing (L).

Schizosiphon cinctus Nägeli [in Kützing, loc. cit. 1849] ex Bornet & Flahault, loc. cit. 1886. —TYPE specimen from Switzerland:·Zürich, auf Schlamm im Katzensee, *Nägeli 278,* in herb. Kützing (L).

Schizosiphon cataractae Nägeli [in Kützing, ibid., p. 330. 1849] ex Bornet & Flahault, ibid. 3: 376. 1886. —TYPE specimen from Switzerland: Rheinfall, an Felsen, *Nägeli 177,* in herb. Kützing (L).

Schizosiphon radians Kützing & De Brébisson [in Kützing, ibid., p. 331. 1849] ex Bornet & Flahault, ibid. 4: 348. 1887. *Rivularia radians* Thuret, Ann. Sci. Nat. VI. Bot. 1: 382. 1875. —TYPE specimen from Calvados, France: sur les Chara, Falaise, *De Brébisson 446,* in herb. Kützing (L).

Physactis villosa β *major* Kützing [Sp. Algar., p. 333. 1849] ex Forti, Syll. Myxoph., p. 654. 1907. [*P. villosa* var. *major* Kützing, Tab. Phyc. 2: 19. 1852.] —TYPE specimen from Oldenburg, Germany: in fossis, Jever, *Koch 155,* Jul. 1844, in herb. Kützing (L).

Physactis gelatinosa Nägeli [in Kützing, Sp. Algar., p. 333. 1849] ex Forti, ibid., p. 656. 1907. *Rivularia Pisum* e *gelatinosa* Rabenhorst, Fl. Eur. Algar. 2: 207. 1865. *R. Pisum* var. *gelatinosa* Forti, loc. cit. 1907. —TYPE specimen from Switzerland: Zürich, Torfgräben, *Nägeli 106,* in herb. Kützing (L).

Physactis cerasum Kützing [ibid., p. 334. 1849] ex Forti, ibid., p. 651. 1907. *Rivularia Pisum* d *cerasum* Rabenhorst, ibid. 2: 206. 1865. —TYPE specimen from Germany: Torfgruben, Hanau, *Theobald 75,* 19 Jun. 1845, in herb. Kützing (L).

Limnactis dura Kützing [Sp. Algar., p. 335. 1849] ex Bornet & Flahault, Ann. Sci. Nat. VII. Bot. 4: 347. 1887. *Rivularia radians* c *dura* Kirchner, Krypt.-Fl. Schles. 2(1): 223. 1878. *R. radians* var. *dura* Kirchner ex Cooke, Brit. Fresh-w. Alg., p. 279. 1884. —TYPE specimen from Calvados, France: in stagnis ad *Charam hispidam* etc., Falaise, *De Brébisson 181,* in herb. Kützing (L).

Limnactis parvula Kützing [loc. cit. 1849] ex Forti, Syll. Myxoph., p. 656. 1907. [*L. minutula* b *parvula* Rabenhorst, Fl. Eur. Algar. 2: 210. 1865.] —TYPE specimen from Baden, Germany: Freiburg, auf Potamogeton, *A. Braun 134,* Aug. 1848, in herb. Kützing (L).

[*Limnactis minutula* β *Orsiniana* Kützing, ibid., p. 336. 1849.] *Rivularia Orsiniana* Meneghini *pro synon.* in Kützing, loc. cit. 1849. —TYPE specimen from Italy: Monte Corno, *Meneghini,* in herb. Kützing (L).

Rivularia Boryana β *flaccida* Kützing, loc. cit. 1849. [*Gloeotrichia Boryana* β *flaccida* Ralfs ex Rabenhorst, ibid. 2: 202. 1865.] —TYPE specimen from Wales: Swansea, *J. Ralfs,* in herb. Kützing (L).

Rivularia rigida Kützing, loc. cit. 1849. [*Gloeotrichia Brauniana* b *rigida* Rabenhorst, ibid. 2: 203. 1865.] —TYPE specimen from Oldenburg, Germany: in piscina prope Wifels, *Koch 32,* in herb. Kützing (L).

Rivularia Brebissoniana Kützing, Sp. Algar., p. 337. 1849. *Gloeotrichia Brebissoniana* Rabenhorst [loc. cit. 1865] ex Bornet & Flahault, Ann. Sci. Nat. VII. Bot. 4: 370. 1887. *Potarcus Brebissonianus* Kuntze (as "Portacus"), Rev. Gen. Pl.

2: 911. 1891. —TYPE specimen from Calvados, France: Falaise, *De Brébisson 466*, in herb. Kützing (L).

Rivularia durissima Kützing, loc. cit. 1849. *Gloeotrichia durissima* Rabenhorst [Fl. Eur. Algar. 2: 204. 1865] ex Bornet & Flahault, ibid. 4: 367. 1887. [*G. Pisum* c *durissima* Kirchner, Krypt.-Fl. Schles. 2(1): 222. 1878.] *G. Pisum* var. *durissima* Kirchner ex Hansgirg, Beih. z. Bot. Centralbl. 18(2): 491. 1905. *Rivularia Pisum* var. *durissima* Kirchner ex Forti, Syll. Myxoph., p. 655. 1907. —TYPE specimen from Germany: Neckarau bei Mannheim, *Mettenius* (comm. A. Braun no. 131), in herb. Kützing (L).

Rivularia parvula Kützing, loc. cit. 1849. *Gloeotrichia parvula* Rabenhorst [ibid. 2: 205. 1865] ex Bornet & Flahault, ibid. 4: 370. 1887. *Potarcus parvulus* Kuntze (as "Portacus"), Rev. Gen. Pl. 2: 912. 1891. —TYPE specimen from Baden, Germany: Kork, Rheinebene, *A. Braun 11*, Sept. 1845, in herb. Kützing (L).

Rivularia minor Kützing, loc. cit. 1849. *Gloeotrichia minor* Rabenhorst [ibid. 2: 204. 1865] ex Bornet & Flahault, loc. cit. 1887. *Potarcus minor* Kuntze (as "Portacus"), loc. cit. 1891. —TYPE specimen from Germany: Hanau, *Theobald*, 15 Aug. 1841, in herb. Kützing (L).

Dasyactis rivularis Nägeli [in Kützing, Sp. Algar., p. 339. 1849] ex Bornet & Flahault, Ann. Sci. Nat. VII. Bot. 4: 350. 1887. —TYPE specimen from Switzerland: Zürich, auf Steinen in Bächen, *Nägeli 255*, in herb. Kützing (L).

Dasyactis Naegeliana Kützing [loc. cit. 1849] ex Bornet & Flahault, loc. cit. 1887. *Zonotrichia Naegeliana* Rabenhorst (as "Negeliana") [Fl. Eur. Algar. 2: 216. 1865] ex Bornet & Flahault, ibid. 4: 351. 1887. —TYPE specimen from Switzerland: Zürich, auf Steinen, *Nägeli 381*, in herb. Kützing (L).

Euactis rufescens Nägeli [in Kützing, ibid., p. 342. 1849] ex Forti, Syll. Myxoph., p. 671. 1907. [*Zonotrichia fluviatilis* d *rufescens* Rabenhorst, ibid. 2: 215. 1865.] *Rivularia rufescens* Nägeli ex Bornet & Flahault, ibid. 4: 349. 1887. —TYPE specimen from Switzerland: Zürich, *Nägeli 180*, in herb. Kützing (L). Fig. 44.

Euactis rivularis Nägeli [in Kützing, loc. cit. 1849] ex Bornet & Flahault, Ann. Sci. Nat. VII. Bot. 4: 350. 1887. *Zonotrichia rivularis* Nägeli [ex Rabenhorst, ibid. 2: 214. 1865] ex Bornet & Flahault, loc. cit. 1887. —TYPE specimen from Switzerland: Zürich, auf Felsen, *Nägeli 179*, in herb. Kützing (L).

[*Euactis rivularis* β *mollis* Kützing, loc. cit. 1849.] [*E. mollis* Kützing, Tab. Phyc. 2: 25. 1852.] [*E. rivularis* var. *mollis* Kützing ex Jack, Leiner & Stizenberger, Krypt. Badens 3: 106. 1855.] [*Zonotrichia fluviatilis* b *mollis* Rabenhorst, Fl. Eur. Algar. 2: 215. 1865.] —TYPE specimen from Switzerland: Zürich, auf Steinen in Bächen, *Nägeli 225*, in herb. Kützing (L).

Euactis rivularis γ *fluviatilis* Kützing [Sp. Algar., p. 342. 1849] ex Bornet & Flahault, loc. cit. 1887. *E. fluviatilis* Kützing [Tab. Phyc. 2: 25. 1852] ex Bornet & Flahault, loc. cit. 1887. *E. rivularis* var. *fluviatilis* Stizenberger [in

Rabenhorst, Alg. Sachs. 33–34: 332. 1853] ex Forti, Syll. Myxoph., p. 671. 1907. *Zonotrichia fluviatilis* Rabenhorst [Fl. Eur. Algar. 2: 214. 1865] ex Bornet & Flahault, ibid. 4: 351. 1887. *Isactis fluviatilis* Kirchner [Krypt.-Fl. Schles. 2(1): 223. 1878] ex Forti, ibid., p. 670. 1907. *Rivularia haematites* var. *fluviatilis* Kirchner ex Hansgirg, Sitzungsber. K. Böhm. Ges. Wiss., Math.-Nat. Cl. 1890(2): 126. 1891. *R. haematites* fa. *fluviatilis* Poliansky in Elenkin, Monogr. Alg. Cyanoph., Pars Spec. 2: 1131. 1949. –TYPE specimen from Switzerland: Schaffhausen, am Rhein, *Nägeli 226*, in herb. Kützing (L).

Euactis Regeliana Nägeli [in Kützing, Sp. Algar., p. 343. 1849] ex Bornet & Flahault, ibid. 4: 350. 1887. *Zonotrichia Regeliana* Rabenhorst [ibid. 2: 216. 1865] ex Bornet & Flahault, ibid. 4: 351. 1887. –TYPE specimen from Switzerland: Zürich, auf Felsen in Bergbächen, *Nägeli 309*, in herb. Kützing (L).

Euactis Heeriana Nägeli [in Kützing, loc. cit. 1849] ex Bornet & Flahault, ibid. 4: 350. 1887. *Zonotrichia Heeriana* Rabenhorst [loc. cit. 1865] ex Bornet & Flahault, ibid. 4: 351. 1887. –TYPE specimen from Switzerland: Zürich, auf Felsen in Bergbächen, *Nägeli 310*, in herb. Kützing (L).

Schizosiphon sabulicola A. Braun [in Kützing, ibid., p. 894. 1849] ex Bornet & Flahault, ibid. 3: 366. 1886. *Calothrix sabulicola* Kirchner [Krypt.-Fl. Schles. 2(1): 220. 1878] ex Bornet & Flahault, loc. cit. 1886. *C. parietina* var. *sabulicola* Hansgirg, Prodr. Algenfl. Böhm. 2: 49. 1892. *C. parietina* fa. *sabulicola* Hansgirg ex Forti, Syll. Myxoph., p. 623. 1907. –TYPE specimen from Baden, Germany: Freib·urg, *Braun 123*, in herb. Kützing (L).

Mastigonema pluviale A. Braun (as "Mastichonema") [in Kützing, loc. cit. 1849] ex Bornet & Flahault, loc. cit. 1886. *Calothrix parietina* var. *pluvialis* Hansgirg, Physiol. & Algol. Stud., p. 162. 1887. *C. parietina* fa. *pluvialis* Hansgirg ex Forti, Syll. Myxoph., p. 623. 1907. –TYPE specimen presumably from Germany: auf Mauern wo die Regenwasser bleibt, *Braun 128*, Oct. 1848, in herb. Kützing (L).

Limnactis rivularis Kützing [Sp. Algar., p. 894. 1849] ex Forti, ibid., p. 671. 1907. –TYPE specimen from Germany: ad saxa in rivulis Hercyniae, *Hampe*, in herb. Kützing (L).

Euactis Shuttleworthiana A. Braun [in Kützing, ibid., p. 895. 1849] ex Forti, loc. cit. 1907. [*Zonotrichia fluviatilis* c *Shuttleworthiana* Rabenhorst, Fl. Eur. Algar. 2: 215. 1865.] –TYPE specimen from Switzerland: auf Steinen in Bache bei Bern, *Shuttleworth*, Aug. 1843 (comm. Braun no. 138), in herb. Kützing (L).

Euactis lacustris Nägeli [in Kützing, loc. cit. 1849] ex Bornet & Flahault, Ann. Sci. Nat. VII. Bot. 4: 351. 1887. [*Zonotrichia lacustris* Rabenhorst, loc. cit. 1865.] –TYPE specimen from Switzerland: Zürichsee, an Steinen und Mauern, *Nägeli 446*, in herb. Kützing (L).

Dasyactis torfacea Nägeli [in Kützing, Sp. Algar., p. 895. 1849] ex Bornet &

Flahault, ibid. 4: 352. 1887. *Limnactis torfacea* Rabenhorst [ibid. 2: 212. 1865] ex Forti, Syll. Myxoph., p. 667. 1907. —TYPE specimen from Switzerland: im Torfmoor von Dübindorf, Zürich, *Nägeli 528,* in herb. Kützing (L).

Dasyactis brunnea Nägeli [in Kützing, loc. cit. 1849] ex Bornet & Flahault, ibid. 4: 349. 1887. *Zonotrichia brunnea* Rabenhorst [ibid. 2: 217. 1865] ex Bornet & Flahault, loc. cit. 1887. —TYPE specimen from Switzerland: Zürich, Sihlwald, an Felsen in Bächen, *Nägeli 463,* in herb. Kützing (L).

[*Coenocoleus cirrhosus* Berkeley & Thwaites in Smith & Sowerby, Suppl. Engl. Bot. 4: 2940. 1849] —TYPE specimen, attached to no. 2940 in the collection of Sowerby's original plates for *English Botany,* presumably from Snowdon lakes, Wales (BM).

[*Mastigonema subulatum* Kützing (as "Mastichonema"), Tab. Phyc. 2: 14. 1851.] [*M. caespitosum* fa. *tenuius* Rabenhorst (as "tenuior"), Fl. Eur. Algar. 2: 226. 1865.] —TYPE specimen, here designated, since it is the only material labeled thus in Kützing's herbarium, from Germany: auf Steinen im Torfmoor, Schwaben, *Schnurmann* (L).

Physactis spirifera Kützing [ibid. 2: 20. 1852] ex Bornet & Flahault, Ann. Sci. Nat. VII. Bot. 4: 370. 1887. *Rivularia Pisum* f *spirifera* Rabenhorst (as "sperifera"), ibid. 2: 207. 1865. —TYPE specimen from Germany: Neudamm, *Itzigsohn,* in herb. Kützing (L).

[*Physactis terebralis* Kützing, loc. cit. 1852.] *Rivularia terebralis* Kützing ex Rabenhorst, loc. cit. 1865. *Potarcus terebralis* Kuntze (as "Portacus"), Rev. Gen. Pl. 2: 912. 1891. —TYPE specimen from Germany: Weissensee bei Berlin, *A. Braun,* Aug. 1851, in herb. Kützing (L).

Limnactis flagellifera Kützing [Tab. Phyc. 2: 21. 1852] ex Bornet & Flahault, ibid. 4: 348. 1887. [*L. Schnurmannii* b *flagellifera* Rabenhorst, Fl. Eur. Algar. 2: 210. 1865.] *Rivularia minutula* var. *flagellifera* Hansgirg, Prodr. Algenfl. Böhm. 2: 46. 1892. —TYPE specimen from Calvados, France: Falaise, *Lenormand 100,* in herb. Kützing (L).

Limnactis Schnurmannii A. Braun (as "Schnurmanni") [in Kützing, loc. cit. 1852] ex Bornet & Flahault, loc. cit. 1887. —TYPE specimen from Baden, Germany: in Neustadter Torfmoor auf Steinen, *Schnurmann,* in herb. Kützing (L).

Rivularia microscopica Dickie in Sutherland, Journ. Voy. Baffin's Bay & Barrow Str. 2: cxciii. 1852.—TYPE specimen with *Enteromorpha compressa* from Parry islands, Franklin district, Canada: Assistance bay, *P. C. Sutherland 58/1,* Aug. 1851 (BM).

Limnochlide flos-aquae var. *fulva* Auerswald [in Rabenhorst, Alg. Sachs. 41–42: 410. 1854] ex Bornet & Flahault, Ann. Sci. Nat. VII. Bot. 7: 241. 1888. [*Sphaerozyga flos-aquae* var. *fulva* Auerswald ex Rabenhorst, Krypt.-Fl. Sachs. 1: 100. 1863.] —TYPE specimen from Germany: Leipzig, *Auerswald,* 24 Oct. 1854 (PH).

Symphyosiphon Castellii Massalongo [Flora, N. R. 13(1:16): 243. 1855] ex Forti, Syll. Myxoph., p. 539. 1907. *Scytonema Castellii* Rabenhorst [Fl. Eur. Algar. 2: 261. 1865] ex Bornet & Flahault, ibid. 5: 113. 1887. *Calothrix Castellii* Bornet & Flahault [Mém. Soc. Nat. Sci. Nat. & Math. Cherbourg 25: 203. 1885] ex Bornet & Flahault, Ann. Sci. Nat. VII. Bot. 3: 369. 1886. —TYPE specimen from Italy: aquis thermalibus opp. Calderani (Caldiero), prov. Veronensis, *A. Massalongo* (PC).

Calothrix Cesatii Rabenhorst [Alg. Sachs. 43–44: 428. 1855] ex Bornet & Flahault, ibid. 3: 371. 1886. *Schizosiphon Cesatianus* Rabenhorst [Fl. Eur. Algar. 2: 237. 1865] ex Bornet & Flahault, ibid. 3: 372. 1886. —TYPE specimen from Piemonte, Italy: Felsen und Steinen der Sturzbäche um Biella, *Cesati,* Aug. 1854, in Rabenhorst, Algen Sachsens no. 428 (PH).

Lyngbya bugellensis Rabenhorst [Alg. Sachs. 43–44. 1855] ex Gomont, Ann. Sci. Nat. VII. Bot. 16: 153. 1892. —TYPE specimen from Piemonte, Italy: Oropa, an Felsen und Steinen der Sturzbäche, *Cesati,* Aug. 1854 (FH).

Schizosiphon hirundinosus Cesati [in Rabenhorst, ibid. 53–54: 534. 1856] ex Bornet & Flahault, ibid. 3: 366. 1886. —TYPE specimen from Piemonte, Italy: Vercelli, in Kiesgruben, *Cesati,* Mai 1856, in Rabenhorst, Algen Sachsens no. 534 (FH).

Sclerothrix Rousseliana Montagne [Ann. Sci. Nat. IV. Bot. 6: 184. 1856] ex Bornet & Flahault, Ann. Sci. Nat. VII. Bot. 4: 367. 1887. —TYPE specimen from France: Belle-Croix, apud fontem Bellaqueum, *Roussel,* 29 Jul. 1856, in herb. Montagne (PC).

Rivularia Boryana fa. *congesta* Suringar, Obs. Phycol., p. 35. 1857. [*Gloeotrichia durissima* c *congesta* Rabenhorst, Fl. Eur. Algar. 2: 205. 1865.] *Rivularia pygmaea* c *congesta* Rabenhorst ex Forti, Syll. Myxoph., p. 654. 1907. —TYPE specimen from the Netherlands: aan waterplanten, Horten om Leeuwarden, *Suringar DDD94b,* Aug. 1854 (L).

Physactis villosa fa. *obsoleta* Suringar [ibid., p. 37. 1857] ex Forti, ibid., p. 656. 1907. —TYPE specimen from the Netherlands: prope Leeuwarden (Miedum), *Suringar DDD93b,* Jul. 1854 (L).

Schizosiphon sociatus Suringar [ibid., p. 38. 1857] ex Bornet & Flahault, ibid. 3: 375. 1886. —TYPE specimen from the Netherlands: Giekerk, *Suringar DDD92,* Jul. 1854 (L).

Sirosiphon Heppii Rabenhorst [Alg. Sachs. 61–62: 610. 1857] ex Bornet & Flahault, ibid. 5: 78. 1887. [*Scytonema leptosiphon* Stizenberger ex Rabenhorst, Alg. Eur., Alphab. Verz. d. Gatt. & Art., p. 13. 1860.] *Scytonema Heppii* Rabenhorst ex Bornet & Flahault, ibid. 5: 113. 1887. —TYPE specimen from Switzerland: auf einem Kalkschieferblocke in einer feuchten Wiese bei Kiffurchweil, *Hepp,* in Rabenhorst, Algen Sachsens no. 610 (FH).

Rivularia angulosa fa. *dura* Hilse in Rabenhorst, Alg. Sachs. 65–66: 648. 1857. *R. angulosa* var. *dura* Hilse ex Rabenhorst, Fl. Eur. Algar. 2: 201. 1865. —TYPE specimen from Silesia, Poland: bei Strehlen, *Hilse,* in Rabenhorst, Algen Sachsens no. 648 (FH).

Physactis mexicana Kützing [Ostern-progr. Realsch. Nordhausen 1862–63: 9. 1863] ex Forti, Syll. Myxoph., p. 676. 1907. *Rivularia mexicana* Rabenhorst, ibid. 2: 222. 1865. —TYPE specimen from Mexico: in lagunis, Orizaba, *F. Müller,* in herb. Kützing (L).

Inomeria fusca Kützing [loc. cit. 1863] ex Forti, ibid., p. 358. 1907. *Inactis fusca* Kützing ex Forti, loc. cit. 1907. —TYPE specimen from Germany: Quellbach in Baiern, *Martens,* 1854, in herb. Kützing (L).

Schizosiphon Vieillardii Kützing [loc. cit. 1863] ex Bornet & Flahault, Ann. Sci. Nat. VII. Bot. 4: 356. 1886. *Rivularia Vieillardii* Bornet & Flahault, Mém. Soc. Nat. Sci. Nat. & Math. Cherbourg 25: 206. 1885. —TYPE specimen from New Caledonia: *Vieillard 2008,* in herb. Kützing (L).

Rivularia insignis Sprée in Rabenhorst, Alg. Eur. 145–146: 1452. 1863. —TYPE specimen from the Netherlands: in fossa prope Rhenen, *T. Sprée,* in Rabenhorst, Algen Europas no. 1452 (FH).

Rivularia Pisum b *saccata* Rabenhorst, Krypt.-Fl. Sachs. 1: 102. 1863. [*Gloeotrichia Pisum* b *saccata* Rabenhorst ex Kirchner, Krypt.-Fl. Schles. 2(1): 222. 1878.] —TYPE specimen from Sachsen, Germany: in Gräben an Wasserpflanzen, in Rabenhorst, Algen Sachsens no. 36 (FH).

Ainactis gothica Areschoug [Alg. Scand. Exs., ser. 2, 5: 234. 1864] ex Bornet & Flahault, Ann. Sci. Nat. VII. Bot. 4: 350. 1887. [*Zonotrichia haematites* b *gothica* Rabenhorst, Fl. Eur. Algar. 3: 422. 1868.] —TYPE specimen in Areschoug, Algae Scandinavicae Exsiccatae, ser. 2, no. 234 (FH).

Rivularia lacustris Cramer ex Rabenhorst, Hedwigia 3(4): 60. 1864; in Wartmann & Schenk, Schweiz. Krypt. 7: 347. 1864. —TYPE specimen from Switzerland: Katzensee, Zürich, *C. Cramer,* 31 Jul. 1863, in herb. Bornet–Thuret (PC).

[*Gloeotrichia durissima* b *minuta* Rabenhorst, Fl. Eur. Algar. 2: 204. 1865.] —TYPE specimen from Switzerland: St. Gallen, *Wartmann,* Oct. 1859, 1860, in Rabenhorst, Algen Europas no. 1095 (PH).

Gloeotrichia parvula b *Westendorpii* Rabenhorst [ibid. 2: 205. 1865] ex Bornet & Flahault, Rev. Nostocac. Hétéroc., Table, p. 6. 1888.—TYPE specimen from Belgium: fossé à Grimberghe près de Termonde, in Westendorp & Wallays, Herbier Cryptogamique Belge no. 1345 (FH).

Rivularia villosa fa. *minor* Rabenhorst, Fl. Eur. Algar. 2: 207. 1865. —TYPE specimen from Calvados, France: prope Falaise, *A. De Brébisson,* in Rabenhorst, Algen Europas no. 2184 (F).

Zonotrichia pulchra Rabenhorst [Fl. Eur. Algar. 2: 217. 1865] ex Bornet & Flahault, Ann. Sci. Nat. VII. Bot. 4: 348. 1887. *Dasyactis pulchra* Nägeli *pro synon.* [in Rabenhorst, loc. cit. 1865] ex Forti, Syll. Myxophyc., p. 673. 1907.

—TYPE specimen from Switzerland: Engadin, St. Moriz, *Nägeli,* in herb. Bornet–Thuret (PC).

[*Mastigonema caespitosum* fa. *gracillimum* Rabenhorst (as "gracillima"), ibid. 2: 226. 1865.] —TYPE specimen from Silesia, Poland: Felsen im Steinbruch auf dem Galgenberge bei Strehlen, *Hilse,* in Rabenhorst, Algen Sachsens no. 871 (PH).

Mastigonema pluviale var. *Kemmleri* Rabenhorst [Fl. Eur. Algar. 2: 227. 1865] ex Bornet & Flahault, ibid. 3: 366. 1886. —TYPE specimen from Württemberg, Germany: bei Eschenau (Hall) auf Kalkgesteinplatten neben der Bühler, *Kemmler,* Apr. 1858, in Rabenhorst, Algen Sachsens no. 733 (F).

Mastigonema Bauerianum Grunow [in Rabenhorst, Fl. Eur. Algar. 2: 227. 1865] ex Forti, Syll. Myxoph., p. 640. 1907. —TYPE specimen from Germany: ad Berolinum, *Bauer,* in herb. Grunow (W).

Schizosiphon intertextus Grunow [in Rabenhorst, ibid. 2: 236. 1865] ex Bornet & Flahault, Ann. Sci. Nat. VII. Bot. 3: 376. 1886. *Calothrix intertexta* Kirchner [Krypt.-Fl. Schles. 2(1): 220. 1878] ex Hansgirg, Prodr. Algenfl. Böhm. 2: 52. 1892. *C. Orsiniana* var. *intertexta* Hansgirg, loc. cit. 1892. *Dichothrix Orsiniana* var. *intertexta* Hansgirg ex Forti, ibid., p. 642. 1907. —TYPE specimen from Silesia, Poland: in Aupegrunde im Riesengebirge, *Hilse,* Aug. 1859, in Rabenhorst, Algen Europas no. 1177 (W).

Schizosiphon Bauerianus Grunow [in Rabenhorst, Fl. Eur. Algar. 2: 238. 1865] ex Bornet & Flahault, loc. cit. 1886. *Dichothrix Baueriana* Bornet & Flahault [Mém. Soc. Nat. Sci. Nat. & Math. Cherbourg 25: 204. 1885] ex Bornet & Flahault, Ann. Sci. Nat. VII. Bot. 3: 375. 1886. *Calothrix Baueriana* Hansgirg, Sitzungsber. K. Böhm. Ges. Wiss., Math.-Nat. Cl. 1890(2): 126. 1891. *Dichothrix orbelica* Petkoff *pro synon.* ex. Hollerbach, Kossinskaia & Poliansky, Opred. Presnov. Vodor. SSSR 2: 371. 1953. —TYPE specimen from Germany: Plötzensee bei Berlin, *Bauer,* in herb. Grunow (W). Fig. 41.

Schizosiphon gracilis Hilse [Jahres-Ber. Schles. Ges. Vaterl. Cultur 1864: 94. 1865; in Rabenhorst, Alg. Eur. 177–178: 1770. 1865] ex Bornet & Flahault, ibid. 3: 366. 1886. —TYPE specimen from Silesia, Poland: am Rande einer Lache am Fuchsberge von Schwoitsch bei Breslau, *Hilse,* 20 Nov. 1864, in Rabenhorst, Algen Europas no. 1770 (F).

Euactis Beccariana De Notaris [Erb. Critt. Ital., ser. 1, 27–28: 1332. 1866] ex Bornet & Flahault, ibid. 4: 356. 1887. *Zonotrichia Beccariana* Rabenhorst [Fl. Eur. Algar. 3: 422. 1868] ex Forti, Syll. Myxoph., p. 663. 1907. *Rivularia Beccariana* Bornet & Flahault, Mém. Soc. Nat. Sci. Nat. & Math. Cherbourg 25: 206. 1885. —TYPE specimen from Emilia, Italy: nel rio Gemese presso il Sasso, nel Bolognese, *Beccari,* Apr. 1864, in Erbario Crittogamico Italiano, ser. 1, no. 1332 (FH). Fig. 59.

Rivularia Marcucciana De Notaris, Erb. Critt. Ital., ser. 1, 27–28: 1331. 1866.

Gloeotrichia Marcucciana Thuret [in Bornet & Thuret, Notes Algol. 2: 170. 1880] ex Bornet & Flahault, Ann. Sci. Nat. VII. Bot. 4: 368. 1887. —TYPE specimen from Toscana, Italy: in una pozza alla spiaggia del mare, presso la bocca di fiume morto, nella Selva Pisana, *Savi & Marcucci,* Sept. 1865, in Erbario Crittogamico Italiano, ser. 1, no. 1331 (FH).

Schizosiphon Rabenhorstianus Hilse [in Rabenhorst, Alg. Eur. 183–184: 1836. 1866] ex Bornet & Flahault, ibid. 4: 348. 1887. —TYPE specimen from Silesia, Poland: in saxis prope Prieborn ad Strehlen, *Hilse,* Sept. 1861, in Rabenhorst, Algen Europas no. 1836 (F).

Schizosiphon Kuehneanus Rabenhorst (as "Kühneanus") [ibid. 185–186: 1851. 1866] ex Bornet & Flahault, ibid. 3: 377. 1886. —TYPE specimen from Rheinland, Germany: Hammerstein am Rhein, *J. Kühn,* in Rabenhorst, Algen Europas no. 1851 (F).

Mastigonema paludosum Crouan (as "Mastichonema paludosa") [Fl. Finistère, p. 116. 1867] ex Forti, Syll. Myxoph., p. 630. 1907. —TYPE specimen from Finistère, France: sur pierres dans un marais montueux, côte nord de Plougastel, 24 Mar. 1854, in herb. Crouan (Lab. Marit., Concarneau).

Gloeotrichia incrustata Wood [Proc. Amer. Philos. Soc. 11: 127. 1869] ex Bornet & Flahault, Ann. Sci. Nat. VII. Bot. 4: 370. 1887. *Rivularia incrustata* Forti, ibid., p. 656. 1907. —TYPE specimen here designated, from near the original locality in Pennsylvania: in pool, Susquehanna river, Wyoming county, *R. R. Grant Jr. 9,* Sept. 1965 (PH).

Dasyactis mollis Wood [ibid. 11: 128. 1869] ex Wolle, Fresh-w. Alg. U. S., p. 249. 1887.—TYPE specimen here designated, from near the original locality in Michigan: Pigeon river at Mullet Lake, Cheboygan county, *H. K. Phinney 27M41/7,* 13 Aug. 1941 (D).

Rivularia cartilaginea Wood, loc. cit. 1869. —TYPE specimen here designated, from near the original locality in Michigan: in lake near the beach pool, Harrison's Landing, Black Lake, Cheboygan county, *Phinney 212,* 11 Aug. 1942 (D).

Phormidium oryzetorum Martens [in Kurz, Proc. Asiatic Soc. Bengal 1870: 12. 1870] ex Gomont, Ann. Sci. Nat. VII. Bot. 16: 191. 1892. —TYPE specimen from Bengal: *S. Kurz 1932,* in herb. Agardh (LD).

[*Mastigonema granulatum* Martens in Kurz, ibid. 1870: 258. 1870.] —TYPE specimen from India: Calcutta, *Kurz 2666,* in herb. Weber (L).

Mastichothrix longissima Crouan [in Mazé & Schramm, Essai Class. Alg. Guadeloupe, ed. 2, p. 31. 1870–77] ex Bornet & Flahault, Ann. Sci. Nat. VII. Bot. 3: 364. 1886. —TYPE specimen from Guadeloupe: dégorgement de l'étang des Pères blancs, *Conquérant,* 25 Feb. 1859, in herb. Bornet–Thuret (PC).

Calothrix submarina Crouan [in Mazé & Schramm, ibid., p. 36. 1870–77] ex

Bornet & Flahault, ibid. 3: 372. 1886. —TYPE specimen from Guadeloupe: Vieux-Fort, anse de la Petite-Fontaine près le Gouffre, *Mazé & Schramm 1076*, 28 Oct. 1860, in herb. Bornet-Thuret (PC).

Schizosiphon Nordstedtianus Rabenhorst [Alg. Eur. 225–226: 2246. 1871] ex Bornet & Flahault, ibid. 3: 377. 1886. —TYPE specimen from Sweden: in rupibus humidis prope Trollhättan, *O. Nordstedt*, in Rabenhorst, Algen Europas no. 2246 (FH).

Zonotrichia mollis Wood [Smithson. Contrib. Knowl. 241: 48. 1872] ex Forti, Syll. Myxoph., p. 675. 1907. —TYPE specimen here designated from the original locality in New York: rocks near the base of the cliff at the west end of Goat island, Niagara Falls, *F. Drouet, H. B. Louderback & A. Owen 14982*, 24 Aug. 1969 (PH).

Zonotrichia parcezonata Wood [ibid. 241: 49. 1872] ex Forti, ibid., p. 670. 1907. —TYPE specimen, with that of the preceding, here designated from the original locality in New York: Niagara Falls, *Drouet, Louderback & Owen 14982* (PH).

Mastigonema sejunctum Wood [ibid. 241: 53. 1872] ex Forti, ibid., p. 631. 1907. —TYPE specimen, with that of *Dasyactis mollis* Wood above, from near the original locality in Michigan: Pigeon river at Mullet lake, Cheboygan county, *H. K. Phinney 27M41/7* (D).

Rivularia echinulus Areschoug, Alg. Scand. Exs., ser. 2, no. 375. 1872. —TYPE specimen from Sweden: in fonte prope angulum septent. lacus Trehörningen, inter Hallqued et Rasbo, *A. Areschoug*, in Areschoug, Algae Scandinavicae Exsiccatae no. 375 (S).

[*Schizosiphon Hendersonii* Dickie (as "Hendersoni") in Henderson & Hume, Lahore to Yarkand, p. 344. 1873.] —TYPE specimen from Tibet: from hot saline springs above Gogra (Nurla), *G. Henderson 5*, 1870 (BM).

Scytonema Rhizophorae Zeller [Journ. Asiatic Soc. Bengal 42(2:3): 183. 1873] ex Bornet & Flahault, Ann. Sci. Nat. VII. Bot. 5: 112. 1887. —TYPE specimen from Burma: Pegu, Elephant Point, on Kambala trees in mangrove swamps, *Kurz 3267* (BM).

Calothrix decipiens Bornet & Thuret [Notes Algol. 1: 12. 1876] ex Bornet & Flahault, ibid. 3: 348. 1886. —TYPE specimen from France: dans le ruisseau de la Brague, Antibes, *G. Thuret*, 18 Mar. 1875, in herb. Bornet–Thuret (PC).

Mastichothrix minuta Reinsch [Journ. Linn. Soc. Bot. 15: 207. 1877] ex Forti, Syll. Myxoph., p. 618. 1907. *Calothrix minuta* Forti, loc. cit. 1907. *C. Borziana* J. de Toni, Noter. Nomencl. Algol. 3: 1. 1936. —TYPE specimen from Kerguelen island: in folia muscorum, *Eaton* (BM).

Schizosiphon kerguelenensis Reinsch [ibid. 15: 211. 1877] ex Forti, ibid., p. 633. 1907. *S. kerguelensis* Reinsch ex Forti, loc. cit. 1907. —TYPE specimen from Kerguelen island: in muscis aquaticis, *Eaton* (BM).

Zonotrichia paradoxa Wolle [Bull. Torrey Bot. Club 6(26): 138. 1877] ex Forti, ibid., p. 672. 1907. *Rivularia paradoxa* Forti, loc. cit. 1907. —TYPE specimen from Pennsylvania: culms of Sagittaria, *F. Wolle* (NY).

Mastigonema violaceum Wolle (as "violacea") [loc. cit. 1877] ex Forti, ibid., p. 619. 1907. *Calothrix violacea* Forti, loc. cit. 1907. *C. Fortii* J. de Toni, Noter. Nomencl. Algol. 1: 6. 1934. —TYPE specimen presumably from Pennsylvania: parasitic on *Lyngbya Wollei* (D).

Mastigonema fuscum Wolle (as "fusca") [loc. cit. 1877] ex Tilden, Minn. Alg. 1: 266. 1910. —TYPE specimen presumably from Pennsylvania: parasitic, *F. Wolle* (D).

Mastigonema luteum Wolle (as "lutea") [ibid. 6(26): 139. 1877] ex Drouet, Field Mus. Bot. Ser. 20(2): 57. 1939. —TYPE specimen presumably from Pennsylvania: stones, limestone springs, *F. Wolle* (D).

Mastichothrix turgida Wolle (as "Mastigothrix") [ibid. 6(35): 184. 1877] ex Forti, Syll. Myxoph., p. 632. 1907. —TYPE specimen, labeled with this name by F. Wolle, presumably from Pennsylvania (D).

Aulosira laxa Kirchner [Krypt.-Fl. Schles. 2(1): 238. 1878] ex Bornet & Flahault, Ann. Sci. Nat. VII. Bot. 7: 256. 1888. —TYPE specimen, here designated in the absence of material labeled thus by O. Kirchner, from Upper Austria: in einem kleinen Wege in einer Wiese, Regauer Wald Umg. Vöcklabruck, *v. Mörl*, 1864, det. E. Bornet in herb. Grunow (W). Fig. 37.

Rivularia pelagica Gobi, Hedwigia 17: 35. 1878. *R. flos-aquae* Gobi *pro synon.*, ibid. 17: 37. 1878. —TYPE specimen from Esthonia: "Rivularia flos-aquae Gobi, misit Gobi 1883," in herb. Bornet—Thuret (PC).

Rivularia fluitans Cohn, Hedwigia 17: 4. 1878; in Rabenhorst, Alg. Eur. 253–255: 2540. 1878. *Gloeotrichia fluitans* Richter, Forschungsber. Biol. Sta. Plön 2: 46. 1894. —TYPE specimen from Pommern, Germany: Lauenburg, *A. Schmidt*, 19–21 Jul. 1877, in Rabenhorst, Algen Europas no. 2540 (PH).

Zonotrichia Marcucciana Rabenhorst [ex Piccone, N. Giorn. Bot. Ital. 10: 314. 1878] ex Bornet & Flahault, Ann. Sci. Nat. VII. Bot. 4: 351. 1887. —TYPE specimen from Sardinia: Tacquitara (Barbargia), dans eaux calcarifères, *Marcucci*, 1866, in Un. Itin. Crypt. no. XI (F).

Gloeotrichia punctulata Thuret [in Bornet & Thuret, Notes Algol. 2: 168. 1880] ex Bornet & Flahault, ibid. 4: 369. 1887. [*G. pustulata* Thuret in Bornet & Thuret *pro synon.*, ibid. 2: 169. 1880.] *Rivularia punctulata* Forti, Syll. Myxoph., p. 652. 1907; Lemmermann, Krypt.-fl. Mark Brandenb. 3: 254. 1907. —TYPE specimen from Manche, France: Cherbourg, mare de Tourlaville, *G. Thuret*, 30 Aug. 1874, in herb. Bornet–Thuret (PC). Fig. 32.

Gloetrichia Rabenhorstii Bornet & Thuret [ibid. 2: 171. 1880] ex Bornet & Flahault, ibid. 4: 368. 1887. *Rivularia Rabenhorstii* Forti, ibid., p. 653. 1907; Lemmermann, ibid. 3: 252. 1907. —TYPE specimen from Brandenburg,

Germany: Neudammer Teiche, *Itzigsohn & Rothe,* Sept. 1856, as "Limnactis dura" in Rabenhorst, Algen Sachsens no. 554 (PH). Fig. 31.

Inactis obscura Dickie [Journ. Linn. Soc. Bot. 18: 126. 1880] ex Gomont, Ann. Sci. Nat. VII. Bot. 15: 329. 1892. —TYPE specimen from Amazonas, Brazil: on sandstone beneath the fall, Cachoeira Grande, Manaos, *Trail,* 27 Dec. 1874 (BM). See Drouet in Amer. Journ. Bot. 25: 660. 1938.

Calothrix Hosfordii Wolle [Bull. Torrey Bot. Club 8(4): 1881] ex Bornet & Flahault, Ann. Sci. Nat. VII. Bot. 3: 370. 1886. *Dichothrix Hosfordii* Bornet in Setchell, Erythaea 1896: 190. 1896. —TYPE specimen from Vermont: wet rocks, La Platte river, Charlotte, *F. H. Hosford 71,* in herb. Bornet–Thuret (PC).

Leptochaete crustacea Borzi [N. Giorn. Bot. Ital. 14(4): 298. 1882] ex Bornet & Flahault, ibid. 3: 342. 1886. *Homoeothrix crustacea* Margalef, Collect. Bot. (Barcelona) 3(3): 231. 1953. *Homoeothrix Margalefii* Komárek & Kalina, Österr. Bot. Zeitschr. 112(3): 424. 1965. —TYPE specimen, here designated, the only material found determined thus by A. Borzi, from Patagonia: på stenen i Ch. Malpasos utlopp, *O. Borge 397,* 23 Feb. 1899 (S).

Microchaete diplosiphon Gomont [Bull. Soc. Bot. France 32: 212. 1885] ex Bornet & Flahault, ibid. 5: 84. 1887. *Coleospermum diplosiphon* Elenkin, Izv. Imp. Bot. Sada Petra Velik. 25(1): 5. 1915. *Fremyella diplosiphon* Drouet, Field Mus. Bot. Ser. 20(2): 32. 1939. —TYPE specimen from Seine-et-Oise, France: cultures d'Oedogoniées rapportée de Lardy, *Bornet & Bonnier,* 1884, prep. 23 Jul. 1885, in herb. Gomont (PC).

Calothrix balearica Bornet & Flahault [Mém. Soc. Nat. Sci. Nat. & Math. Cherbourg 25: 200. 1885] ex Bornet & Flahault, Ann. Sci. Nat. VII. Bot. 3: 348. 1886. *Homoeothrix balearica* Lemmermann, Krypt.-Fl. Mark Brandenb. 3:239. 1907. —TYPE specimen from Minorca, Balearic islands: Pozo de Santa Magdalene, *Feminias 316,* 1 May 1878, in herb. Bornet–Thuret (PC). Fig. 60.

Calothrix stellaris Bornet & Flahault [Mém. Soc. Nat. Sci. Nat. & Math. Cherbourg 25: 202. 1885] ex Bornet & Flahault, Ann. Sci. Nat. VII. Bot. 3: 365. 1886. —TYPE specimen from Uruguay: in paludibus prope Montevideo, *J. Arechaveleta,* Mar. 1884, in herb. Bornet–Thuret (PC).

Dichothrix Nordstedtii Bornet & Flahault [Mém. Soc. Nat. Sci. Nat. & Math. Cherbourg 25: 203. 1885] ex Bornet & Flahault, Ann. Sci. Nat. VII. Bot. 3: 374. 1886. *Homoeothrix caespitosa* Kirchner in Engler & Prantl, Natürl. Pflanzenfam. 1(1A): 87. 1900. *Diplothrix Nordstedtii* Bornet & Flahault *pro synon.* ex Kirchner, loc. cit. 1900. —TYPE specimen from Norway: ad saxa in rivulo ad Odde in Hardanger, *O. Nordstedt,* Jul. 1872, as "Calothrix cespitosa" in Rabenhorst, Algen Europas no. 2315, in herb. Bornet–Thuret (PC). Fig. 50.

Calothrix Braunii Bornet & Flahault [Mém. Soc. Nat. Sci. Nat. & Math. Cherbourg 25: 203. 1885] ex Bornet & Flahault, Ann. Sci. Nat. VII. Bot. 3:

368. 1886. —TYPE specimen from Baden, Germany: Donaueschingen, *A. Braun*, Oct. 1849, in herb. Bornet—Thuret (PC). Fig. 35.

Calothrix adscendens Bornet & Flahault [Mém. Soc. Nat. Sci. Nat. & Math. Cherbourg 25: 203. 1885] ex Bornet & Flahault, Ann. Sci. Nat. VII. Bot. 3: 365. 1886. *Mastigonema adscendens* Nägeli *pro synon.* ex Bornet & Flahault, loc. cit. 1886. *M. parasiticum* Wolle *pro synon.* ex Setchell, Erythaea 7(5): 46. 1899. —TYPE specimen from Berlin, Germany: in *Hypno scorpioide* in paludosis prope Moabit, *A. Braun*, Aug. 1851, in herb. Bornet—Thuret (PC). Fig. 47.

Isactis caespitosa fa. *tenuior-viridis* Rabenhorst ex Wolle, Fresh-w. Alg. U. S., p. 245. 1887.—TYPE specimen here designated in the absence of the material indicated by Wolle, from Pennsylvania: *I. caespitosa*, from stones, large spring, Nazareth, *F. Wolle*, 10 Jul. 1887 (D).

Gloeotrichia solida Richter in Hauck & Richter, Phyk. Univ. 2: 83. 1887. *G. Pisum* var. *solida* Hansgirg, Prodr. Algenfl. Böhm. 2: 44. 1892. *Rivularia Pisum* var. *solida* Hansgirg ex Forti, Syll. Myxoph., p. 653. 1907. —TYPE specimen from Saxony, Germany: prope Lipsiam, *H. Reichel*, 26 Sept. 1886, in Hauck & Richter, Phykotheka Universalis no. 83B (PH).

Scytonema Hofmannii var. *calcicola* Hansgirg (as "calcicolum"), Sitzungsber. K. Böhm. Ges. Wiss., Math.-Nat. Cl. 1890(2): 97. 1891. *S. Hofmannii* fa. *calcicola* Kossinskaia in Elenkin, Monogr. Alg. Cyanoph., Pars Spec. 1: 913. 1938. —TYPE specimen from Bohemia, Czechoslovakia: auf feuchten Kalkstein-felsen an der Prag-Duxer Bahn bei Nová Ves nächst St. Procop, *A. Hansgirg*, May 1890 (W).

Tolypothrix rivularis Hansgirg, ibid. 1891: 337. 1891. —TYPE specimen from Steiermark, Austria: in einem Bächlein bei Rudersdorf nächst Graz, *Hansgirg*, Sept. 1890 (W).

Calothrix Baueriana var. *minor* Hansgirg, Prodr. Algenfl. Böhm. 2: 53. 1892. *Dichothrix Baueriana* var. *minor* Hansgirg ex Forti, Syll. Myxoph., p. 640. 1907. *Calothrix Baueriana* fa. *minor* Poliansky in Elenkin, ibid. 2: 1118. 1949. —TYPE specimen from Bohemia, Czechoslovakia: auf Plänerkalkfelsen am Rande des Žehuner-Teiches nächst Chlumec, *Hansgirg*, Jul. 1885 (W).

Dichothrix Nordstedtii var. *salisburgensis* Beck, Ann. K. K. Naturh. Hofmus. Wien 9: 137. 1894. *Homeoethrix caespitosa* fa. *salisburgensis* Poliansky ex Star-mach, Fl. Slodkow. Polski 2: 623. 1966. —TYPE specimen from Salzburg, Austria: in fontibus et aquis nivalibus prope lacum Palfnersee supra Wild-bad-Gastein, *G. Beck*, in Museum Palatinum Vindobonense, Kryptogamae Exsiccatae, Algae no. 73 (W).

Rivularia Bornetiana Setchell, Bull. Torrey Bot. Club 22(10): 426. 1895. —TYPE specimen from Rhode Island: on Ruppia, brackish water, Watch Hill pond, *W. A. Setchell 526*, 3 Sept. 1892 (UC).

Calothrix stagnalis Gomont, Journ. de Bot. 9(11): 201. 1895. —TYPE specimen from Maine-et-Loire, France: étang de St.-Nicolas près Angers, *M. Gomont*, 11 Jul. 1894, in herb. Gomont (PC).

Dichothrix calcarea Tilden, Amer. Alg. 2: 165. 1896. *Calothrix calcarea* Starmach. Fl. Slodkow. Polski 2: 612. 1966. —TYPE specimen from Minnesota: sides of wooden tank, Minneapolis, *J. E. Tilden*, 1 Oct. 1895, in Tilden, American Algae no. 165 (MIN).

Aphanizomenon holsaticum Richter, Hedwigia 35: 273, 274. 1896; in Hauck & Richter, Phyk. Univ. 15: 746. 1896. *A. flos-aquae* fa. *holsaticum* Elenkin, Monogr. Alg. Cyanoph., Pars Spec. 1: 853. 1938. —TYPE specimen from Holstein, Germany: Oldesloe in einem Teiche, *C. Sonder*, Dec. 1895, in Hauck & Richter, Phykotheka Universalis no. 746 (NY). Fig. 34.

Ammatoidea Normanii W. & G. S. West, Journ. R. Microsc. Soc. London 1897: 506. 1897 (as "Hammatoidea" in Lemmermann, Krypt-Fl. Mark Brandenb. 3: 256. 1910). —TYPE specimen from Devon, England: Dartmoor, *T. Norman*, in the slide collection (BM). Terminal heterocysts are present on some trichomes on this slide. Fig. 48.

Gloeotrichia aethiopica W. & G. S. West, Journ. of Bot. 35: 240. 1897. *Rivularia aethiopica* Forti, Syll. Myxoph., p. 651. 1907. —TYPE specimen from Angola: Mossâmedes, in rivo Caroca prope Cabo Negro, *F. Welwitsch 19*, Sept. 1859 (UC).

Calothrix epiphytica W. & G. S. West, loc. cit. 1897. —TYPE specimen from Angola: Mossâmedes, in stagnis ad ripas flum. Bero, *Welwitsch 190*, Aug. 1859 (UC).

Dichothrix utahensis Tilden, Amer. Alg. 3: 288. 1898. —TYPE specimen from Utah: stream into Great Salt lake one mile northeast of Black Rock, Garfield Beach, *Tilden*, 8 Aug. 1897, in Tilden, American Algae no. 288 (MIN).

Rivularia compacta Collins in Collins, Holden & Setchell, Phyc. Bor.-Amer. 11: 508. 1898. —TYPE specimen from Massachusetts: Spot pond, Middlesex Fells, *F. S. Collins*, Jul. 1890 (NY).

Tolypothrix calcarata fa. *minor* Schmidle in Simmer, Allg. Bot. Zeitschr. 1899(12): 193. 1899. *Hassallia calcarata* fa. *minor* Schmidle ex Forti, Syll. Myxoph., p. 554. 1907. *H. calcarata* var. *minor* Schmidle ex Migula, Krypt.-Fl. 2(1): 135. 1907.—TYPE specimen from Kärnten, Austria: Kreutzeckgebiet, im Kolbitsch, 900 m., Kalk and Tuff, *H. Simmer*, 29 Aug. 1898 (PC).

Gloeotrichia indica Schmidle, Allg. Bot. Zeitschr. 1900(3): 35. 1900. *Rivularia indica* Forti, ibid., p. 652. 1907; Lemmermann, Krypt.-Fl. Mark Brandenb. 3: 254. 1907. —TYPE specimen from India: in lacu prope Igatpuri, *A. Hansgirg*, in Museum Palatinum Vindobonense Kryptogamae Exsiccatae, Algae no. 221 (W).

Dichothrix montana Tilden, Amer. Alg. 6: 572. 1902. *Calothrix montana* Poliansky in Elenkin, Monogr. Alg. Cyanoph., Pars Spec. 2: 1111. 1949.

—TYPE specimen from Montana: on rocks in hot water, Lo Lo hot springs, *D. Griffiths,* 17 Sept. 1898, in Tilden, American Algae no. 572 (MIN).

Calothrix rupestris Royers, Jahres-Ber. Naturw. Ver. Elberfeld 12: 40. 1906. —TYPE specimen from Rheinland, Germany: Ostufer des Laacher-Sees, *H. Royers,* 20 Jun. 1905, mixed sparingly with *Scytonema Hofmannii* Ag. (F).

Calothrix adscendens fa. *culta* Teodorescu, Ann. Sci. Nat. IX. Bot. 5: 17. 1907. —TYPE specimen from Ilfov district, Romania: e lacu non procul a monasterio Caldarusani, in cubiculo colui, *E. C. Teodorescu,* Apr. 1901–3 Feb. 1902 (W).

Calothrix brevissima G. S. West, Journ. Linn. Soc. Bot. 38: 180. 1907. —TYPE specimen from Tanzania: in plankton of Victoria Nyanza near Bukoba, *W. A. Cunnington 252,* Apr. 1905 (UC).

Calothrix cartilaginea G. S. West, ibid. 38: 181. 1907. *Homoeothrix cartilaginea* Lemmermann, Krypt.-Fl. Mark Brandenb. 3: 684. 1907. —TYPE specimen from Tanzania: scrapings from canoe, Lake Tanganyika, *Cunnington 205,* 6 Jan. 1905, in the slide collection (BM). Heterocysts are well developed in some trichomes of this material.

Rivularia globiceps G. S. West, ibid. 38: 182. 1907. *R. aquatica* fa. *globiceps* Geitler, Arch. f. Hydrobiol. Suppl.-Bd. 12(6): 441. 1935. —TYPE specimen from Zambia: in Kituta bay, Lake Tanganyika, *Cunnington 74,* 24 Aug. 1904 (UC).

Gloeotrichia longiarticulata G. S. West, ibid. 38: 183. 1907. *Rivularia longiarticulata* Lemmermann, loc. cit. 1907. —TYPE specimen from Malawi: floating, Anchorage bay, Lake Nyasa, *Cunnington 1,* 10 Jun. 1904, in the slide collection (BM).

Rivularia intermedia Lemmermann, Krypt.-Fl. Mark Brandenburg 3: 253. 1907. *Gloeotrichia intermedia* Geitler in Pascher, Süsswasserfl. 12: 233. 1925. —TYPE specimen, here designated in the absence of material in the Lemmermann herbarium, from the original locality in Berlin, Germany: with the Type of *Calothrix adscendens* Born. & Flah., in *Hypno scorpioide* in paludosis prope Moabit, *A. Braun,* Aug. 1851, in herb. Bornet–Thuret (PC).

Calothrix scytonemicola Tilden, Minn. Alg. 1: 265. 1910. —TYPE specimen from Hawaii: on Scytonema on stagnant water on beach, Meheiwa, Makao, Koolauloa, Oahu, *J. E. Tilden,* 20 Jun. 1900, in Tilden, American Algae no. 480 (MIN).

Calothrix intricata Fritsch, Nat. Antarct. Exped., Nat. Hist. 6(Fresw. Alg.): 36. 1912. —TYPE specimen from Antarctica: Alg. Discovery 23, in the slide collection (BM).

Calothrix gracilis Fritsch, ibid. 6(Fresw. Alg.): 37. 1912. *C. Fritschii* J. de Toni, Noter. Nomencl. Algol. 1: 6. 1934. —TYPE specimen from Antarctica: Alg. Discovery 23, in slide collection (BM).

Calothrix minuscula Weber-van Bosse, Liste d. Alg. d. Siboga, p. 42. 1913.

—TYPE specimen from Celebes, Indonesia: op draagalg., Meer van Tempé, *A. A. Weber-van Bosse 862,* 1888 (L).

Rivularia oolithica Bremekamp, Teysmannia 25: 71. 1914. —TYPE specimen from eastern Java, Indonesia: in fonte Tjeding in monte Idjen, *C. E. B. Bremekamp,* in Mineralogisch-Geologisch Instituut, Utrecht.

Campylonema lahorense Ghose, New Phytol. 19: 35. 1920. *Schmidleinema lahorense* J. de Toni, Noter. Nomencl. Algol. 8: [4]. 1936. *Camptylonemopsis lahorensis* Desikachary, Proc. Indian Acad. Sci. 28(2B): 43. 1948. —TYPE specimen, erroneously interpreted by me in Monogr. Acad. Nat. Sci. Philadelphia 15: 102 (1968) as *Schizothrix Friesii* (Ag.) Gom., from Lahore, Pakistan, in the slide collection (BM).

Dichothrix minima Setchell & Gardner in Gardner, Univ. Calif. Publ. Bot. 6(17): 474. 1918. *Calothrix subminima* Kossinskaia, Opred. Morsk. Sinez. Vodor., p. 160. 1948. —TYPE specimen from Washington: near Fairhaven near Bellingham, *W. A. Setchell & N. L. Gardner 544,* Jul. 1899 (UC).

Microchaete naushonensis Collins, Rhodora 20: 141. 1918. *Fremyella naushonensis* J. de Toni, Diagn. Alg. Nov. 1(3): 233. 1938. —TYPE specimen from Naushon island, Massachusetts: on Sphagnum and Oedogonium, Tarpaulin pond, *J. M. Farber,* Aug. 1917 (NY).

Aulosira fertilissima Ghose, Journ. Linn. Soc. Bot. 46: 342. 1923. *Ghosea fertilissima* Cholnoki, Bol. Soc. Portug. Ciênc. Nat., ser. 2, 4(1): 92. 1952. —TYPE specimen from Shalamar gardens, Lahore, Pakistan: no. 326 in the collection of F. E. Fritsch (BM).

Calothrix gelatinosa Fritsch & Rich, Trans. R. Soc. South Africa 11: 373. 1924. —TYPE specimen from South Africa: in dripping water, Sweetwaters Bush, Maritzburg, *J. W. Bews 92,* 2 Aug. 1915 (BM).

Calothrix scytonemicola var. *brasiliensis* Borge, Ark. f. Bot. 19(17): 5. 1925. —TYPE specimen from Mato Grosso, Brazil: Cáceres, *F. C. Hoehne 52,* Jan. 1914 (S).

Calothrix linearis Gardner, Rhodora 28: 3. 1926. —TYPE specimen from Fukien, China: vicinity of University of Amoy, *H. H. Chung A54* (FH).

Microchaete tapahiensis Setchell, Univ. Calif. Publ. Bot. 12(4): 66. 1926. *Fremyella tapahiensis* J. de Toni, Noter. Nomencl. Algol. 8: [4]. 1936. —TYPE specimen from Tahiti: on leaves wet with drip from cliffs, Tapahi, *W. A. Setchell 5045,* 23 May 1922 (UC).

Anabaena unispora Gardner, Mem. New York Bot. Gard. 7: 59. 1927. *A. Volzii* var. *unispora* Bourrelly, Bull. Inst. Franç. Afr. Nord, sér. A, 19(4): 1049. 1957. —TYPE specimen from Puerto Rico: pool near the park, Santurce, *N. Wille 49a,* 28 Dec. 1914 (NY). Fig. 36.

Nodularia epiphytica Gardner, ibid. 7: 65. 1927. —TYPE specimen from Puerto Rico: on limestone, Hato Arriba, Arecibo, *Wille 1415a,* 1 Mar. 1915 (NY).

Calothrix simplex Gardner, ibid. 7: 66. 1927. —TYPE specimen from Puerto Rico: stream near San Lorenzo, *Wille 501,* 18 Jan. 1915 (NY).

Calothrix parietina var. *torulosa* Gardner, ibid. 7: 67. 1927. —TYPE specimen from Puerto Rico: in Río Grande near Sabana Grande, *Wille 914,* 9 Feb. 1915 (NY).

Calothrix Braunii var. *mollis* Gardner, ibid. 7: 68. 1927. *C. Braunii* fa. *mollis* Poliansky in Elenkin, Monogr. Alg. Cyanoph., Pars Spec. 2: 1068. 1949. —TYPE specimen from Puerto Rico: on earth, Maricao, *Wille 1037b,* 13 Feb. 1915 (NY).

Calothrix linearis Gardner, loc. cit. 1927. *C. intermedia* Gardner, New York Acad. Sci., Sci. Surv. Porto Rico 8(2): 305. 1932. *C. Gardneri* J. de Toni, Noter. Nomencl. Algol. 1: 6. 1934. —TYPE specimen from Puerto Rico: water pump in Maricao, *Wille 1276a,* 21 Feb. 1915 (NY).

Calothrix juliana var. *tenuior* Gardner, Mem. New York Bot. Gard. 7: 69. 1927. *Homoeothrix juliana* var. *tenuior* J. de Toni, Gen. & Sp. Myxoph. 1(D–L): 280. 1949. —TYPE specimen from Puerto Rico: in Arroyo de los Corchos, *Wille 1696,* 13 Mar. 1915 (NY).

Calothrix evanescens Gardner, loc. cit. 1927. —TYPE specimen from Puerto Rico: in water, Jayuya, *Wille 1776a,* 15 Mar. 1915 (NY).

Dichothrix Willei Gardner, ibid. 7: 70. 1927. —TYPE specimen from Puerto Rico: in ditch 10 km. north of Utuado, *Wille 1554,* 6 Mar. 1915 (NY).

Calothrix simulans Gardner, loc. cit. 1927. —TYPE specimen from Puerto Rico: west of Experiment Station, Río Piedras, *Wille 1946,* 23 Mar. 1915 (NY).

Rivularia flagelliformis Gardner, ibid. 7: 71. 1927. *Gloeotrichia flagelliformis* Gardner ex Geitler, Rabenh. Krypt.-Fl., ed. 2, 14: 636. 1932. —TYPE specimen from Puerto Rico: in a water reservoir near Río Piedras, *Wille 126,* 28 Dec. 1914 (NY).

Microchaete tenera var. *tenuior* Gardner, Mem. New York Bot. Gard. 7: 71. 1927. *Fremyella tenera* var. *tenuior* J. de Toni, Archivio Bot. 15(3–4): 290. 1939. —TYPE specimen from Puerto Rico: by the road near Adjuntas, *E. G. Britton 1571* (NY).

Dichothrix Chungii Gardner, Univ. Calif. Publ. Bot. 14(1): 6. 1927. —TYPE specimen from Fukien, China: rocks in a mountain ravine, Kushan near Foochow, *H. H. Chung A329a* (FH).

Calothrix gypsophila fa. *saccoideo—fruticulosa* Poliansky, Trudy Bot. Inst. Akad. Nauk SSSR, ser. 2, 2: 26. 1935. —TYPE specimen from Massachusetts: in Suntaug lake, Lynnfield, *W. A. Setchell,* 8 Sept. 1892, in Wittrock & Nordstedt, Algae Aquae Dulcis Exsiccatae no. 1309 (PH).

Calothrix aequalis Skuja in Handel-Mazzetti, Symb. Sinic. 1:19. 1937. —TYPE specimen from Szechwan, China: ad orientem Ningyüen prope

Lemoka ad rupes fonte calido irrigatas, *H. Handel-Mazzetti 1620*, 23 Apr. 1914 (W).

Calothrix micromeres Skuja in Handel-Mazzetti, ibid. 1: 20. 1937. —TYPE specimen from Yunnan, China: in lacu ad Yungning, in lapidibus submersis, *H. Handel-Mazzetti 3096*, 18 Jun. 1914 (W).

Dichothrix Handelii Skuja in Handel-Mazzetti, ibid. 1: 21. 1937. —TYPE specimen from Yunnan, China: prope Bödö (Peti) ad austro-orient. Dschungdien (Chungtien) topho fontis supra Guto, *H. Handel-Mazzetti 4471*, 4 Aug. 1914 (W).

Sacconema homoiochlamys Skuja in Handel-Mazzetti, ibid. 1: 23. 1937. —TYPE specimen from Szechwan, China: prope Ningyüen plankton lacus, *H. Handel-Mazzetti 1794*, 3 May 1914 (W).

Dichothrix Baueriana var. *crassa* Godward, Journ. of Ecol. 25(2): 562. 1937. *Calothrix Baueriana* fa. *crassa* Godward ex Starmach, Fl. Slodkow. Polski 2: 612. 1966. —TYPE specimen, here designated, from the original locality: in England: scrapings from rocks along the shore of Lake Windermere at Watbarrow Point, *J. W. G. Lund*, 1971 (D).

Homoeothrix Pearsallii Godward, ibid. 25(2): 564. 1937. —TYPE specimen, here designated, from the original locality in England: scrapings from rocks along the shore of Lake Windermere at Watbarrow Point, *J. W. G. Lund*, 1971 (D).

Rivularia coadunata fa. *pseudogypsophila* Poliansky, Bot. Zhurn. SSSR 22(2): 156. 1937. —TYPE specimen from Baden, Germany: auf Steinen in Bächen des Torfmoors bei Neustadt, *Schnurmann*, in Rabenhorst, Algen Sachsens no. 771 (PH).

Anabaena parva Philson, Journ. Elisha Mitchell Sci. Soc. 55(1): 99. 1939. —TYPE specimen from South Carolina: on Lake Chapin near Myrtle Beach, *P. J. Philson SC42*, 4 Jul. 1933 (F).

Calothrix genuflexa Philson, ibid. 55(1): 112. 1939. —TYPE specimen from South Carolina: on a log in Muldrow's Mill pond, Florence, *Philson SC/31*, 3 Jul. 1933 (F).

Calothrix Braunii var. *maxima* Philson, ibid. 55(1): 113. 1939. —TYPE specimen from South Carolina: in lily pool, Gaffney, Cherokee county, *Philson 58*, 18 Jul. 1933 (F).

Dichothrix inyoensis Drouet, Field Mus. Bot. Ser. 20(7): 159. 1943. —TYPE specimen from California: pool in a salt playa in Death Valley 35.7 miles south of Furnace Creek on the east highway, *M. J. Groesbeck 3*, Feb. 1940 (F).

Calothrix omeiensis Chu, Sinensia 15(1-6): 154. 1944. —TYPE specimen from Szechwan, China: on wet rock under swift stream at Niu-sä Temple, Omei, *H.-J. Chu 1366*, 23 Aug. 1942 (F).

Rivularia Jaoi Chu, ibid. 15(1-6): 155. 1944. —TYPE specimen from Szechwan, China: on wet rock under swift stream at Niu-sä Temple, Omei, *H.-J. Chu 1366,* 23 Aug. 1942 (F).

Calothrix dolichomeres Skuja, N. Acta R. Soc. Sci. Upsal., ser. 4, 14(5): 25. 1949. —TYPE specimen from Burma: Royal lake, Rangoon, *L. P. Khanna 521,* 2 Dec. 1935 (D).

Calothrix geitonos Skuja, ibid. 14(5): 26. 1949. —TYPE specimen from Burma: on a floating piece of bamboo, Kauadaugyi, Rangoon, *Khanna 507,* 22 Nov. 1935 (D).

Fortiea incerta Skuja, ibid. 14(5): 30. 1949. —TYPE specimen from Burma: Royal lake, Rangoon, *Khanna 479,* 8 Nov. 1935 (D).

Calothrix thermalis fa. *Tildenii* Poliansky in Elenkin, Monogr. Alg. Cyanoph., Pars Spec. 2: 1045. 1949. —TYPE specimen from Wyoming: overflow of channel of Spasmodic geyser, Upper Geyser Basin, Yellowstone national park, *W. H. Weed,* 1897, in Tilden, American Algae no. 287 (D).

Calothrix fusca fa. *minutissima* Poliansky in Elenkin, ibid. 2: 1057. 1949. —TYPE specimen from Sweden: in Krokträsket insulae Värmdö prope Stockholmiam, *K. Bohlin,* 23 Aug. 1896, in Wittrock, Nordstedt & Lagerheim, Algae Aquae Dulcis Exsiccatae no. 1304 (PH).

Calothrix australiensis Scott & Prescott, Rec. Amer.-Austral. Sci. Exped. to Arnhem Land 3: 14. 1958. —TYPE specimen from Arnhem Land, Australia: freshwater marsh at Yirrkalla, *R. L. Specht A80,* 27 Aug. 1948 (EMC).

Calothrix desertica Schwabe, Österr. Bot. Zeitschr. 107(3–4): 282. 1960. —TYPE specimen a culture from Antofagasta, Chile: aus Feinsand nahe La Portada etwa 1 km. landenwärts von der Steilküstenkante, *G. H. Schwabe,* 7 May 1960 (D).

Gloeotrichia Raciborskii var. *kaylanaensis* Goyal, Journ. Bombay Nat. Hist. Soc. 61(1): 72. 1964. —TYPE specimen from Rajasthan, India: ad Kaylana prope Jodhpur, *S. K. Goyal Myx-J19,* 8 Oct. 1959 (IARI).

Calothrix anomala Mitra in Pandey, N. Hedwigia 10(1–2): 203. 1965. —TYPE specimen from Uttar Pradesh, India: from liquid and agar cultures, Indian soil algae, *A. K. Mitra 56,* in the slide collection (BM).

Dichothrix elongata Compère, Bull. Inst. Franç. Afr. Nord 22(A): 31. 1970. —TYPE specimen from Massif de l'Ennedi, North Chad: Versant Sud, Aoué, sources, *T. Monod 13772,* 24 Dec. 1966 (BR).

Original specimens have not been available to me for the following names; their original descriptions must serve as Types until the specimens are found:

Tremella utriculata Hudson [Fl. Angl., ed. altera 2: 564. 1778] ex Agardh, Syst. Algar., p. 26. 1824.

Tremella natans Hedwig [Theoria Generat. & Fructif. Pl. Crypt. Linn., Retract. & Aucta, p. 218. 1798] ex Agardh, loc. cit. 1824. *Linckia natans* Lyngbye [Tent. Hydroph. Dan., p. 196. 1819] ex Bornet & Flahault, Ann. Sci. Nat. VII. Bot. 4: 369. 1887. *Gaillardotella natans* Bory de Saint-Vincent [in Mougeot & Nestler, Stirp. Crypt. Vogeso-rhen. 8: 796. 1823] ex Bornet & Flahault, loc. cit. 1887. [*Linckiella natans* Gaillon, Mém. Soc. d'Émul. d'Abbeville 1833: 473. 1833.] *Raphidia natans* Carmichael [in Harvey in Hooker, Brit. Fl. 2(Engl. Fl. 5): 394. 1833] ex Bornet & Flahault, ibid. 4: 370. 1887. *Rivularia natans* Hedwig ex Fries, Fl. Scan., p. 323. 1835. *R. Lyngbyana* Kützing, Phyc. Gener., p. 238. 1843. *Gloeotrichia natans* Rabenhorst [Deutschl. Krypt.-Fl. 2(2): 90. 1847] ex Bornet & Flahault, ibid. 4: 369. 1887. *Rivularia Hedwigiana* Kützing, Sp. Algar., p. 337. 1849. *R. Boryana* Kützing, ibid., p. 336. 1849. *Gloeotrichia Boryana* Rabenhorst [Fl. Eur. Algar. 2: 201. 1865] ex Bornet & Flahault, ibid. 4: 370. 1887. *Rivularia globosa* Grateloup *pro synon.* ex Bornet & Thuret, Notes Algol. 2: 170. 1880. *Cylindrospermum hepaticum* Opiz *pro synon.* ex Hansgirg, Prodr. Algenfl. Böhm. 2: 72. 1892.

Potarcus bicolor Rafinesque [Journ. de Phys. 89: 107. 1819] ex Kuntze (as "Portacus"), Rev. Gen. Pl. 2: 911. 1891.

Listia crustacea Meyen [Verh. K. Leop.-Carol. Akad. Naturf. 14(2): 469, fig. 1. 1829] ex Forti, Syll. Myxoph., p. 632. 1907. *Schizosiphon Listeanus* Rabenhorst [Fl. Eur. Algar. 2: 234. 1865] ex Hansgirg, Beih. z. Bot. Centralbl. 18(2): 494. 1905. *Rivularia Listia* Meneghini *pro synon.* ex Hohenbühel-Heufler, Abh. K. Zool.-Bot. Ges. Wien 21: 314. 1871. *Calothrix Listeana* Hansgirg, loc. cit. 1905. —Meyen's (loc. cit.) fig. 2 appears to represent trichomes of *Scytonema Hofmannii* Ag.

[*Siradium intricatum* Liebmann, Naturh. Tidskr. 2: 487. 1838–39.]

[*Calothrix contorta* Biasoletto, Viaggio di S. M. Federico Augusto, Re di Sassonia, per l'Istria, Dalmazia e Montenegro, p. 232. 1841.]

[*Calothrix nova* Biasoletto, ibid., p. 233. 1841.]

Agonium centrale Oersted [De Region, Marin., p. 44. 1844] ex Forti, Syll. Myxoph., p. 684. 1907.

[*Raphidia viridis* Hassall, Hist. Brit. Freshw. Alg. 1: 265. 1845.]

[*Raphidia viridis* var. *marginata* Hassall, loc. cit. 1845.]

Rivularia Duriaei Montagne, Explor. Sci. de l'Algérie, pt. 1, 1: 184. 1846–49.

Rivularia minuta a *scirpina* Ainé, Pl. Crypt.-cell. Saône-et-Loire, p. 254. 1863.

Amphithrix papillosa Rabenhorst [Krypt.-Fl. Sachs. 1: 105. 1863] ex Forti, ibid., p. 603. 1907. *Arthrotilum papillosum* Rabenhorst [Fl. Eur. Algar. 2: 230. 1865] ex Forti, loc. cit. 1907.

Oscillatoria dissiliens Fiorini-Mazzanti (as "Oscillaria") [Atti Accad. Pontif. N. Lincei 16(5): 631. 1863] ex Gomont, Ann. Sci. Nat. VII. Bot. 16: 237. 1892.

[*Zonotrichia brunnea* fa. *aequalis* Rabenhorst, Fl. Eur. Algar. 2: 217. 1865.]

Zonotrichia saxicola Rabenhorst [loc. cit. 1865] ex Bornet & Flahault, Ann. Sci. Nat. VII. Bot. 4: 351. 1887.

Zonotrichia Lindigii Rabenhorst [ibid. 2: 222. 1865] ex Forti, Syll. Myxoph., p. 675. 1907.

Schizosiphon gracilis Rabenhorst [ibid. 2: 237. 1865] ex Bornet & Flahault, ibid. 3: 366. 1886. *Calothrix gracilis* Rabenhorst ex Wolle, Fresh-w. Alg. U. S., p. 237. 1887.

Mastigonema elongatum Wood [Proc. Amer. Philos. Soc. 11: 128. 1869] ex Wolle, ibid., p. 243. 1887. *

Mastichothrix fibrosa Wood [ibid. 11: 129. 1869] ex Wolle, ibid., p. 244. 1887. *Mastigonema fibrosum* Wood (as "fibrosa") ex Wolle, loc. cit. 1887.

Zonotrichia minutula Wood [Smithson. Contrib. Knowl. 241: 50. 1872] ex Bornet & Flahault, ibid. 4: 354. 1887.

Mastigonema fertile Wood [ibid. 241: 51. 1872] ex Wolle, loc. cit. 1887.

Rivularia peguana Zeller, Journ. Asiatic Soc. Bengal 42(2: 3): 181. 1873.

Gloeotrichia Kurziana Zeller [loc. cit. 1873] ex Bornet & Flahault, Ann. Sci. Nat. VII. Bot. 4: 371. 1887.

Mastichothrix articulata Reinsch (as "Mastigothrix") [Journ. Linn. Soc. Bot. 15: 207. 1877] ex Forti, Syll. Myxoph., p. 618. 1907. *Calothrix articulata* Forti, loc. cit. 1907.

Lophopodium sandvicense Nordstedt [Minneskr. K. Fysiogr. Sållsk. Lund 1878(7): 5. 1878] ex Nordstedt, K. Svensk. Vet. Akad. Handl. 22(8): 80. 1888. *Calothrix sandvicensis* Schmidle, Flora 84: 170. 1897.

Calothrix lacucola Wolle [Bull. Torrey Bot. Club 8(4): 39. 1881] ex Bornet & Flahault, ibid. 3: 370. 1886.

Leptochaete fonticola Borzi [N. Giorn. Bot. Ital. 14(4): 298. 1882] ex Bornet & Flahault, ibid. 3: 342. 1886.

Tolypothrix Nostoc Zopf [Zur Morphol. d. Spaltpfl., p. 55, Taf. VI, fig. 19. 1882] ex Bornet & Flahault, Ann. Sci. Nat. VII. Bot. 5: 125. 1887. —Zopf's (loc. cit.) figs. 20–31 appear to represent a species of Nostoc.

Leptochaete stagnalis Hansgirg, Notarisia 3: 399. 1888.

Calothrix minuta Bennett, Journ. R. Microsc. Soc. London 1888: 4. 1888.

Microchaete diplosiphon var. *cumbrica* W. West, Journ. R. Microsc. Soc. London 1892: 739. 1892. *Fremyella diplosiphon* var. *cumbrica* J. de Toni (as "cambrica"), N. Notarisia, N. S. 1(1): 7. 1947.

Dichothrix interrupta W. & G. S. West, Journ. R. Microsc. Soc. London 1894: 16. 1894. *Calothrix Westiana* Poliansky in Elenkin, Monogr. Alg. Cyanoph., Pars Spec. 2: 1088. 1949.

Dichothrix Baueriana var. *hibernica* W. & G. S. West, ibid. 1894: 17. 1894.

Calothrix endophytica Lemmermann, Forschungsber. Biol. Sta. Plön 4: 184. 1896. *Homoeothrix endophytica* Lemmermann, Krypt.-Fl. Mark Brandenb. 3: 240. 1907.

Calothrix Kawraiskyi Schmidle, Trudy Tiflis Bot. Sada 2: 273. 1897.

Calothrix javanica De Wildeman, Ann. Jard. Bot. Buitenzorg, Suppl. 1: 34. 1897.

Rivularia aquatica De Wildeman, ibid. 1: 40. 1897.

Calothrix calida Richter in Kuntze, Rev. Gen. Pl. 3(2): 388. 1898.

Calothrix Kuntzei Richter in Kuntze, loc. cit. 1898.

Aphanizomenon flos-aquae var. *gracile* Lemmermann (as "gracilis"), Forschungsber. Biol. Sta. Plön 6(2): 204. 1898. *A. gracile* Lemmermann, Krypt.-Fl. Mark Brandenb. 3(2): 193. 1907. *Anabaena flos-aquae* var. *gracilis* Lemmermann (as "gracile") ex Geitler *pro synon.*, Rabenh. Krypt.-Fl., ed. 2, 14: 825. 1932. *Aphanizomenon flos-aquae* fa. *gracile* Elenkin, Monogr. Alg. Cyanoph., Pars Spec. 1: 853. 1938.

Calothrix balearica var. *tenuis* W. & G. S. West, Journ. of Bot. 36: 337. 1898. *Homoeothrix balearica* var. *tenuis* W. & G. S. West ex Geitler, ibid. 14: 575. 1932.

Calothrix Weberi Schmidle, Hedwigia 38: 173. 1899.

Gloeotrichia natans var. *aequalis* Schmidt, Dansk Bot. Tidsskr. 22: 402. 1899.

Rivularia borealis Richter, Biblioth. Bot. 42: 4. 1899.

Calothrix wembaerensis Hieronymus & Schmidle in Kirchner in Engler & Prantl, Natürl. Pflanzenfam. 1(1A): 87. 1900.

Rivularia Hansgirgii Schmidle (as "Hansgirgi"), Allg. Bot. Zeitschr. 1900(3): 34. 1900.

Gloeotrichia longicauda Schmidle, Hedwigia 40: 51. 1901. *Rivularia longicauda* Schmidle *pro synon.*, loc. cit. 1901.

Gloeotrichia Pilgeri Schmidle, ibid. 40: 52. 1901. *Calothrix Pilgeri* Schmidle *pro synon.*, loc. cit. 1901. *Rivularia Pilgeri* Schmidle *pro synon.*, loc. cit. 1901.

Calothrix membranacea Schmidle, Bot. Jahrb. 30: 61. 1901.

Calothrix africana Schmidle, ibid. 30(2): 249. 1901.

Calothrix Goetzei Schmidle, ibid. 30(2): 248. 1901.

Calothrix Fuellebornii Schmidle, ibid. 32: 62. 1902.

Calothrix parietina var. *hibernica* W. & G. S. West, Trans. R. Irish Acad. 32(B:1): 71. 1902.

Calothrix parietina var. *thermalis* G. S. West, Journ. of Bot. 40: 243. 1902. *C. parietina* fa. *thermalis* Poliansky in Elenkin, Monogr. Alg. Cyanoph., Pars Spec. 2: 1074. 1949.

Anabaena Volzii Lemmermann, Abh. Nat. Ver. Bremen 18(1): 153. 1904.

Gloeotrichia echinulata fa. *brevispora* W. & G. S. West, Ann. R. Bot. Gard. Calcutta 6(2): 241. 1907.

Aphanizomenon flos-aquae var. *Klebahnii* Elenkin, Izv. Imper. S.-Peterburg Bot. Sada 9: 147, 151. 1909. *A. Klebahnii* Elenkin, ibid. 9: 148. 1909. *A. flos-aquae* fa. *Klebahnii* Elenkin, Monogr. Alg. Cyanoph., Pars Spec. 1: 852. 1938.

Calothrix antarctica Fritsch, Nat. Antarct. Exped., Nat. Hist. 6(Freshw. Alg.): 36. 1912.

Homoeothrix africana G. S. West, Ann. So. Afr. Mus. 9: 68. 1912.

Rivularia laurentiana Klugh, Rhodora 15: 91. 1913.

Dichothrix austrogeorgica Carlson, Wiss. Ergebn. Schwed. Südpolar-Exped. 1901—03, 4(14): 9. 1913.

Gloeotrichia Raciborskii Woloszyńska, Bull. Int. Acad. Sci. Cracovie, Cl. Sci., Math. & Nat., sér. B, 1912: 687. 1913. *Rivularia Raciborskii* Woloszyńska ex van Oye, Hedwigia 63(3–4): 197. 1922.

Gloeotrichia Lilienfeldiana Woloszyńska, ibid. 1912: 689. 1913. *Rivularia Lilienfeldiana* Woloszyńska ex van Oye, loc. cit. 1922. *Gloeotrichia Raciborskii* var. *Lilienfeldiana* Geitler in Pascher, Süsswasserfl. 12: 233. 1925.

Rivularia kamtschatica Elenkin, Kamchatsk. Eksped. vod. P. Riabuchinskago, Bot. Otd. 2(Sporov. Rast. Kamchat. 1): 210. 1914. *Gloeotrichia kamtschatica* Poliansky in Elenkin, Monogr. Alg. Cyanoph., Pars Spec. 2: 1165. 1949.

Calothrix clavata G. S. West, Mém. Soc. Sci. Nat. Neuchâtel 5(2): 1019. 1914.

Calothrix columbiana G. S. West, loc. cit. 1914.

Calothrix marchica Lemmermann, Abh. Naturw. Ver. Bremen 23(1): 247. 1914.

Calothrix fusca var. *michailovskoensis* Elenkin, Estestv.-Istor. Koll. E. P. Sheremetevoi, Mikhailovskoe, Moskov. Gub., VI, Vodor., 1: 15. 1915. *C. fusca* fa. *michailovskoensis* Poliansky (as "michailovskoense") in Elenkin, Monogr. Alg. Cyanoph., Pars Spec. 2: 1061. 1949.

Dichothrix fusca Fritsch, Ann. So. Afr. Mus. 9: 581. 1918.

Dichothrix spiralis Fritsch, loc. cit. 1918.

Rivularia planctonica Elenkin, Izv. Glavn. Bot. Sada R. S. F. S. R. 20(1): 16. 1921.

Aphanizomenon platense Seckt (as "platensis"), Bol. Acad. Nac. Cienc. Córdoba (Argentina) 25: 420. 1921.

Calothrix gracilis fa. *flexuosa* Fritsch & Stephens, Trans. R. Soc. So. Afr. 9: 67. 1921. *C. Fritschii* fa. *flexuosa* J. de Toni, Archivio Bot. 15(3–4): 289. 1939.

Calothrix Ramenskii Elenkin, Bot. Mater. Inst. Sporov. Rast. Glavn. Bot. Sada Petrograd 1(1): 9. 1922.

Calothrix fusca fa. *minor* Wille in Hedin, So. Tibet 6(3, Bot.): 169. 1922. *C. fusca* var. *minor* Wille ex Geitler in Pascher, Süsswasserfl. 12: 221. 1925.

Rivularia dura fa. *viridis* Wille in Hedin, ibid. 5: 185. 1922.

Microchaete spirulina Steinecke, Bot. Arch. 3(5): 272. 1923. *Leptobasis spirulina* Geitler, Beih. z. Bot. Centralbl. 41(2): 275. 1925. *Fortiea spirulina* J. de Toni, Noter. Nomencl. Algol. 8: [3]. 1936.

Rivulariopsis floccosa Voronikhin, Bot. Mater. Inst. Sporov. Rast. Glavn. Bot. Sada Petrograd 2(8): 115. 1923. *Calothrix floccosa* Geitler in Pascher, Süsswasserfl. 12: 228. 1925. *C. wembaerensis* fa. *floccosa* Poliansky in Elenkin, Monogr. Alg. Cyanoph., Pars Spec. 2: 1026. 1949.

Calothrix aeruginosa Voronikhin, loc. cit. 1923. *C. Woronichinii* J. de Toni, Archivio Bot. 15(3–4): 289. 1939.

Dichothrix compacta var. *calcarata* Voronikhin, ibid. 2(8): 116. 1923. *Rivularia calcarata* Poliansky in Hollerbach, Kossinskaia & Poliansky, Opred. Presn. Vodor. SSSR 2: 379. 1953.

Ammatoidea simplex Voronikhin (as "Hammatoidea"), loc. cit. 1923.

Dichothrix subdichotoma Voronikhin, loc. cit. 1923. *Calothrix subdichotoma* Poliansky in Elenkin, Monogr. Alg. Cyanoph., Pars Spec. 2: 1088. 1949.

Calothrix Elenkinii Kossinskaia, Bot. Mater. Inst. Sporov. Rast. Glavn. Bot. Sada Petrograd 3(1): 11. 1924.

Calothrix cylindrica Frémy, Rev. Algol. 1(1): 37. 1924.

Calothrix minima Frémy, loc. cit. 1924.

Dichothrix Orsiniana var. *africana* Frémy, ibid. 1(1): 38. 1924. *D. Orsiniana* fa. *africana* Poliansky in Elenkin, Monogr. Alg. Cyanoph., Pars Spec. 2: 1107. 1949. *Calothrix gypsophila* fa. *africana* Poliansky in Elenkin, ibid. 2: 1109. 1949.

Gloeotrichia Le-Testui Frémy, loc. cit. 1924.

Calothrix parietina fa. *nodosa* Ercegović, Acta Bot. Inst. Bot. R. Univ. Zagreb. 1: 93. 1925.

Calothrix parietina fa. *crassior* Ercegović, loc. cit. 1925.

Calothrix parietina fa. *brevis* Ercegović, ibid. 1: 94. 1925.

Calothrix parva Ercegović, loc. cit. 1925. *C. fusca* fa. *parva* Poliansky, Bot. Mater. Otd. Sporov. Rast. Bot. Inst. Akad. Nauk SSSR 5: 110. 1941.

Leptochaete nidulans var. *major* Gaidukov, Zap. Belorussk. Gos. Inst. Selsk. i Pesnogo Khozyaistva, Minsk, 4: 81. 1925. *L. nidulans* fa. *major* Elenkin, Monogr. Alg. Cyanoph., Pars Spec. 2: 1823. 1949.

Microchaete calothrichoides var. *tenuis* Liebetanz, Bull. Int. Acad. Polon. Sci. & Lettr., Cl. Sci. & Math., sér. B, 1925: 110. 1926. *Fremyella calothrichoides* var. *tenuis* J. de Toni, N. Notarisia, N. S. 1(1): 7. 1947.

Microchaete diplosiphon var. *tenuis* Hodgetts, Trans. R. Soc. So. Afr. 13: 93. 1926. *Fremyella diplosiphon* var. *tenuis* J. de Toni, Archivio Bot. 20: 2. 1946.

Microchaete capensis Hodgetts, ibid. 13: 94. 1926. *Fremyella capensis* J. de Toni, loc. cit. 1946.

Microchaete loktakensis Brühl & Biswas, Mem. Asiatic Soc. Bengal 8(5): 265. 1926. *Fremyella loktakensis* J. de Toni, Noter. Nomencl. Algol. 8: [4]. 1936.

Calothrix Kossinskajae Poliansky, Russk. Arkh. Protistol. 6: 63. 1927.

Calothrix brevissima var. *moniliforma* Ghose, Journ. Burma Res. Soc. 17(3): 242. 1927.

Calothrix clavatoides Ghose, ibid. 17(3): 253. 1927.

Calothrix Flahaultii Frémy (as "Flahaulti"), Arch. de Bot. Caen, Bull. Mens. 1(1): 5. 1927.

Spelaeopogon Kashyapii Bharadwaja (as "Kashyapi"), Ann. of Bot. 42: 69. 1928. *Scytonematopsis Kashyapii* Geitler, Arch. f. Hydrobiol., Suppl.-Bd. 12(6): 444. 1935.

Leptochaete Capsosirae Frémy, Arch. de Bot. Caen 3(Mém. 2): 240. 1929. —In his fig. 216, Frémy appears to imply that cells presumably referable to *Entophysalis rivularis* (Kütz.) Dr. are produced by these trichomes.

Calothrix Viguieri Frémy, ibid. 3(Mém. 2): 252. 1929.

Calothrix Bossei Frémy, ibid. 3(Mém. 2): 255. 1929.

Calothrix atricha Frémy, ibid. 3(Mém. 2): 261. 1929.

Microchaete violacea Frémy, ibid. 3(Mém. 2): 284. 1929. *Fremyella violacea* J. de Toni, Noter. Nomencl. Algol. 8: [4]. 1936.

Aulosira africana Frémy, ibid. 3(Mém. 2): 380. 1929.

Calothrix minima-clavata Budde, Arch. f. Hydrobiol. 20: 456. 1929.

Homoeothrix costaricensis Kufferath (as "costaricense"), Ann. Crypt. Exot. 2(1): 34. 1929.

Lyngbya striata Kufferath, ibid. 2(1): 48. 1929.

Dichothrix catalaunica González Guerrero, Cavanillesia 3(1—5): 55. 1930.

Gloeotrichia natans var. *zujaris* González Guerrero, Bol. R. Soc. Españ. Hist. Nat. 30: 223. 1930.

Microchaete setcasasii González Guerrero, ibid. 30: 412. 1930. *Fremyella setcasasii* J. de Toni (as "setcacasii"), Noter. Nomencl. Algol. 8: [4]. 1936.

Calothrix Candelii González Guerrero, ibid. 30: 413. 1930.

Calothrix Contarenii var. *sancti-nectarii* Frémy, Bull. Soc. Bot. France 77: 676. 1930.

Scytonematopsis Woronichinii Kisseleva, Zhurn. Russk. Bot. Obshch. 15: 169. 1930. *Scytonema Woronichinii* Kisselev ex Starmach, Fl. Slodkow. Polski 2: 660. 1966.

Rivularia Manginii Frémy (as "Mangini"), Trav. Déd. à L. Mangin, p. 103. 1931.

Anabaena Volzii fa. *recta* Kisselev, Tr. Sred.-Aziat. Gos. Univ., ser. XIIa, Geogr., Fasc. 9, p. 74. 1931.

Gloeotrichia natans fa. *bucharica* Kisselev, ibid., p. 75. 1931.

Calothrix inserta Nygaard, Trans. R. Soc. So. Afr. 20(2): 118. 1932.

Aulosira Fritschii Bharadwaja, Ann. of Bot. 47(185): 123. 1933.

Aulosira prolifica Bharadwaja, ibid. 47(185): 131. 1933.

Calothrix ramosa Bharadwaja, Arch. f. Protistenk. 81(2): 268. 1933.

Homoeothrix juliana var. *lyngbyoides* Geitler, Arch. f. Hydrobiol., Suppl.-Bd. 12(4): 630. 1933.

Calothrix turfosa Geitler, loc. cit. 1933.

Calothrix reptans Geitler, ibid. 12(4): 631. 1933.

Scytonematopsis calothrichoides Geitler, loc. cit. 1933.

Scytonematopsis incerta Geitler, loc. cit. 1933.

Calothrix Ghosei Bharadwaja, Proc. Indian Acad. Sci. 2(1): 99. 1935.

Calothrix Fritschii Bharadwaja, ibid. 2(1): 100. 1935. *C. Bharadwajae* J. de Toni, Diagn. Alg. Nov. 1(6): 501. 1939.

Microchaete tenera var. *tenuis* Bharadwaja, ibid. 2(1): 101. 1935. *Fremyella tenera* var. *tenuis* J. de Toni, ibid. 1(6): 503. 1939.

Microchaete grisea var. *brevis* Bharadwaja, loc. cit. 1935. *Fremyella grisea* var. *brevis* J. de Toni, Archivio Bot. 15(3—4): 289. 1939.

Homoeothrix juliana fa. *brevicellularis* Geitler, Arch. f. Hydrobiol., Suppl.-Bd. 12(6): 431. 1935.

Aulosira cylindrica El-Nayal, Bull. Fac. Sci. Egypt. Univ. (Cairo) 4: 101. 1935.

Dichothrix bivaginata Frémy, Bull. Soc. d'Hist. Nat. Afr. Nord 26: 89. 1935.

Scytonematopsis hydnoides Copeland, Ann. New York Acad. Sci. 36: 103. 1936.

Ammatoidea yellowstonensis Copeland (as "Hammatoidea"), ibid. 36: 105. 1936.

Calothrix Geitleri Copeland, ibid. 36: 110. 1936.

Calothrix charicola Copeland, ibid. 36: 112. 1936.

Calothrix coriacea Copeland, ibid. 36: 117. 1936.

Calothrix cavernarum Copeland, ibid. 36: 120. 1936.

Calothrix gigas Copeland, ibid. 36: 122. 1936.

Calothrix Baileyi Copeland, ibid. 36: 124. 1936.

Microchaete bulbosa Copeland, ibid. 36: 128. 1936. *Fremyella bulbosa* J. de Toni, N. Notarisia, N. S. 1(1): 6. 1947.

Gloeotrichia intermedia var. *kanwaensis* C. B. Rao (as "kanwaense"), Proc. Indian Acad. Sci. 3(2B): 168. 1936.

Gloeotrichia Raciborskii var. *bombayensis* Dixit (as "bombayense"), Proc. Indian Acad. Sci. 3(1B): 97. 1936.

Gloeotrichia Raciborskii var. *conica* Dixit, loc. cit. 1936.

Gloeotrichia Raciborskii var. *salsettensis* Dixit (as "salsettense"), loc. cit. 1936.

Calothrix longissima Hirose, Journ. Jap. Bot. 13(11): 798. 1937.

Rivularia globiceps var. *longissima* Hirose, ibid. 13(11): 801. 1937.

Rivularia sphaerica Hirose, ibid. 13(11): 802. 1937.

Isactis nipponica Hirose, ibid. 13(11): 803. 1937. *Rivularia nipponica* Bourrelly, Les Alg. d'Eau Douce 3: 412. 1970.

Calothrix marchica var. *crassa* C. B. Rao, Proc. Indian Acad. Sci. 6(6B): 349. 1937.

Calothrix marchica var. *intermedia* C. B. Rao, ibid. 6(6B): 350. 1937.

Gloeotrichia Raciborskii var. *kashiensis* C. B. Rao (as "kashiense"), ibid. 6(6B): 351. 1937.

Aulosira fertilissima var. *tenuis* C. B. Rao, ibid. 6(6B):353. 1937.

Anabaena unispora var. *crassa* C. B. Rao, ibid. 6(6B): 362, 1937. *A. Volzii* var. *crassa* Fritsch, Journ. Indian Bot. Soc. 28(3): 155. 1949.

Aulosira sinensis Li, Bull. Fan Mem. Inst. Biol. (Peiping), Bot. Ser. 8(1): 25. 1937.

Gloeotrichia heterocystosa Lillick, Papers Michigan Acad. Sci., Arts & Lett. 22: 146. 1937.

Microchaete striata Elenkin, Monogr. Alg. Cyanoph., Pars Spec. 1: 884. 1938.

Calothrix violacea Jaag, Mitt. Naturf. Ges. Schaffhausen 14: 116. 1938.

Gloeotrichia echinulata var. *berhampurensis* C. B. Rao (as "berhampurense"), Proc. Indian Acad. Sci. 8(3B): 161. 1938.

Calothrix Castellii var. *samastipurensis* C. S. Rao (as "samastipurense"), ibid. 9(3B): 145. 1939.

Calothrix fusca var. *crassa* C. S. Rao, loc. cit. 1939.

Gloeotrichia Raciborskii var. *longispora* C. S. Rao, ibid. 9(3B): 146. 1939.

Calothrix compacta Jao, Sinensia 10(1—6): 223. 1939.

Calothrix intorta Jao, loc. cit. 1939.

Calothrix subsimplex Jao, ibid. 10(1—6): 224. 1939.

Calothrix subantarctica Jao, ibid. 10(1—6): 225. 1939.

Calothrix hunanica Jao, ibid. 10(1-6): 227. 1939. *C. humanica* Jao ex Starmach, Fl Slodkow. Polski 2: 589. 1966.

Dichothrix hamata Jao, ibid. 10(1—6): 228. 1939. *Calothrix hamata* Starmach, ibid. 2: 617. 1966.

Dichothrix sinensis Jao, ibid. 10(1—6): 229. 1939. *Calothrix sinensis* Starmach, ibid. 2: 613. 1966.

Gloeotrichia seriata Jao, ibid. 10(1—6): 230. 1939.

Homoeothrix Zavattarii Marchesoni, R. Accad. d'Ital. (Roma), Centro Studi per l'Afr. Orient. Ital. 4(Missione Biol. nel Paese dei Borana, 4): 396. 1939.

Homoeothrix juliana fa. *tenuis* Singh, Indian Journ. Agric. Sci. 9(1): 58. 1939, *H. juliana* var. *tenuis* Singh ex Srinivasan, Bull. Bot. Surv. India 7: 199. 1965.

Gloeotrichia Ghosei Singh, Proc. Indian Acad. Sci. 9(2): 64. 1939.

Aulosira major Emoto & Hirose, Bot. & Zool. (Syokubutu Oyobi Doluta) 8(1): 38. 1940.

Homoeothrix thermalis Emoto & Hirose, ibid. 8: 1726. 1940.

Microchaete thermalis Emoto & Yoneda, Ecol. Rev. 6: 269. 1940.

Calothrix fonticola Brabez, Beih. z. Bot. Centralbl. 61A: 213. 1941.

Calothrix obtusa Brabez, loc. cit. 1941.

Gloeotrichia Tuzsonii Palik (as "Tuzsoni"), Arch. f. Protistenk. 95(1): 45. 1941.

Aphanizomenon Morrenii Kufferath (as "Morreni"), Bull. Soc. R. Bot. Belg. 74: 95. 1942.

Rivularia dura var. *confluens* Frémy, Blumea, Suppl. 2: 25. 1942.

Rivularia Manginii var. *confluens* Frémy, ibid. 2: 26. 1942.

Fremyella aequalis Frémy, ibid. 2: 32. 1942. *Microchaete aequalis* Desikachary, Cyanoph., p. 513. 1959.

Homoeothrix fluviatilis Jao, Sinensia 15(1—6): 69. 1944.

Calothrix subsimplex var. *kwangsiensis* Jao, ibid. 15(1—6): 80. 1944.

Dichothrix hemisphaerica Jao, loc. cit. 1944.

Ammatoidea sinensis Ley (as "Hammotoidea"), Sinensia 15(1—6): 101. 1944.

Calothrix conjuncta Schwabe, Mitt. Deutsch. Ges. f. Natur. & Völkerk. Ostasiens (Shanghai), Suppl. Bd. 21: 144. 1944.

Homoeothrix dicax Schwabe, loc. cit. 1944.

Gloeotrichia salina fa. *epiphytica* Schwabe, ibid. 21: 143. 1944.

Leptobasis nipponica Schwabe, ibid. 21: 145. 1944.

Homoeothrix caespitosa var. *argentinensis* González Guerrero, Anal. Jard. Bot. Madrid 5: 322. 1945.

Calothrix Allorgei Frémy, Bull. Mus. Nat. d'Hist. Nat. Paris, sér. 2, 17: 73. 1945.

Gloeotrichia Juignetii Frémy (as "Juigneti"), ibid. 17: 75. 1945.

Dichothrix gelatinosa Böcher, K. Danske Videnskab. Selsk. Skr. 4(4): 1. 1946. *Calothrix gelatinosa* Poliansky in Hollerbach, Kossinskaia & Poliansky, Opred. Presn. Vodor. SSSR 2: 367. 1953.

Rivularia Cavanillesiana González Guerrero, Anal. Jard. Bot. Madrid 6: 255. 1946.

Aulosira bombayensis Gonzalves, Indian Bot. Soc. Prof. M.O. P. Iyengar Commem. Vol., p. 381. 1947.

Homoeothrix kwangtungensis Ley, Bot. Bull. Acad. Sinica 1: 79. 1947.

Montanoa castellana González Guerrero, ibid. 8: 268. 1948.

Aulosira Godoyana González Guerrero, ibid. 8: 270. 1948.

Aulosira confluens Jao, Bot. Bull. Acad. Sinica 2(1): 45. 1948.

Camptylonemopsis pulneyensis Desikachary, Proc. Indian Acad. Sci. 28(2B): 40. 1948.

Camptylonemopsis minor Desikachary, ibid. 28(2B): 42. 1948.

Camptylonemopsis Iyengarii Desikachary, ibid. 28(2B): 43. 1948.

Calothrix brevissima fa. *angustior-et-longior* G. S. West ex Poliansky in Elenkin, Monogr. Alg. Cyanoph., Pars Spec. 2: 1029. 1949.

Calothrix intricata fa. *dnjeprensis* Poliansky in Elenkin, ibid. 2: 1036. 1949. *C. dnjeprensis* Kondratieva, Viznach. Prisnov. Vodor. Ukrainsk. RSR 1(2): 404. 1968.

Calothrix Braunii fa. *major* Poliansky in Elenkin, ibid. 2: 1066. 1949.

Calothrix Braunii fa. *Schirschovii* Poliansky in Elenkin, loc. cit. 1949.

Rivularia coadunata fa. *pavlovskoensis* Elenkin in Poliansky in Elenkin, ibid. 2: 1148. 1949.

Calothrix gloeocola Skuja, N. Acta R. Soc. Sci. Upsal., ser. 4, 14(5): 27. 1949.

Calothrix Ramenskii var. *minor* Voronikhin, Bot. Mater. Otd. Sporov. Rast. Bot. Inst. Komarova Akad. Nauk SSSR 6(1–6): 39. 1949. *C. Ramenskii* fa. *minor* Poliansky in Hollerbach, Kossinskaia & Poliansky, Opred. Presn. Vodor. SSSR 2: 357. 1953.

Aphanizomenon flos-aquae fa. *incurvatum* Tschernov (as "incurvata"), Bot. Mater. Otd. Sporov. Rast. Bot. Inst. Komarova Akad. Nauk SSSR 6(7–12): 114. 1950.

Gloeotrichia natans fa. *thaumastospora* Tschernov, ibid. 6(7–12): 123. 1950.

Scytonematopsis ambigua Emoto & Hirose, Journ. Balneol. Soc. Japan 5(2): 20. 1952.

Calothrix Feldmannii Bourrelly in Bourrelly & Manguin, Alg. d'Eau Douce Guadeloupe, p. 152. 1952.

Calothrix Rodriguezii Bourrelly in Bourrelly & Manguin, ibid., p. 153. 1952.

Aulosira Schweickerdtii Cholnoki, Bol. Soc. Portug. Ciênc. Nat., ser. 2, 4(1): 92. 1952. *Ghosea Schweickerdtii* Cholnoky *pro synon.*, loc. cit. 1952.

Calothrix aberrans Cholnoki, ibid. 4(1): 93. 1952.

Calothrix Schweickerdtii Cholnoki, loc. cit. 1952.

Calothrix scytonemicola var. *africana* Cholnoki, ibid. 4(1): 94. 1952.

Dichothrix Baueriana var. *africana* Cholnoki, ibid. 4(1): 96. 1952.

Dichothrix Schweickerdtii Cholnoki, ibid. 4(1): 97. 1952.

Microchaete Schweickerdtii Cholnoki, ibid. 4(1): 104. 1952.

Homoeothrix flagelliformis Vozzhennikova, Bot. Mater. Otd. Sporov. Rast. Bot. Inst. Komarova Akad. Nauk SSSR 9: 75. 1953.

Calothrix aeruginosa fa. *minor* Melnikova, Bot. Mater. Otd. Sporov. Rast. Bot. Inst. Akad. Nauk SSSR 9: 65. 1953.

Calothrix Elenkinii fa. *edaphica* Melnikova, loc. cit. 1953.

Pseudanabaena edaphyca Melnikova, ibid. 9: 67. 1963.

Gloeotrichia spiroides Kondratieva, Bot. Zhurn. Akad. Nauk Ukrainsk. RSR 11(1): 106. 1954.

Calothrix bugensis Poliansky, Bot. Mater. Otd. Sporov. Rast. Bot. Inst. Leningrad 10: 17. 1955.

Microchaete brunescens Komárek, Preslia 28: 373. 1956.

Gloeotrichia Andreanszkyana Claus, Verh. Zool.-Bot. Ges. Wien 97: 95. 1957. *Vindobona Andreanszkyana* Claus *pro synon.*, loc. cit. 1957.

Calothrix gypsophila fa. *turfosa* Tarnavschi & Mitroiu, Bul. Stiint. Sect. Biol. & Stiinte Agric., Ser. Bot., Acad. Rep. Pop. Romine, 9(1): 51. 1957.

Calothrix fusca fa. *ampliusvaginata* Starmach, Fragm. Florist. & Geobot. 3(2): 137. 1958.

Calothrix fusca fa. *durabilis* Starmach, ibid. 3(2): 138. 1958.

Calothrix wembaerensis var. *fusca* Behre in Quézel, Univ. d'Alger, Inst. Rech. Sahar., Mission Bot. au Tibesti, p. 17. 1958.

Calothrix inaequabilis Čado, Zbornik na Rabotite, Filoz. Fac. Univ. Skopje, Khidrobiol. Zavod., Okhrid, 6(5): 8, 39. 1958.

Campylonema umidum Čado (as "Camptylonema"), ibid. 6(5): 11, 40. 1958. *Schmidleinema umidum* Čado ex Starmach, Fl. Slodkow. Polski 2: 713. 1966.

Rivularia lapidosa Čado, ibid. 6(6): 3, 11. 1958.

Calothrix estonica Kukk, Bot. Mater. Otd. Sporov. Rast. Bot. Inst. Akad. Nauk SSSR 12: 25. 1959.

Rivularia calcarata fa. *bistrata* Kukk, ibid. 12: 26. 1959.

Scytonematopsis Woronichinii fa. *minor* Muzaffarov, Bot. Mater. Otd. Sporov. Rast. Bot. Inst. Akad. Nauk SSSR 12: 34. 1959.

Homoeothrix gloeophila Starmach, Acta Hydrobiol. Kraków 2(3—4): 230. 1960.

Homoeothrix articulata Starmach, ibid. 2(3—4): 233. 1960.

Calothrix Santapaui Gonzalves & Kamat, Journ. Bombay Nat. Hist. Soc. 57(2): 454. 1960.

Calothrix karnatakensis Gonzalves & Kamat, Journ. Univ. Bombay 28(5): 29. 1960.

Calothrix karnatakensis var. *major* Gonzalves & Kamat, loc. cit. 1960. *C. karnatakensis* fa. *major* Gonzalves & Kamat ex Starmach, Fl. Slodkow. Polski 2: 587. 1966.

Calothrix karnatakensis var. *major* fa. *varians* Gonzalves & Kamat, Journ. Univ. Bombay 28(5): 30. 1960.

Calothrix wembaerensis var. *minor* Gonzalves & Kamat, ibid. 28(5): 31. 1960.

Calothrix columbiana var. *constricta* Gonzalves & Kamat, ibid. 28(5): 32. 1960.

Gloeotrichia Pisum fa. *strialuta* Gonzalves & Kamat, ibid. 28(5): 33. 1960.

Gloeotrichia echinulata var. *epiphytica* Gonzalves & Kamat, ibid. 28(5): 35. 1960.

Gloeotrichia Raciborskii var. *major* Gonzalves & Kamat, loc. cit. 1960.

Gloeotrichia Raciborskii var. *kashiensis* fa. *intermedia* Vasishta, Journ. Bombay Nat. Hist. Soc. 58(1): 144. 1961.

Gloeotrichia Raciborskii var. *kashiensis* fa. *hoshiarpurensis* Vasishta, ibid. 58(1): 145. 1961.

Rivularia maharastrensis Kamat, Journ. Univ. Bombay, N. S. 30(3—5B): 24. 1961.

Homoeothrix Santolii Wawrik (as "Santoli"), Rev. Algol., N. S. 6(2): 95. 1962.

Fortiella Subaiana Claus, N. Hedwigia 4(1—2): 63. 1962.

Fortiella Subaiana var. *simplex* Claus, ibid. 4(1—2): 65. 1962.

Plectonema tatricum Starmach, Acta Hydrobiol. Kraków 10(4): 427. 1963.

Calothrix Debii Bharadwaja, Proc. Indian Acad. Sci. 57(4B): 251. 1963.

Gloeotrichia Raciborskii var. *longispora* fa. *major* Bharadwaja, ibid. 57(4B): 254. 1963.

Rivularia Joshii Vasishta, Journ. Indian Bot. Soc. 41(4): 516. 1963.

Calothrix Desikacharyensis Vasishta, ibid. 42: 579. 1963.

Aulosira fertilissima var. *hoshiarpurensis* Vasishta, Journ. Bombay Nat. Hist. Soc. 60(3): 677. 1963.

Aulosira terrestris Subba Raju (as "terrestre"), Phykos 3: 33. 1964.

Calothrix sphaerospora Prasad & Srivastava, Phykos 4(2): 83. 1965.

Aulosira implexa var. *tenuis* Pandey & Mitra, N. Hedwigia 10(1—2): 91. 1965.

Calothrix wembaerensis var. *epiphytica* Pandey & Mitra, ibid. 10(1—2): 93. 1965.

Calothrix Braunii fa. *major* Pandey, N. Hedwigia 10(1—2): 202. 1965.

Homoeothrix auraria Obuchova (as "aurarius"), Bot. Mater. Herb. Inst. Bot. Akad. Nauk Kazakhst. (Alma-Ata) 3: 69. 1965.

Raphidiopsis longisetae Eberly, Trans. Amer. Microsc. Soc. 85(1): 134. 1966.

Aphanizomenon flos-aquae var. *Haerdtlii* Fjerdingstad (as "Härdtlii"), Schweiz. Zeitschr. f. Hydrol. 28(2): 144. 1966.

Aphanizomenon flos-aquae var. *minus* Nair (as "minor"), Hydrobiologia 30(1): 149. 1967.

Calothrix submarchica Archibald, N. Hedwigia 13(3—4): 393. 1967.

Calothrix Galpinii Cholnoky-Pfannkuche, N. Hedwigia 15(2—4): 432. 1968.

Aphanizomenon Ussaczevii Proshkina-Lavrenko in Proshkina-Lavrenko & Makarova, Vodor. Plankt. Kaspisk. Mor., p. 111. 1968.

Aphanizomenon ovalisporum var. *caspicum* Usaczev (as "caspica") ex Proshkina-Lavrenko & Makarova, loc. cit. 1968.

Gloeotrichia murgabica Kogan & Yazkulieva, Izv. Akad. Nauk Turkmen. SSR, Ser. Biol. Nauk No. 5: 37. 1970; Nov. Syst. Nizsh. Rast. Bot. Inst. Komarova Akad. Nauk SSSR 9: 8. 1972.

Aphanizomenon flos-aquae fa. *macrosporum* Fedorov, Nov. Sist. Nizsh. Rast. Bot. Inst. Komarova Akad. Nauk SSSR 1969, 6: 20. 1970.

Trichomata aeruginea, luteo-viridia, olivacea, fusca, rosea, violacea, vel cinereo-viridia, cylindrica vel extrema longe et sensim attenuantia, vel ad unum extremum tumida, ad dissepimenta constricta vel non-constricta, diametro 3—24μ crassa, partim aut passim increscentia partim decrescentia, ambitu recta vel curvantia vel spiralia, longitudine indeterminata, per destructionem cellulae intercalaris vel per constrictionem ad dissepimentum frangentia, ad unum extremum vel utrinque extrema per paucas vel plures cellulas angustas perlongas decolorantes terminantia. Cellulae diametro trichomatis longiores vel breviores, 3—20μ longae, protoplasmate homogeneo vel granuloso saepe pseudovacuolato, dissepimentis non granulatis; heterocystae saepe ad solum extremum nonnumquam ad utrinque extrema terminales, vel intercalares, interdum seriatae, hemisphaericae vel quasisphaericae vel discoideae vel cylindricae, diametro 4—25μ crassae; sporae cylindricae vel ovoideae raro sphaericae, solitariae vel seriatae, muris hyalinis vel lutescentibus vel fuscescentibus; cellulae vegetativae terminales primum hemisphaericae demum obtuse conicae vel cylindricae ad extrema rotundae. Materia vaginalis primum hyalina demum lutea vel fusca. Planta trichomata longa vel brevia nuda aut vulgo in vaginis discretis cylindricis saepe ramosis, vulgo laminosis, aut in gelatina distributa comprehens. Fig. 28—60.

Trichomes blue-green, yellow-green, olive, brown, red, violet, or gray-green, cylindrical or toward the tips long and noticeably attenuating, or at one tip swollen, constricted or not constricted at the cross walls, $3-24\mu$ in diameter, partly or here and there increasing or decreasing in diameter, straight or curving or spiraled, indeterminate in length, breaking by the destruction of an intercalary cell or by constriction at a cross wall, often terminating at one end or at both ends in few to many long, narrow cells which lose their pigments. Cells longer or shorter than their diameters, $3-20\mu$ long, the protoplasm homogeneous or granulose, often pseudovacuolate, the cross walls not granulated; heterocysts often terminal at one end or at both ends of the trichome, or intercalary, sometimes seriate, hemispherical or almost spherical or discoid or cylindrical, $4-25\mu$ in diameter; spores cylindrical or ovoid, rarely spherical, solitary or seriate, the walls hyaline or becoming yellow or brown; terminal vegetative cells at first hemispherical, becoming obtuse-conical or cylindrical with rotund tips. Sheath material at first hyaline, becoming yellow or brown. Plant consisting of long or short naked trichomes or trichomes in discrete cylindrical, often branched sheaths, or distributed in gelatinous matrices. Figs. 28–60.

Calothrix parietina is distributed all over the world, on rocks, soil, wood, larger plants and animals, shells, and other substrates, in places influenced by fresh water, seldom by brackish or slightly marine water. This species can possibly grow under circumstances where the amount of water is so restricted as that in which *Scytonema Hofmannii* can grow, but it is generally found on substrates frequently or permanently wet.

The common ecophene (Fig. 40) consists of short trichomes within thin, often discrete sheath material, growing and multiplying without special order in the mass, frequently in association with *Schizothrix calcicola* (Ag.) Gom. and many times with *Anacystis montana* (Lightf.) Dr. & Daily or *Entophysalis rivularis* (Kütz.) Dr. This growth form, often uncalcified, is found in fresh-water seepage and on almost any substrate in temporary or permanent bodies of fresh water. As the plant ages, the trichomes proliferate; and the sheath material becomes single- or twin-branched, often profusely. In water containing silt or moderate to high concentrations of dissolved salts, the plants become silted or the sheath material becomes impregnated with or surrounded by calcium carbonate (thus forming marl and tufa), calcium sulfate, or (in certain hot springs) silica. These plants become mealy or stony, the trichomes growing in a parallel or radiate fashion toward the source of light. Such complex plants, as well as those not impregnated with minerals, have been described in the past under the generic names *Dichothrix* Zanard. and *Rivularia* Roth, even though, as in my collection no. 12860 (D) from Beaver Island, Michigan, one commonly observes all stages of transition from the common ecophene to complicated soft or stony cushions present within a radius of a few centimeters of substrate. Very often one finds the entire or peripheral bottoms of ponds, lakes, and streams covered with calcified or uncalcified growths of this species, mixed with *Schizothrix calcicola* and *Microcoleus vaginatus* (Vauch.) Gom.; in very deep bodies of water, broad peripheral areas of benthic algae consist principally of these species.

The trichomes, by mechanical abrasion, browsing of animals, or other means, may lose either the terminal heterocysts or the attenuated colorless end cells, or both. The cell or cells next to a lost heterocyst may mature as heterocysts, but seldom if ever are the colorless cells regenerated. A fragment of a trichome may develop attenuated colorless cells at both ends, and it is not surprising to find heterocysts at both ends. Numerous ecophenes have been described as species (Figs. 48, 52) of *Homoeothrix* Thur. and *Ammatoidea* W. & G. S. West, where trichomes have not yet developed heterocysts. Repeated collections of these over a period of time, as in a large series of samples (PH) taken annually from the Potomac river in Frederick and Montgomery counties, Maryland, by Dr. M. H. Hohn, Dr. P. J. Halicki, Dr. C. W. Reimer, and Mr. R. R. Grant, 1957–1970, show that heterocysts do develop as external conditions change. Similar observations have been made in my cultures of these plants. Pearson & Kingsbury (1966) have likewise noted such changes. Also, in most cases, meticulous examination of the Type specimens of species described as being devoid of heterocysts shows that heterocysts in various stages of differentiation are indeed present (Figs. 48, 52).

In other aquatic ecophenes, the trichomes become surrounded with gelatinous or mucous, often calcified, sheath materials of various shapes and consistencies; they grow radiately and multiply. Eventually the one or few cells next to a terminal heterocyst enlarge, elongate, and mature as spores (Figs. 29, 31, 32, 33). The inner layers of the sheath material, often calcified in part, adhere to the spore walls. These spores, in the laboratory, germinate *in situ* (Figs. 39, 53), as in my culture (PH) of materials collected by Dr. H. S. Forest in Conesus lake, New York; and the germlings develop precisely as do the trichomes of the common ecophene. I have been unsuccessful in germinating spores of dried material. Some plants produce pseudovacuoles in the protoplasts and grow as water blooms (Maxwell, 1971). All these ecophenes have been assigned in the recent past to species of a special genus, *Gloeotrichia* J. Ag. The same kind of spores often develops in solitary naked trichomes and in trichomes inclosed in firm cylindrical sheaths attached to larger plants in bodies of fresh water.

In shallow and usually temporary bodies of water, especially during hot sunny weather, segments of trichomes issue from the sheaths (Fig. 54) of the common ecophene and, by continuous growth to indeterminate lengths, often with a minimum of breakage and of branching of sheaths, form more or less gelatinous sheets over masses of Cladophoraceae and other larger plants. Cylindrical intercalary heterocysts, and but few terminal heterocysts, may be differentiated. Where heterocysts are rare, the trichomes may be difficult to distinguish from those of *Schizothrix Friesii* (Ag.) Gom. Colorless attenuated cells may or may not develop at one end or at both ends of each trichome. Almost every vegetative cell, often every other one, or perhaps only one here and there, may enlarge and mature into a cylindrical, ovoid, or spheroid spore (Fig. 38). Such spores in laboratory culture, even after many years of storage in the herbarium—as in my no. 2359 (D) from Miller, Indiana; my no. 3036 (D) from Hermosillo, Mexico; and my no. 14602 (D) from Madera Canyon, Arizona—germinate (Figs. 38, 55, 56) and grow into typical trichomes of the common ecophene (Fig. 40). Such plants have customarily been placed in species of *Aulosira* Kirchn.

In deeper and more permanent bodies of water, segments of trichomes issue in prodigious numbers from sheaths of the common benthic ecophene, develop series of colorless cells at both ends, produce pseudovacuoles in all other cells including often the few intercalary heterocysts, and each under appropriate external conditions matures one or few long cylindrical spores (Figs. 34, 58). These ecophenes, with sheath material dispersed or in the form of slime, are the water blooms commonly designated as "species" of *Aphanizomenon* Morr. They occur chiefly in ponds and lakes of the North Temperate Zone, where benthic algal populations contain excellent growths of the common ecophene. The spores do not, apparently, survive desiccation; although most laboratory cultures of the dried plants which I have made have developed the common ecophene (Fig. 40) in profusion. When the spores germinate in water, the germlings (Rose, 1934) are recognizable trichomes of the common ecophene. The precise role of the spores in the production of such vast numbers of trichomes in water blooms has not as yet been determined.

Another ecophene, described as *Anabaena unispora* Gardn. (Fig. 36), develops as slimy, pale blue-green masses of trichomes in temporary bodies of water in tropical regions and in temperate areas where the summers are hot. The large ovoid spores commonly mature singly or seriately on one side of a heterocyst.

Many plants (ecophenes) which one encounters in the field and in laboratory culture will, of course, be morphologically transitional between or among those mentioned above.

My attempts to grow this species, from Pottstown, Pennsylvania, and from Lake Itasca, Minnesota, in marine waters in the laboratory have been unsuccessful.

Geitler (1934) reported isolation of this species from the body of the lichen *Placynthium nigrum* (Huds.) S. F. Gray. Trichomes of *Calothrix parietina*, more or less infested with fungi, inhabit the cephalodia of certain lichens. Fungi are often seen growing in the sheaths of otherwise healthy trichomes. Also, bacteria, *Schizothrix calcicola*, and uni- or pauci-cellular algae thrive within, and contribute to the degradation or augmentation of, the sheath materials.

The following specimens, selected from among the some 5,500 studied, are representative of *Calothrix parietina* as described above:

Cultures of Uncertain Origin: Cambridge Botany School 1410/1, 1482/3 (D); Indiana University 382 (D), 484 (PH), 583 (D), 590 (PH); Hopkins Marine Station (as *Fremyella diplosiphon* and *F. tenera*, D); Kaiser Research Foundation, M-13.1.1 (PH); University of Wisconsin (as *F. diplosiphon*, *Aphanizomenon flos-aquae*, and *Gloeotrichia echinulata*, D).

Russia: St. Petersburg (LD).

Finland: Höckböleträsk (Alandia), *K. E. Hirn* (as *Aphanizomenon flos-aquae* in Wittr. & Nordst., Alg. Exs. 1341c, D, PH).

Norway: Lomsfjelderne, *O. Nordstedt* (as *Scytonema myochrous* in Aresch., Alg. Scand. Exs., ser. nov. 377, D, F); Govelid and Memurutungen, *Nordstedt* (as *S. figuratum* and *Rivularia radians* in Wittr. & Nordst., Alg. Exs. 275, 878a, D, PH); Odde in Hardanger, *Nordstedt* (TYPE of *Dichothrix Nordstedtii* Born. & Flah. as *Calothrix cespitosa* in Rabenh., Alg. Eur. 2315, PC; duplicates: W. R. Taylor; in Wittr. & Nordst., Alg. Exs. 857, D D, PH).

Sweden: Stockholm and Lund, *C. A. Agardh* (TYPES of *Scytonema compactum*

Ag., *S. pulverulentum* Ag., and *Rivularia Pisum* Ag., LD); Gotland, *P. T. Cleve*
(TYPE of *Ainactis gothica* Aresch. in Aresch., Alg. Scand. Exs., ser. nov. 234, S;
duplicates: D, FH); Uppsala, *A. Areschoug, V. Wittrock* (as *Gloeotrichia Boryana*
and *G. Brauniana* in Aresch., Alg. Scand. Exs., ser. nov. 373, 374, D, F); inter
Hallqved et Rasbo, *A. Areschoug* (TYPE of *Rivularia echinulus* Aresch. in
Aresch., Alg. Scand. Exs., ser. nov. 375, S; duplicates: D, FH); Trollhättan,
Nordstedt (TYPE of *Schizosiphon Nordstedtianus* Rabenh. in Rabenh., Alg. Eur.
2246, D); Lerdala, *Nordstedt* (as *Zonotrichia rivularis* in Rabenh., Alg. Eur. 2287,
D, UC; as *Rivularia rivularis* in Wittr. & Nordst., Alg. Exs. 189, D, F, PH);
Ǫrtofta, *Nordstedt* (as *Gloeotrichia natans* and *G. Pisum* in Wittr. & Nordst., Alg.
Exs. 187, 188, D, NY); pr. Grebbesta, *Nordstedt* (as *Rivularia torfacea* in Wittr. &
Nordst., Alg. Exs. 190, D, PH); Lötsjön, Valloxen, Nacka pr. Stockholm, and
Kålungen, *V. Wittrock* (as *Aphanizomenon flos-aquae, Hypheothrix cataractarum,* and
Gloeotrichia Pisum in Wittr. & Nordst., Alg. Exs. 278a,b, 392, 660, D, PH);
Gotlandia, *Cleve* (as *Rivularia haematites* in Wittr. & Nordst., Alg. Exs. 665a, D,
S); Danviken and Björbolund, *G. Lagerheim* (as *Calothrix parietina* and *Gloeo-
trichia natans* in Wittr. & Norst., Alg. Exs. 751a,b, 753a, D, F, PH); Österslöf, *J.
Braun* (as *G. Pisum* in Wittr. & Nordst., Alg. Exs. 864, D, NY); Krokträsket pr.
Stockholm, *K. Bohlin* (TYPE of *Calothrix fusca* fa. *minutissima* Polj. in Wittr.,
Nordst. & Lagerh., Alg. Exs. 1304, PH); Mjölhatteträsk Gotlandiae, *Wittrock*
(as *Dichothrix gypsophila* in Wittr., Nordst. & Lagerh., Alg. Exs. 1308, D, PH);
Åkarp par. Borlöf, *Hirn* (as *Aphanizomenon flos-aquae* in Wittr., Nordst. &
Lagerh., Alg. Exs. 1342, D); Klågerup, *H. G. Simmons* (as *Calothrix Braunii* in
Wittr., Nordst. & Lagerh., Alg. Exs. 1606, NY).

Denmark: Soeborghuus, *Liebmann* (TYPE of *C. brevis* Liebm., C); Donse
Mölle, *O. Nordstedt* (as *C. Braunii* in Wittr. & Nordst., Alg. Exs. 856, D, F, PH);
ad lacum Lyngbye, *Lyngbye* (TYPE of *Linckia hypnicola* Lyngb., C); Fuursöe,
Lyngbye (TYPE of *L. dura* var. *lutescens* Lyngb., C); Hallenslev Mose, culture, *T.
Christensen 4993* (D, L).

Estonia: *Gobi* (TYPE of *Rivularia pelagica* Gobi, PC).

Poland: Strehlen, *Bleisch* (as *Mastigonema pluviale* in Rabenh., Alg. Sachs.
647, D, F), *Hilse* (TYPE of *Rivularia angulosa* fa. *dura* Hilse in Rabenh., Alg.
Sachs. 648, PH; TYPE of *Mastigonema caespitosum* fa. *gracillimum* Rabenh. in
Rabenh., Alg. Sachs. 871, D; as *Limnactis flagellifera, Rivularia durissima,* and
Leptothrix rosea in Rabenh., Alg. Sachs. 928, 976, 1467, D; TYPE of *Schizosiphon
Rabenhorstianus* Hilse in Rabenh., Alg. Eur. 1836, PH); Breslau, *Hilse* (TYPE of
S. gracilis Hilse in Rabenh., Alg. Eur. 1770., PH; as *Rivularia rigida* in Rabenh.,
Alg. Eur. 1837, D, FH), *B. Schröder* (as *Gloeotrichia natans* in Wittr., Nordst. &
Lagerh., Alg. Exs. 1311, D, F); Aupegrund im Riesengebirge, *Hilse* (TYPE of
Schizosiphon intertextus Grun., as *Symphyosiphon intertextus* in Rabenh., Alg. Eur.
1177, D); Lauenburg, *A. Schmidt* (TYPE of *Rivularia fluitans* Cohn in Rabenh.,
Alg. Eur. 2540, PH); Garder See, *W. Krieger* (as *Gloeotrichia echinulata* in Mig.,

Crypt. Exs. Alg. 285, D, F, NY, OC); Peterwitz, *Bleisch* (as *Mastichothrix fusca* in Rabenh., Alg. Eur. 1499, D, F).

Romania: Siminicea (Botosani), *E. C. Teodorescu 1204* (D, F, W); Peris (Ilfov), *Teodorescu 1338/2* (D, W); Ciurel (Ilfov), *Teodorescu 1333* (D, F, W); e lacu Caldarusano (Ilfov), *Teodorescu 1070* (TYPE of *Calothrix adscendens* fa. *culta* Teod., W; in Mus. Vindob. Krypt. Exs. Alg. 3137, D, F); riv. Colintina (Ilfov), *Teodorescu* (as *Gloeotrichia Pisum* in Mus. Vindob. Krypt. Exs. Alg. 633b, D, F); Comana (Vlasca), *Teodorescu* (as *G. natans* in Mus. Vindob. Krypt. Exs. Alg. 221b, D, F, US).

Germany: (TYPES of *Rivularia Linckia* Roth and *R. Linckia β simplex* Roth, LD; TYPE of *R. angulosa* Roth, PC; TYPE of *Linckia amblyonema* Spreng., F; TYPE of *Mastigonema pluviale* A. Br., L); Baiern, *Martens* (TYPE of *Inomeria fusca* Kütz., L); Hercyniae, *Hampe* (TYPE of *Limnactis rivularis* Kütz., L); Mecklenburg, *Flotow* (TYPE of *Rivularia pygmaea* Kütz., L); Sachsen, *L. Rabenhorst* (as *Physactis Pisum* in Rabenh., Alg. Sachs. 52, D, F, PH, NY); Artern, *F. T. Kützing* (TYPE of *Scytonema salinum* Kütz. with *Chthonoblastus salinus* in Kütz., Alg. Aq. Dulc. Dec. 136, D, F, L, NY, PH); bei Berlin, *Willdenow* (TYPE of *Tremella globulosa* Hedw., W), *A. Braun* (TYPES of *Physactis terebralis* Kütz. and *P. saccata* Kütz., L), *Bauer* (TYPES of *Mastigonema Bauerianum* Grun. and *Schizosiphon Bauerianus* Grun., W); Carlsruhe, *Braun* (TYPES of *Rivularia Brauniana* Kütz. and *R. gigantea* Trentep., L); Clausthal, *Römer* (TYPES of *Mastigonema caespitosum* Kütz. and *Limnochlide hercynica* Kütz., L); Constanz, *E. Stizenberger* (as *Euactis rivularis* in Jack, Lein. & Stizenb., Krypt. Badens 105, 106, D, F, and in Rabenh., Alg. Sachs. 289, 332, 678, D, F, PH; as *Rivularia pygmaea* in Rabenh., Alg. Sachs. 355, D, F, PH); Donaueschingen (Baden), *Braun 49* (TYPE of *Calothrix Braunii* Born. & Flah., PC); Driesen (Neumark), *Lasch* (as *Physactis chalybea* in Rabenh., Alg. Sachs. 245, D, PH); Eilenburg, *Kützing* (TYPE of *Rivularia minuta* Kütz., L); Eisenach, *W. Migula* (as *Aphanizomenon flos-aquae* and *Gloeotrichia Pisum* in Mig., Krypt. Exs. Alg. 78, OC, 260, D, F); Eschenau, *Kemmler* (TYPE of *Mastigonema pluviale* var. *Kemmleri* Rabenh. in Rabenh., Alg. Sachs. 733, D); Flensburg, *R. Häcker* (as *Rivularia Sprengeliana* and *R. Lyngbyana* in Rabenh., Alg. Sachs. 793, 932, D, F, FH); Freiburg (Baden), *Braun* (TYPES of *Schizosiphon sabulicola* A. Br. and *Limnactis parvula* Kütz., L), *Schnurmann* (TYPES of *L. Schnurmannii* A. Br. and *Mastigonema subulatum* Kütz., L); Görlitz, *Peck* (as *Rivularia minuta* in Rabenh., Alg. Sachs. 143, D, F, PH); bei Halle, *Kützing* (TYPES of *R. Sprengeliana* and *Dasyactis salina* Kütz., L), *O. Bulnheim* (as *D. salina* in Rabenh., Alg. Sachs. 570, D, PH); Hammerstein am Rhein, *J. Kühn* (TYPE of *Schizosiphon Kuehneanus* Rabenh. in Rabenh., Alg. Eur. 1851, PH); Hanau, *Theobald* (TYPES of *Physactis cerasum* Kütz. and *Rivularia minor* Kütz., L); Heringen, *Wallroth* (TYPE of *Limnactis minutula* Kütz., L); Salzdettfurth bei Hildesheim, *F. A. Römer* (TYPE of *Schizosiphon curvulus* Kütz., NY); Höchst (Nassau), *A. De Bary* (as *Rivularia angulosa* in Rabenh., Alg. Sachs. 931, D, F, PH); Husbyer Moor

(Schleswig), *Hansen* (TYPE of *Physactis chalybea* Kütz., L); Jever, *Koch 155* (TYPE of *P. villosa β major* Kütz., L); Kork, Strassburg gegenüber, *Braun 11* (TYPE of *Rivularia parvula* Kütz., L); Kreuth (Baiern), *Bausch* (as *Euactis amnigena* in Rabenh., Alg. Sachs. 679, D, F); Laacher-See, *H. Royers* (TYPE of *Calothrix rupestris* Roy., with *Scytonema myochrous*, F); bei Leipzig, *O. Bulnheim* (TYPE of *Rivularia Pisum* b *saccata* Rabenh. as *Physactis Pisum* in Rabenh., Alg. Sachs. 36, PH; as *Schizosiphon salinus* in Rabenh., Alg. Sachs. 609, D, F), *Auerswald* (TYPE of *Limnochlide flos-aquae* var. *fulva* Auersw. in Rabenh., Alg. Sachs. 410, PH; as *L. flos-aquae* in Rabenh., Alg. Sachs. 246, D, F), *P. Richter* (as *Mastichothrix aeruginea* in Rabenh., Alg. Eur. 2155, D, F, UC), *H. Reichel* (TYPE of *Gloeotrichia solida* Richt. in Hauck & Richt., Phyk. Univ. 83, MIN); Neckarau bei Mannheim, *Mettenius* (TYPE of *Rivularia durissima* Kütz., L); bei Mansfeld, *P. Richter* (as *Gloeotrichia Pisum* in Wittr. & Nordst., Alg. Exs. 660a, D, NY), *Kützing* (TYPE of *Rivularia salina* Kütz., L; duplicates as *R. angulosa* in Kütz., Alg. Aq. Dulc. Dec. 90, D, PH; as *R. atra* in Kütz., Alg. Aq. Dulc. Dec. 88, D, PH); Merseburg, *Kützing* (as *R. dura* in Kütz., Alg. Aq. Dulc. Dec. 89, D, PH; TYPES of *Chalaractis villosa* Kütz. and *C. mutila* Kütz., L); Moabit bei Berlin, *Braun* (TYPES of *Calothrix adscendens* Born. & Flah. and *Rivularia intermedia* Lemm., PC), *De Bary* (as *R. minor* in Rabenh., Alg. Sachs. 295, D, F); Mosbach, *Migula* (as *Gloeotrichia natans* and *G. Pisum* in Mig., Crypt. Exs. 134, 260, OC); Müggelsee, *P. Hennings* (as *G. natans* in Henn., Phyk. March. 37, FH); Neudamm, *Itzigsohn & Rothe* (as *Amphithrix incrustata, Rivularia gigantea, Limnactis dura*, and *Physactis Pisum* in Rabenh., Alg. Sachs. 198, 211, 235, 236, D, F, NY); TYPE of *Gloeotrichia Rabenhorstii* Born. & Thur. in Rabenh., Alg. Sachs. 554, PC), *Itzigsohn* (TYPE of *Physactis spirifera* Kütz., L; as *Rivularia minuta* in Rabenh., Alg. Sachs. 416, D, F); Neustadt (Baden), *Schnurmann* (TYPE of *R. coadunata* fa. *pseudogypsophila* Pol., as *Limnactis Schnurmannii* in Rabenh., Alg. Sachs. 771, PH); Nordhausen (TYPE of *Schizosiphon gypsophilus* Kütz., L), *Kützing* (TYPES of *Rivularia lenticula* Kütz., *Amphithrix crustacea* Kütz., *Mastigonema fasciculatum* Kütz., *M. paradoxum* Kütz., and *Mastichothrix fusca* Kütz., L); Oldesloe (Holstein), *C. Sonder* (TYPE of *Aphanizomenon holsaticum* Richt. in Hauck & Richt., Phyk. Univ. 746, NY; duplicate: MIN); Pfohren, *F. Brunner* (as *Gloeotrichia Sprengeliana* in Jack, Lein. & Stizenb., Krypt. Badens 803, D, F); Plöner Seen, *O. Zacharias* (as *G. echinulata* in Hauck & Richt., Phyk. Univ. 587, NY); Potsdam, *A. Braun* (as *Physactis Pisum* in Rabenh., Alg. Sachs. 870, D, NY); Schnepfenthal (Thüringen) (as *Physactis spirifera* in Rabenh., Alg. Sachs. 316, FH); Schwäbisch Hall, *Kemmler* (as *Rivularia lenticula* in Rabenh., Alg. Sachs. 975, D, FH); Wanzleben bei Mansfeld, *Kunze* (TYPE of *Dasyactis Kunzeana* Kütz., L); Weissenfels, *Kützing* (TYPE of *Mastichothrix aeruginea* Kütz., L); Wifels (Oldenburg), *Koch 32* (TYPE of *Rivularia rigida* Kütz., L); Wurzen, *Bulnheim* (as *R. Sprengeliana* in Rabenh., Alg. Eur. 1125, D, F, FH).

Czechoslovakia: Moravia: Eisgrub, *H. Zimmermann* (as *Aphanizomenon flos-*

aquae in Mus. Vindob. Krypt. Exs. Alg. 1005, D, F); Studenec, *F. Nováček* (D); Mohelno, *Nováček* (D). Slovakia: Pozsony, *J. A. Bäumler* (as *Aphanizomenon flos-aquae* in Fl. Hung. Exs. Alg. 722, D, F); Lersch-villa (Magas-Tatra), *E. Kol* (as *Rivularia haematites* in Fl. Hung. Exs. Alg. 921, D, F). Bohemia: Carlsbad, *Schwabe* (TYPE of *Oscillatoria subulata* Corda, PC); St. Prokop, *A. Hansgirg* (as *Calothrix parietina* in Wittr. & Nordst., Alg. Exs. 751c and in Fl. Exs. Austro-Hung. 2394, D, F; TYPE of *Scytonema Hofmannii* var. *calcicola* Hansg., W), *E. Bauer* (as *Calothrix parietina* in Mus. Vindob. Krypt. Exs. Alg. 1006, D, F); Chlumec, *Hansgirg* (TYPE of *C. Baueriana* var. *minor* Hansg., W); Chotzen, *Hansgirg* (as *C. salina* in Fl. Exs. Austro-Hung. 2395, D, F); Elbe-Kostelec, *Hansgirg* (as *Gloeotrichia Pisum* in Wittr. & Nordst., Alg. Exs. 754, D, F, PH); Böhmisch-Leipa, Budějovice, Dobříš, Doksy, Eichwald, Eisenbrod, Hluboká, Hohenfurth, Karlstein, Königgrätz, Kostelec na Labi, Polná, Přelouč, Roud-nice, Rožd'álovic, Šárka, Smichov, Trojský Ostrov, Veselí, Vrané, and Zás-muky, *Hansgirg* (D, F, W); Čimelic, *Hansgirg* (as *Gloeotrichia Pisum* in Fl. Exs. Austro-Hung. 2800I, FH).

Hungary: Soltvadkert (Pest), *F. Filarszky & J. B. Kümmerle* (as *G. natans* in Fl. Hung. Exs. Alg. 723, D, F, FH); Ó-Buda, *Filarszky* (as *Rivularia dura* in Mus. Vindob. Krypt. Exs. Alg. 74, D, F); ad Olaszinum, *Kalchbrenner* (as *Euactis rivularis* in Rabenh., Alg. Sachs. 931, D, F).

Austria: Lower Austria: Wien, *G. de Beck* (as *Microchaete tenera* in Mus. Vindob. Krypt. Exs. Alg. 334, D, F), *S. Stockmayer* (as *Calothrix solitaris* in Hauck & Richt., Phyk. Univ. 294, K in BM; as *C. adscendens* in Mus. Vindob. Krypt. Exs. Alg. 147, D, F); Frankenfels, *Stockmayer* (as *Rivularia rufescens* in Mus. Vindob. Krypt. Exs. Alg. 332, D, F); Langau, *de Beck* (D, F, W); ad Scheibbs, *Stockmayer* (as *R. haematites* in Mus. Vindob. Krypt. Exs. Alg. 331, D, F; also in Mig., Crypt. Exs. Alg. 43, OC); Vöslau—Oyenhausen, *K. H. Rechinger* (D, W). Burgenland: Podersdorf, *C. Rechinger* (as *Calothrix parietina* in Mus. Vindob. Krypt. Exs. Alg. 1006b, D, F). Upper Austria: Vöcklabruck, *von Mörl* (TYPE of *Aulosira laxa* Kirchn., W); Leonstein, *A. Grunow* (D, W); Almsee, *Schiedermayr* (as *Euactis fluviatilis* and *E. mollis* in Rabenh., Alg. Eur. 1304 and Suppl., UC); Hallstatt, *K. & L. Rechinger* (as *Rivularia haematites* in Mus. Vindob. Crypt. Exs. Alg. 3935, D, W). Salzburg: Wildbad-Gastein, *Beck* (TYPE of *Dichothrix Nordstedtii* var. *salisburgensis* Beck in Mus. Vindob. Krypt. Exs. Alg. 73, W); Salzburg (TYPE of *Euactis chrysocoma* Kütz., L); Hallein, *Sauter* (as *Schizosiphon rufescens* in Rabenh., Alg. Sachs. 579, D, UC; as *S. crustiformis* in Rabenh., Alg. Eur. 1121, D, F); Saalfelden, *Sauter* (as *Zonotrichia chrysocoma* in Rabenh., Alg. Sachs. 145, D, F). Steyermark: Rudersdorf nächst Graz, *A. Hansgirg* (TYPE of *Tolypothrix rivularis* Hansg., W; duplicates: D, F); Grundlsee, *K. H. Rechinger* (D, W); Tobelbad, *Hansgirg* (D, F, W); Mürzsteg, *G. de Beck* (D, F, W). Carinthia: Kolbitsch, Kreutzeckgebiet, *H. Simmer* (TYPE of *T. calcarata* fa. *minor* Schmidle, PC); Pontafel, *Hansgirg* (as *Rivularia rufesens* in Wittr. &

Nordst., Alg. Exs. 756, D, F); Friesach and Klagenfurt, *Hansgirg* (D, F, W). Tyrol: Kufstein, Jenbach, and Patsch, *Hansgirg* (D, F, W).

Yugoslavia: Slovenia: Franzdorf, Laibach, Lupoglava, Pirano, Podnart, Pola, Steinbrück, Tüffer, and Veldes, *Hansgirg* (D, F, W); Kronau, *F. Hauck* (as *R. haematites* in Wittr. & Nordst., Alg. Exs. 666, D, F), *C. Keissler* (as *R. haematites* in Mus. Vindob. Krypt. Exs. Alg. 331b,c, D, F). Croatia: Fiume and Cherso, *Hansgirg* (D, F, W). Bosnia: Nieder-Tuzla and Doboj, *Hansgirg* (D, F, W). Dalmatia: in flumine Scardona, *Meneghini* (TYPE of *Euactis scardonitana* Kütz., L); Castelnuovo, Ragusa, Sebenico, Spalato, and Zara, *Hansgirg* (D, F, W).

Albania: Paraboar-Lilai, *G. de Toni* (D, de Toni).

Greece: Panormos (Mykonos), *K. H. Rechinger* (D, W).

Switzerland: Gletsch, Innertkirchet, and Rhonegletscher, *H. Royers* (D, F); Albula pr. Bergün, *Hepp* (as *Rivularia haematites* in Wittr. & Nordst., Alg. Exs. 860, D, F); Bern, *Shuttleworth* (TYPE of *Euactis Shuttleworthiana* A. Br., L); Dübendorf, *Nägeli 528* (TYPE of *Dasyactis torfacea* Näg., L); Gratsch bei Meran, *J. Milde* (as *Zonotrichia chrysocoma* in Rabenh., Alg. Eur. 1288, D, F); Horgen (Zürich), *C. Cramer* (as *Euactis Regeliana* and *E. Heeriana* in Rabenh., Alg. Sachs. 555 and 556, D, F); Kiffurchweil, *Hepp* (TYPE of *Sirosiphon Heppii* Rabenh. in Rabenh., Alg. Sachs. 610, FH; duplicates: D, F); Lake of Neuchâtel, *A. Braun 142* (TYPE of *Euactis calcivora* Kütz., L), *Hepp* (as *E. calcivora* and *Zonotrichia calcivora* in Rabenh., Alg. Sachs. 680, D, and Alg. Eur. 1811, UC); Couvet (Neuchâtel), *Bulnheim* (as *Dasyactis torfacea* in Rabenh., Alg. Sachs. 651, D, F); Oberstdorf, *W. Migula* (as *Leptochaete Rivulariarum* in Mig., Crypt. Exs. Alg. 201, OC); St. Gallen, *B. Wartmann* (as *Schizosiphon cinctus* in Wartm. & Wint., Schweiz. Krypt. 857, MIN; as *S. Kuetzingianus* in Rabenh., Alg. Sachs. 816, D, F; TYPE of *Gloeotrichia durissima* b *minuta* Rabenh. as *Rivularia minuta* in Rabenh., Alg. Eur. 1095, PH; duplicates: D, F, FH); St. Moritz, *Nägeli* (TYPE of *Zonotrichia pulchra* Rabenh., PC); Schaffhausen, *Nägeli* (TYPES of *Euactis rivularis* γ *fluviatilis* Kütz. and *Schizosiphon cataractae* Näg., L); Schindellegi (Schwyz), *Nägeli 100* (TYPE of *S. crustiformis* Näg., L); Zürich, *Nägeli* (TYPES of *Dasyactis brunnea* Näg., *D. Naegeliana* Kütz., *D. rivularis* Näg., *Euactis Heeriana* Näg., *E. lacustris* Näg., *E. Regeliana* Näg., *E. rivularis* Näg., *E. rivularis* β *mollis* Kütz., *E. rufescens* Näg., *Hypheothrix turicensis* Näg., *Physactis gelatinosa* Näg., *Schizosiphon cinctus* Näg., *S. decoloratus* Näg., *S. Kuetzingianus* Näg., and *S. parietinus* Näg., L), *H. Schinz* (as *Gloeotrichia Pisum* in Mus. Vindob. Krypt. Exs. Alg. 633, D, F), *G. Winter* (as *Rivularia terebralis* in Wartm. & Wint., Schweiz. Krypt. 758, MIN, and in Rabenh., Alg. Eur. 2563, D, F), *C. Cramer* (TYPE of *R. lacustris* Cram., PC, and in Wartm. & Schenk, Schweiz. Krypt. 347, FH; as *Dasyactis Naegeliana* in Rabenh., Alg. Sachs. 993, D, F).

Italy: Monte Corno, *Meneghini* (TYPE of *Limnactis minutula* β *Orsiniana*

Kütz., L); Fonti del Coppo, *Meneghini* (TYPE of *Mastigonema Orsinianum* Kütz., L). Emilia: Sasso, *Beccari* (TYPE of *Euactis Beccariana* De Not. in Erb. Critt. Ital. ser. 1, no. 1332, FH). Piemonte: Valle d'Andorno, *Cesati* (as *Schizosiphon cinctus* in Rabenh., Alg. Sachs. 732, D, UC); Bardonecchia, *G. Arcangeli* (as *Zonotrichia fluviatilis* in Erb. Critt. Ital., ser. 2, no. 1134, D, F); Biella, *Cesati* (TYPE of *Calothrix Cesatii* Rabenh. in Rabenh., Alg. Sachs. 428, PH); Oropa, *Cesati* (TYPE of *Lyngbya bugellensis* Rabenh. in Rabenh., Alg. Sachs. 436, FH; duplicates: D, F, PH); Vercelli, *Cesati* (TYPE of *Schizosiphon hirundinosus* Ces. in Rabenh., Alg. Sachs. 534, FH; duplicates: D, F, PH). Sardegna: *E. Marcucci* (as *Zonotrichia Marcucciana* in Erb. Critt. Ital., ser. 2, no. 1046, D, F); Tacquitara (Barbargia), *Marcucci* (TYPE of *Z. Marcucciana* Rabenh. in Marc., Un. Itin. Crypt. XI, F); Calabona (Alghero), *Marcucci* (as *Schizosiphon crustiformis* in Marc., Un. Itin. Crypt. XXIIIb, D, F). Liguria: Noli, *Piccone & Cesati* (as *Euactis rivularis* in Erb. Critt. Ital., ser. 2, no. 333, D, F). Toscana: in thermis Julianis, *Meneghini* (TYPES of *Lyngbya juliana* Menegh. and *Schizosiphon julianus* Kütz., L); selva Pisana, *Savi & Marcucci* (TYPE of *Rivularia Marcucciana* De Not. in Erb. Critt. Ital., ser. 1, no. 1331, FH); bagni Ferrari presso Livorno, *Arcangeli* (as *Schizosiphon crassus* in Erb. Critt. Ital., ser. 2, no. 1132, D, F). Venezia Euganea: Colli Euganei, *Meneghini* (TYPES of *S. aponinus* Menegh., *Amphithrix incrustata* Kütz., *Rivularia rudis* Menegh., and *Scytonema melanopleurum* Menegh., L; TYPE of *S. furcatum* Menegh., PC); Forojulii (Fruili), *Meneghini* (TYPE of *Ainactis alpina* Kütz., L); Padova, Meneghini (TYPE of *Rivularia Lens* Menegh., FI); Pontebba, *A. Hansgirg* (as *R. rufescens* in Fl. Exs. Austro-Hung. 3187, MO); in flumine Tartaro, *Meneghini* (TYPE of *Schizosiphon Meneghinianus* Kütz., L); Caldiero, *A. Massalongo* (TYPE of *Symphyosiphon Castellii* Mass., PC; duplicates in Rabenh., Alg. Sachs. 589, D, F). Venezia Giulia: Boliunz, *F. Hauck* (as *Rivularia haematites* in Wittr. & Nordst., Alg. Exs. 665b, D, F); Contovello and Görz, *Hansgirg* (D, F, W); Miramar, *F. Krasser* (as *R. mesenterica* in Mus. Vindob. Krypt. Exs. Alg. 747, D, F); Monfalcone, *F. T. Kützing* (TYPE of *Geocyclus oscillarinus* Kütz., L), *Hauck* (as *Rivularia Biasolettiana* in Wittr. & Nordst., Alg. Exs. 577, D, F); Trieste, *Biasoletto* (TYPE of *R. Biasolettiana* Menegh., FI); Nabresina, *Hansgirg* (D, F, W). Venezia Tridentina: Auer, Branzoll, Kardaun bei Bozen, Neumarkt, Roveredo, and Trento, *Hansgirg* (D, F, W); Bozen, *Heufler* (as *Zonotrichia fluviatilis* in Rabenh., Alg. Eur. 2157, D, F).

Netherlands: Giekerk, *W. F. R. Suringar DDD92* (TYPE of *Schizosiphon sociatus* Sur., L); Leeuwarden, *Suringar* (TYPES of *Physactis villosa* fa. *obsoleta* Sur. and *Rivularia Boryana* fa. *congesta* Sur., L); Nieuwkoopsche plassen, *J. T. Koster 141* (D, L); Rhenen, *T. Sprée* (TYPE of *R. insignis* Sprée in Rabenh., Alg. Eur. 1452, FH; duplicates: D, F, PH).

Belgium: Grimberghe près de Termonde, *C. D. Westendorp & H. C. J. Wallays* (TYPE of *Gloeotrichia parvula b Westendorpii* Rabenh. in Westend. & Wall., Herb. Crypt. Belg. 1345, FH).

Faeroes: Näs (Österö), *H. G. Simmons 394* (MO).

Scotland: (TYPE of *Rivularia crustacea* Carm., BM); Appin (Argyll), *Carmichael* (TYPES of *R. granulifera* Carm. and *Calothrix Berkeleyana* Carm., K in BM); Pentland Hills, Edinburgh, *R. K. Greville* (BM); Lake Dungeon, Kirkcudbrightshire, *G. S. West 2* (D, UC).

England: (TYPE of *Coenocoleus cirrhosus* Berk. & Thw., BM); stream, Brotherswater Hotel near Ambleside, *F. Drouet, J. Talling & H. B. Louderback 15024* (D); Lake Windermere, *J. W. G. Lund* (TYPES of *Homoeothrix Pearsallii* Godw. and *Dichothrix Baueriana* var. *crassa* Godw., D); Dartmoor, Devon, *T. Norman* (TYPE of *Ammatoidea Normanii* W. & G. S. West, BM); Moughton Fell, West Yorks, *West & West* (D, UC); Bristol, *C. E. Broome* (BM); Ilfracombe, *J. Ralfs* (BM); Slapton Pool, *E. M. Holmes* (BM); Frensham Great Pond, Surrey, *C. E. Salmon* (BM).

Wales: Swansea, *Ralfs* (TYPE of *Rivularia Boryana* β *flaccida* Kütz., L); Dolgelly, *Ralfs* (TYPE of *R. botryoides* Carm., K in BM); Anglesea, *Davies* (TYPE of *Conferva echinulata* Sm. & Sow., BM).

Ireland: (TYPE of *Rivularia calcarea* Sm. & Sow., LD); Dublin, *Allman* (TYPE of *Trichormus incurvus* Allm., K in BM); "Renoyle, Cunnamara," *R. J. Shuttleworth 61A* (BM).

France: (as *Gaillardotella natans* in Moug. & Nestl., Stirp. Crypt. Vogesorhen. 796, NY; as *Linckiella natans* in Desmaz., Pl. Crypt. France 751, NY; TYPE of *Batrachospermum haematites* Lam. & DC., PC); "lacus Longemar Vogesorum" (TYPE of *Oscillatoria Mougeotiana* Ag., LD); "Belle-Croix . . . apud Fontem Bellaqueum," *Roussel* (TYPE of *Sclerothrix Rousseliana* Mont., PC). Aisne: Berthenicourt-par-Moy, *J. Mabille 5* (D, F). Alpes-Maritimes: Antibes, *G. Thuret* (TYPE of *Calothrix decipiens* Born. & Thur., PC). Anjou: Étang de Chaumont, *C. Flahault* (FH). Basses-Pyrénées: Biarritz, *Thuret* (D, YU). Calvados: Falaise, *A. De Brébisson* (as *Limnactis dura* in Rabenh., Alg. Eur. 1451, D, F; TYPES of *L. dura* Kütz., *L. flagellifera* Kütz., *Merizomyria flagelliformis* Kütz., *Schizosiphon apiculatus* β *rupestris* Kütz., *S. radians* Kütz., and *Rivularia Brebissoniana* Kütz., L; TYPE of *R. villosa* fa. *minor* Rabenh. in Rabenh., Alg. Eur. 2184, F). Finistère: nord de Plougastel, *P. L. & H.M. Crouan* (TYPE of *Mastigonema paludosum* Crouan in Lab. Mar. Biol., Concarneau). Cantal: Thiézac à Aurillac, *M. Gomont 25* (D, F, PC). Hérault: Agde, *Flahault* (as *Rivularia Warreniae* in Wittr. & Nordst., Alg. Exs. 662, D, S); near Montpellier (as *R. haematites, Schizosiphon Bauerianus, Dichothrix Baueriana, Calothrix parietina*, and *C. juliana* in Wittr. & Nordst., Alg. Exs. 581, 659, 752, 755, and 1305, D, F). Lozère: flumen Tarn, *Flahault* (as *Rivularia Biasolettiana* in Wittr. & Nordst., Alg. Exs. 861, D, F). Maine-et-Loire: Angers, *M. Gomont* (TYPE of *Calothrix stagnalis* Gom., PC; duplicate: D), *F. Hy* (as *C. balearica, C. stagnalis, Dichothrix Baueriana*, and *Aphanizomenon flos-aquae* in Wittr. & Nordst., Alg. Exs. 851, 858, 1341b, 1502, D, NY). Manche: Cherbourg, *G. Thuret* (TYPE of *Gloeotrichia punctulata* Thur., PC; duplicates: D, F, FH), *A. Le Jolis 2209, 2215*

(FH); Lafeuillie, *P. Frémy* (as *G. natans* in Hamel, Alg. de France 6, D, F, MICH); Saint-Gilles près Saint-Lô, *Frémy* (as *Calothrix parietina* in Hamel, Alg. de France 9, D, F, MICH). Nord: Bailleul, *Flahault* (as *Gloeotrichia natans* in Wittr. & Nordst., Alg. Exs. 753b, D, F, NY). Savoie: Aix-les-Bains, *H. S. Forest* (D). Seine-et-Oise: Lardy, *Bornet & Bonnier* (TYPE of *Microchaete diplosiphon* Gom., PC), *Gomont* (as *M. diplosiphon* in Wittr. & Nordst., Alg. Exs. 870, D, F, NY), *E. Bornet & C. Flahault* (as *Dichothrix gypsophila* in Wittr. & Nordst., Alg. Exs. 859, D, F); St.-Quentin près Versailles, *Thuret* (TYPE of *Aphanizomenon incurvum* Morr., PC; as *A. flos-aquae* in Wittr., Nordst. & Lagerh., Alg. Exs. 1341a, D, F).

Spain: Minorca, *Feminias* (TYPE of *Calothrix balearica* Born. & Flah., PC).

Portugal: ad Tagum, *F. Welwitsch 67* (BM, D).

Egypt: cultures of Egyptian algae, Alexandria, *M. S. Taha* (D, PH); Mallaha prope Alexandriam, *A. Hansgirg* (with *Enteromorpha salina* in Mus. Vindob. Krypt. Exs. Alg. 1101, D, F).

Kenya: Lake Rudolf near Slur, *I. B. Talling* (D).

Tanzania: Lake Tanganyika, *W. A. Cunnington 205* (TYPE of *Calothrix cartilaginea* G. S. West, BM); Lake Victoria Nyanza, Bukoba, *Cunnington 216* (TYPE of *C. brevissima* G. S. West, UC).

Malawi: Anchorage bay, Lake Nyasa, *Cunnington 1* (TYPE of *Gloeotrichia longiarticulata* G. S. West, BM).

Zambia: Kituta bay, Lake Tanganyika, *Cunnington 74* (TYPE of *Rivularia globiceps* G. S. West, UC).

Rhodesia: Lake Kariba, *A. J. McLachlan 8, 9* (PH).

South Africa: Maritzburg, *J. W. Bews 92* (TYPE of *Calothrix gelatinosa* Fritsch & Rich, BM).

Tchad: massif de l'Ennedi, Aoué, *T. Monod 13772* (TYPE of *Dichothrix elongata* Comp., BR; duplicate: D).

The Congo: *J. J. Symoens 430* (D); Tshopo falls near Stanleyville, *Léonard 1351* (D).

Angola: Mossâmedes, Cabo Negro, *F. Welwitsch 19* (TYPE of *Gloeotrichia aethiopica* W. & G. S. West, UC); Mossâmedes, Bero, *Welwitsch 190* (TYPE of *Calothrix epiphytica* W. & G. S. West, UC).

South-West Africa: Nauwkloof mountains near Rehoboth, *R. J. Rodin* (D).

Tunisia: fontaine du Bou Kornine près de Tunis, *M. Serpette TL35* (D, Serpette).

Algeria: Hammam Salahin près Biskra, *C. Sauvageau* (D, PC).

Nigeria: Ikogosi warm spring and Ile-Ife, *T. F. Allen* (Allen, PH); Shaki, *D. J. Hambler 1118* (BM, D); Ibadan, *M. Fox 132* (D, Fox Nielsen).

Dahomey: Cotonou, *D. M. John 6531* (GC, PH).

Ghana: Akotokyir near Cape Coast, Volta river at Battor, and Tokose near Weija, *J. B. Hall A15, A62, A163, A164* (Hall, PH).

Azores: Lagôa Grande, Sete Cidades, San Miguel, *W. Trelease* (MO).

Greenland: glaciers, *Drygalski* (PC); Nunatarssuaq, Thule district, *W. S. Benninghoff & H. C. Robbins 53-20, 53-25, 53-30, 53-33* (D); Thule, *R. W. Gerdel III, 2, 4* (D).

Bermuda: Shore Hills, St. Georges island, *A. J. Bernatowicz 51-797* (D, F, MICH).

New Brunswick: Murray lake south of Dalhousie, *M. Le Mesurier 6* (D, F); Blue Bell, Cormier's Lake, Lake Edward, Ennishore, Gillespie, Grand Falls, Salmon River, Salt Springs (Sussex), and Nauwigewauk, *H. Habeeb 10346, 10645, 10755, 10902, 11399, 11527, 11556, 11609, 11624, 11670, 11697, 11734, 11746, 13474* (D, F, Habeeb).

Maine: Long pond, Mount Desert, *F. S. Collins 3854* (D, UC); Kittery Point, *R. Thaxter* (FH).

New Hampshire: Chocorua, *W. G. Farlow* (D, FH), *Collins* (as *Calothrix fusca* in Coll., Hold. & Setch., Phyc. Bor.-Amer. 1407, D, F); Enfield, *L. H. Flint* (D); Plainfield, *Flint & H. T. Croasdale* (D); Durham, *R. M. Whelden* (D, FH).

Vermont: La Platte river, Charlotte, *F. H. Hosford 71* (TYPE of *C. Hosfordii* Wolle, PC; duplicate: UC), *C. G. Pringle* (FH); Willoughby lake, Orleans, *Flint* (D, F).

Massachusetts: Brewster, *Collins* (as *C. stellaris* in Coll., Hold. & Setch., Phyc. Bor.-Amer. 1953, D, F); Cambridge, *Collins 1955* (D, UC); Concord, *F. C. Seymour* (D, F, FH); Cuttyhunk, *H. T. Croasdale* (D, W. R. Taylor), *G. T. Moore* (as *Rivularia Bornetiana* in Coll., Hold. & Setch., Phyc. Bor.-Amer. 2261, D, F); Coonamessett river, East Falmouth, *R. N. Webster & F. Drouet 2250* (D); Eastham, *Collins* (as *R. echinulata* and *Dichothrix Hosfordii* in Coll., Hold. & Setch., Phyc. Bor.-Amer. 1408, 2262, D, F); Essex, *Collins* (as *D. Baueriana* in Coll., Hold. & Setch., Phyc. Bor.-Amer. 1721a, D, NY); Falmouth, *F. Drouet 1084, 2171* (D, W), *E. T. Rose* (D, NY), *E. T. Moul* (D, RUT); Lynnfield, *W. A. Setchell* (TYPE of *Calothrix gypsophila* fa *saccoideo-fruticulosa* Polj. as *Sacconema rupestre* in Wittr., Nordst. & Lagerh., Alg. Exs. 1309, PH), *Collins 1913* (as *S. rupestre* in Hauck & Richt., Phyk. Univ. 741, NY); Marblehead, *Collins* (as *Rivularia Biasolettiana* in Coll., Hold. & Setch., Phyc. Bor.-Amer. 860, D, F); Medford, *Collins* (as *Calothrix stagnalis* in Coll., Hold. & Setch., Phyc. Bor.-Amer. 1114, D, NY), *F. B. Lambert* (as *Aphanizomenon flos-aquae* in Coll., Hold. & Setch., Phyc. Bor.-Amer. 1359, D, F); Middlesex Fells, *Collins* (TYPE of *Rivularia compacta* Coll., NY, and in Coll., Hold. & Setch., Phyc. Bor.-Amer. 508, D, F); Nahant, *Collins* (D, NY); Naushon island, *J. M. Ferber* (TYPE of *Microchaete naushonensis* Coll., NY), *G. R. Lyman & W. R. Maxon* (as *Dichothrix Baueriana* in Coll., Hold. & Setch., Phyc. Bor.-Amer. 1721b, D, F); Pasque island, *W. R. Taylor* (D, Taylor); Stoneham, *Collins* (as *Rivularia compacta* in Coll., N. Amer. Alg. 34, D, NY); Saugus, *Collins* (as *Dichothrix Hosfordii* in Coll., Hold. & Setch., Phyc. Bor.-Amer. 215c, D, F); Wakefield, *Collins* (as *Gloeo-*

trichia Pisum in Coll., Hold. & Setch., Phyc. Bor.-Amer. 1310, D, F); Waverley, *A. B. Seymour* (D, F); Wellesley, *T. Lawlor* (D); Williamstown, *W. G. Farlow* (D, F, FH).

Rhode Island: Knightsville, Cranston, *W. J. V. Osterhout 430* (D, UC); Lime Rock, *Osterhout* (as *Dichothrix Baueriana* in Coll., Hold. & Setch., Phyc. Bor.-Amer. 216a, D, F); Lincoln, *Collins* (as *Calothrix parietina* in Coll., Hold. & Setch., Phyc. Bor.-Amer. 1360b, D, F, MICH); Newport, *H. M. Richards* (as *Dichothrix Baueriana* in Coll., Hold. & Setch., Phyc. Bor.-Amer. 216b, D, F); Watch Hill pond, *W. A. Setchell* (TYPE of *Rivularia Bornetiana* Setch., UC; duplicates, D and in Coll., Hold. & Setch., Phyc. Bor.-Amer. 157, D, F), *Osterhout 441* (D, UC).

Connecticut: Bridgeport, *I. Holden* (as *Calothrix parietina, C. Braunii, Dichothrix Hosfordii*, and *D. Orsiniana* in Coll., Hold. & Setch., Phyc. Bor.-Amer. 12, 112, 215b, 405, D, F); Gaylordsville, *Holden* (as *D. gypsophila* in Coll., Hold. & Setch., Phyc. Bor.-Amer. 562, D, NY); Greenville, *Setchell* (D, UC); Guilford, *W. T. Edmondson* (D, F); Hamden, *Setchell* (as *D. Hosfordii* in Coll., Hold. & Setch., Phyc. Bor.-Amer. 215a, D, F); Ledyard and Lisbon, *Setchell 27, 714* (D, UC); Middlebury, *Edmondson* (D, F); Mudge lake, Litchfield county, *H. K. Phinney 1129* (D, F, Phinney); New Haven, *F. Drouet 2058* (D, NY, W); North Bridgeport, *L. N. Johnson 94* (D, F); Norwich, *Setchell* (as *Calothrix juliana* in Coll., Hold. & Setch., Phyc. Bor.-Amer. 113, D, F); Salisbury, *Setchell & Holden* (as *Gloeotrichia Pisum* in Coll., Hold. & Setch., Phyc. Bor.-Amer. 311, D, F, FH, US); West Goshen, *Phinney 1119* (D, F, Phinney); Waterbury and Woodbridge, *Edmondson* (D, F).

Franklin District: Assistance bay, Parry islands, *Sutherland 58/1* (TYPE of *Rivularia microscopica* Dick., BM); Hazen Camp, Ellesmere island, *D. R. Oliver 292* (PH).

Quebec: Mount Albert, Gaspé, *H. Habeeb 1778, 1965* (D, Habeeb); Montreal, Valleyfield, St.-François-de-Sales, Charlemagne, Rivière-Lois, Wakefield, Rivière-Beaudette, and Kegashka river, *J. Brunel 66, 73, 278, 353, 363, 393, 399, 739* (D, F, MT); Ungava bay at Korok river, entre le lac Tashwak et le lac Payne, estuaire de rivière George, and rivière Payne vers 71° 25′ W, *J. Rousseau 483, 656, 1096, 1134* (D, F, MTJB); Mont-Royal and Les Forges, *C. Lanouette 101, 126* (D, F); Fort Chimo (Koksoak river), *H. A. Senn* (D, F); Petit-Pabos (Gaspé), *E. Jacques 425* (D, F); Sorel, *C. G. Pringle* (FH).

New York: Bowmansville, *J. Blum 201* (D, F); Brewerton, *C. H. Peck* (FH); soil cultures, Brookhaven, *E. H. Franz 104* (PH); Buffalo, *F. Wolle* (D), *Blum 171* (D, F); Chautauqua lake, *W. L. Tressler* (D, F); Conesus lake, *H. S. Forest* (PH); Edwards and Falkirk, *Blum 309, 325* (D, F); Cayuga lake, Ithaca, *G. F. Atkinson* (as *Gloeotrichia natans* in Coll., Hold. & Setch., Phyc. Bor.-Amer. 214, D, F, US); Lake George, *S. E. Jelliffe* (UC); Mendon Ponds Park, *Blum 405* (D,

F); Mohansic lake (Westchester county), *H. C. Bold B123* (D, F); Bronx park, New York, *R. Weikert* (D, NY); Niagara Falls, *O. Kuntze 13* (NY), *F. Wolle* (D), *B. M. Davis* (D, F, MICH), *F. Drouet, H. B. Louderback & A. Owen 14982* (TYPES of *Zonotrichia mollis* Wood and *Z. parcezonata* Wood, PH); Sardinia, Portageville, and North Java, *Blum 167, 296, 402* (D, F); Southold, *R. Latham 3554* (NY); Westerleigh, Staten island, *I. C. G. Cooper* (D, F).

New Jersey: Boonton, Middleville, Springdale, Upper Greenwood lake, and Bernardsville, *H. Habeeb 3131, 3826, 3855, 3948, 4146* (D, Habeeb); Morris pond, *F. Wolle* (as *Calothrix Orsiniana* in Wittr. & Nordst., Alg. Exs. 389, D, F); Pleasantville, *J. E. Peters* (as *C. fusca* in Coll., Hold. & Setch., Phyc. Bor.-Amer. 11, D, F, NY); Toms River, *J. C. Bader* (D, F); Andover, *E. T. Moul 7506* (D, F, RUT).

Pennsylvania: Stroudsburg and Pottstown, and cultures, *F. Drouet 14916, 14977* (PH); North Warren, *T. Flanagan 23* (D, F); Indian Caverns, Huntingdon county, *B. Wertz* (PH); Bethlehem, *F. Wolle* (as *Gloeotrichia parvula* in Rabenh., Alg. Eur. 2539, D, F; TYPE of *Zonotrichia paradoxa* Wolle, NY; TYPES of *Mastigonema violaceum* Wolle, *M. fuscum* Wolle, *M. luteum* Wolle, and *Mastichothrix turgida* Wolle, D); Nazareth, *Wolle* (TYPE of *Isactis caespitosa* fa. *tenuior-viridis* Rabenh., D); Derry Church, Dauphin county, *Wolle* (D, F); Susquehanna river, Wyoming county, *R. R. Grant Jr. 9* (TYPE of *Gloeotrichia incrustata* Wood, PH).

Delaware: Rehoboth Beach, *F. Drouet & H. B. Louderback 8539* (D).

Maryland: Plummers island near Cabin John, *Drouet & E. P. Killip 3966, 5570* (D, F, US); Potomac river near mouth of Monocacy river, *M. H. Hohn 9* (D, PH); Point Patience, Calvert county, *H. C. Bold 5* (D, F).

District of Columbia: culture from pond, *F. E. Allison 20* (D).

Virginia: Dot, Flat Rock, Haysi, Kire, Mountain Lake, Mount Rogers, Oakwood, Staunton, Tazewell, and Williamsburg, *J. C. Strickland et al. 130, 447, 931, 1074, 1223, 1271, 1301, 1305, 1341, 1361* (D, F, Strickland); New Baltimore, *A. J. Pieters* (D, F); Richmond, *M. H. Wood* (D, F, Strickland); Mountain Lake, *I. F. Lewis* (D, F, Strickland).

West Virginia: Stollings, *A. T. Cross* (D, F); Shawnee lake, Mercer county, *E. M. McNeill 318* (D, F).

North Carolina: Swift creek (Wake county), and Raleigh, *L. A. Whitford W8* (D), *W1101* (PH); Hiwassee reservoir, Cherokee county, *H. Silva 1742* (D, F); Beaufort, *H. J. Humm 15* (D, F), *W. Culberson & C. S. Nielsen 1749* (D, FSU); Hot Springs and Franklin, *Nielsen & Culberson 1806, 1894* (D, FSU); Highlands, old Highway 106 (Jackson county), Onion Skin Falls, Highway 181 (Burke county), and Toxaway Gorge, *H. C. Bold H39, H173, H177, H373, H408* (D, F).

South Carolina: Santee Canal, *H. W. Ravenel 28* (BM, D); Walhalla, *Bold*

H67 (D, F); Gaffney, *P. J. Philson 58* (TYPE of *Calothrix Braunii* var. *maxima* Phils., F); Florence, *Philson SC31* (TYPE of *C. genuflexa* Phils., F); Lake Chapin, Myrtle Beach, *Philson SC42* (TYPE of *Anabaena parva* Phils., F).

Georgia: Warwoman creek, Rabun county, *Bold H382* (D, F); Forest Falls, Decatur county, *R. M. Harper 1194a* (D, F, MO).

Florida: Florida Caverns state park, Gulf Beach, McIntosh, Steinhatchee, Newport, Port Leon, Wakulla Springs, Riverside, Carrabelle Beach, and Key West, *F. Drouet et al. 10405, 10564, 11100, 11227, 11347, 11460, 11474, 11525, 11672, 14946* (D, F); Fort Myers, Bunnell, and Tampa, *P. C. Standley 73229, 92779, 92657* (D, F); Gainesville, Orlando, and Dunnellon, *M. A. Brannon 367, 350, 379* (D, F, PC); culture, Tallahassee, *C. S. Nielsen 200* (D, F, FSU); Kissimmee, *A. H. Johnston 1936* (D, F, FSU); Aspalaga, *R. C. Phillips & L. R. Almodóvar 610* (D, FSU); Cocoanut Grove, *R. Thaxter* (D, FH); Big Cypress swamp, Collier county, *M. Díaz-Piferrer 10* (D, FPDB); Dunedin, *S. Ericson* (D, FSU).

Ontario: Niagara Glen park and Niagara Falls, *F. Drouet & H. B. Louderback 14978, 15005* (PH); rapids above Niagara falls, *J. Blum 141* (D, F); Almonte, *A. F. Kemp* (D, FH); Clear lake (Kawartha lakes), *W. A. Strow* (D, F); Kingston, *J. H. Wallace 18* (D, PH).

Ohio: Athens, Addyston, Cincinnati, Glendale, Sharon, Sharonville, South Amherst, and Squaws Harbor (Ottawa county), *W. A. Daily et al. 127, 175, 193, 214, 218, 308, 610, 650* (D, Daily, F); Hovey's pond (Lorain county) and Oberlin, *O. S. Curtis* (OC); North Appalachian experimental watershed (Coshocton county), *L. J. King 689* (D, EAR, F); South Bass island, Lake Erie, *C. E. Taft, E. H. Ahlstrom* (D, Daily, F), *J. Blum 300* (D, F); McConnelsville, *A. H. Blickle* (D, Daily, F); Newtown, *J. B. Lackey* (D, F); cultures from Ohio, *C. M. Palmer,* Environmental Health Center no. 22, 25 (D, F).

Kentucky: Slade and Clay City, *W. A. Daily et al. 754, 756* (D, Daily, F); Earlington and Lexington, *B. B. McInteer 16, 592, 1040* (D, Daily, F); Carters Caverns (Carter county), *L. Walp* (D, F); Louisville, *A. T. Hotchkiss* (D), *T. W. Tichenor 25* (D, F).

Tennessee: Porters Prong (Sevier county), Horse Creek (Washington county), Knoxville, Nolichucky Dam, Walnut Log, Neuberts Springs (Knox county), Elkmont, Laurel Lake, Norris Dam, and Abrams Falls (Blount county), *H. Silva 209, 471, 834, 913, 1204, 1533, 1642, 1659, 1871, 1887* (D, F, TENN); Harrogate, Knoxville, Linden, Montgomery Bell state park, and Nashville, *H. C. Bold* (D, F); Daytona, *A. J. Sharp 1987* (D, F, TENN).

Alabama: Tuscaloosa, *L. G. Williams* (PH), *R. L. Caylor 2* (D, F); Fort Morgan, *Caylor 51-7a* (D, F); Marion, *J. Snow 12* (D, Daily); Wilson Lake (Colbert County), *H. Silva 2082* (D, F, TENN).

Michigan: Charlevoix, Beaver island, Falls of Tahquamenon, and Fox Park (Menominee county), *F. Drouet & H. B. Louderback 12357, 12392, 12521,*

12542, 13155 (D); Ann Arbor, *L. N. Johnson 1021* (D, F), *W. R. Taylor* (D, Taylor); Pictured Rocks (Alger county), Black lake (Cheboygan county; TYPE of *Rivularia cartilaginea* Wood, D), Mullet Lake (TYPES of *Dasyactis mollis* Wood and *Mastigonema sejunctum* Wood, D), Douglas Lake, Trout Lake, Alanson, Pointe aux Pins, and Ocqueoc lake (Presque Isle county), *H. K. Phinney 200, 212, 252, 489, 27M39/1, 27M41/1, 9M41/7-1, 13M41/7* (D, F, Phinney); Sturgeon Bay, *J. Blum 366* (D, F); Lake Medora, (Keweenaw county), *W. C. Beckman* (MICH); Lancaster lake (Cheboygan county), *A. H. Gustafson* (D, F); Buchanan, *J. M. Beal* (D, F); Lake Odessa, *W. E. Wade* (D, F); Grand Traverse bay, *J. H. Hoskins* (D, F, Daily).

Indiana: Beverly Shores and Miller, *Drouet 5820, 5821, 5862* (D, F); Columbus, Hartsville, Monticello, Logansport, Terre Haute, Greensburg, Bristol, Fulton, North Vernon, Lake Wawasee, Silver Lake, Cedar Lake, Mill Creek, Indianapolis, Culver, Valparaiso, St. Paul, Knox, Angola, and Riley, *F. K. & W. A. Daily et al. 91, 876, 934, 1040, 1107a, 1142, 1162, 1165, 1482, 1501, 1550, 2193, 2533, 2547, 2563, 2635, 2661, 2679, 2822, 2996* (D, Daily); Richmond, *L. J. King 26, 374* (D, EAR, F); Salem, Winona Lake, and Indianapolis, *C. M. Palmer* (D, Palmer); Myers lake (Marshall county), *W. R. Eberly 4, 5* (D); Bass lake (Starke county) and Lake Maxinkuckee, *H. W. Clark & B. W. Evermann 35, 121, 136, 147, 261* (D, Daily, F, US); Nashville, *R., E. A. & M. Fritsche 4* (D, Daily, F); Notre Dame, *A. T. Cross* (D, F); Shakamak lake (Sullivan county), *F. Geisler 2375* (D, Daily, F); Stroh and Wilmot, *D. G. Frey* (D, Daily); Richmond, *M. S. Markle* (D, EAR, F); Oswego, *H. B. Metcalf 3* (D, F).

Wisconsin: Terry Andrae state park (Sheboygan county), Oshkosh, Madison, and Prairie du Chien, *F. Drouet & H. B. Louderback 5094, 5493, 5500, 12556, 13248* (D); Round and Whitefish lakes (Sawyer county), Boulder and Big Arbor Vitae lakes (Vilas county), and Shell lake (Washburn county), *G. W. Prescott 2W229, 3W156, 3W175, 3W246* (D, EMC, F); Lake Geneva, *P. D. Voth* (D, F), *W. E. Lake* (D, F); Chetek, *G. M. Smith* (as *Aphanizomenon flos-aquae* in Coll., Hold. & Setch., Phyc. Bor.-Amer. 2209, D, F, W. R. Taylor); Wisconsin Dells, *R. D. Wood* (D, F); Lake Nokomis (Lincoln county), *F. C. Seymour* (D); West Superior, *C. Bullard* (D, FH); Granite lake (Barron county), *G. M. Smith* (D, EMC, F); Lake Rotomer (Rock county), *G. P. Fitzgerald* (D, Daily); Milwaukee, *L. G. Smith* (D, F); culture, Madison, *W. J. Hayes Jr.* (D, F, J. C. Strickland).

Illinois: laboratory cultures, *P. D. Voth, K. Damann, A. E. Vatter Jr.,* and *R. McMillan* (D, F); Chicago, Lemont, St. Charles, and Momence, *F. Drouet et al. 2504, 5270, 5312, 5774, 12624, 13350* (D); Chicago, *Kung Chu Fan 10597, 10644, 10662* (D, Fan); Evanston and Robbsville recreational area, *H. K. Phinney 388, 982, 1027* (D, F, Phinney); Charleston, *E. N. Transeau* (as *Calothrix Kawraiskyi* in Coll., Hold. & Setch., Phyc. Bor.-Amer. 1719b, D, F); Barrington, *P. C. Standley 92455* (D, F); Carbondale, *W. B. Welch C02X* (D, F); Reddick, *E. M. Schugman* (D, F).

Mississippi: Gulfport, *F. Drouet & R. L. Caylor 9945, 9946* (D, F); Buffalo river (Wilkinson county), *Caylor 43-33* (D, F); Sandy Hook, *L. H. Flint & Drouet* (D, F, Flint).

Minnesota: Itasca state park, *F. Drouet 11882, 12243* (D, MIN), *F. O. Fortich 17, 25* (D), *K. C. Fan 10253, 10381* (D, Fan); Mud Lake Wild Life Refuge (Marshall county), *Drouet 12104, 12114* (D, MIN), *Fan 10335* (D, Fan); Redby, *Drouet 12019* (D, MIN); Winona, Lake Phelan, Waterville, and Minneapolis, *J. C. Arthur* (FH); Minneapolis, *J. E. Tilden* (as *Gloeotrichia natans, Calothrix parietina*, and the TYPE of *Dichothrix calcerea* Tild. in Tild., Amer. Alg. 80, 164, 165, D, F, MIN, US); Duluth, *Tilden* (as *Gloeotrichia natans* in Tild., Amer. Alg. 569, D, F); Osceola Mills and Winona, *J. M. Holzinger* (D); Lake Minnetonka, *D. T. MacDougal* (as *G. incrustata* in Tild., Amer. Alg. 81, D, F), *W. G. Farlow* (as *Rivularia fluitans* in Wittr. & Nordst., Alg. Exs. 664, D, F); Minneapolis, *A. P. Anderson* (as *Porphyrosiphon Notarisii* in Tild., Amer. Alg. 65, D, F, MO); Long lake (Hennepin county), *B. T. Shaver & J. E. Tilden* (as *Aphanizomenon flos-aquae* in Tild., Amer. Alg. 173, D, F); Remer, *D. Richards 1105* (D, F); Lake Elmo (Washington county), *E. G. Reinhard* (D, F); East Silent lake (Ottertail county), *S. Eddy* (D, F); Okabena lake (Johnson county), *C. B. Reif* (D, F).

Iowa: Iowa City, Fairport, and Dane Ray, *G. W. Prescott Ia2, 265, 321* (D, EMC); East Okoboji lake, *J.C. Arthur* (FH), *J.D. Dodd 1* (D, Daily); Spirit lake, *C. W. Reimer* (PH), *Dodd 5* (D, Daily); West Lake Okoboji, *Reimer* (PH); Silver lake (Dickinson county), *E. Whitehouse 23409* (D, F, SMU); Jemmerson slough (Dickinson county), *D. Fritze* (PH); Ventura, *F. M. Begres* (PH); Clear Lake, *T. E. Jensen 21* (PH).

Missouri: Alley Spring state park and Waynesville, *F. Drouet & H. B. Louderback 13487, 14359a* (D); Columbia, Warrensburg, Gravois Mills, and Choteau Springs (Cooper county), *Drouet 77, 681, 689, 720, 778, 2391, 5647* (D); Center creek (Jasper county), *J. Cairns 4* (PH); Iberia and Zora, *C. Shoop 47, 133* (D, F); St. Louis, *H. H. Iltis 4137* (D, MO); Fayette, *J. R. Hurt NF-11* (D, F); Liberty, *C. J. Elmore* (D).

Arkansas: Salem, *Drouet & Louderback 13480* (D); Boxley, *D. M. Moore & H. H. Iltis* (D, F, UARK); Monticello, *M. Thomason 250* (D, F), *D. Demaree 24552a, 25469* (D, F); Arkansas Post, *Demaree 25467a* (D, F); Stuttgart, *K. L. Olsen 6* (D); Lake Gertrude (Garland county), *N. E. Gray 126* (D, F); Fayetteville, *Iltis* (D, F).

Louisiana: Lake Charles, New Orleans, and Mandeville, *F. Drouet et al. 8758, 9379, 9381, 9537* (D, F); Baton Rouge, Calhoun, Covington, Gonzales, and Hammond, *L. H. Flint* (D, F, Flint); Carville, Crowley, Eunice, and Golden Meadow, *G. W. Prescott La56, La73, La89a, La95, La96* (D, EMC).

North Dakota: Court lake, *M. A. Brannon* (D, F).

South Dakota: Columbia, *D. Griffiths 20* (BKL; as *Anabaena oscillarioides* var. *elongata* in Tild., Amer. Alg. 293, MIN); Aberdeen, *Griffiths 16, 21* (BKL); Big Stone Lake, *Griffiths* (as *Rivularia Biasolettiana* in Tild., Amer. Alg. 166, D, F); Tacoma Park, *D. Saunders 3025* (D, F); Viborg, *P. D. Voth* (D, F).

Nebraska: Ames, Angora, Arthur, Aurora, Benkelman, Champion, Columbus, Eagle Canyon (Keith county), Falls City, Fremont, Gretna, Halsey national forest (Thomas county), Harrisburg, Hayes Center, Johnson lake (Gosper county), Keystone, Kimball, Kingsley dam (Keith county), Lake McConaughy, Lexington, Lincoln, Merriman, North Platte, McCook, Omaha, Oshkosh, Parks, Phillips, Rock Creek Lake state park, Schickley, Stewart, Sutherland, Valentine, and Verdon, *W. Kiener 10622, 11541, 13683b, 13946, 15016, 15525, 15609, 16550, 16743, 17200, 17515, 17635, 17721c, 18815d, 19407, 19609, 19828, 19930, 21023, 21134, 21350, 21385b, 21878a, 21945, 22139b, 22216, 22350a, 22470, 22653, 23100, 23774, 24672b, 24695, 28766* (D, F, NEB-Kiener); Greenwood, *T. A. Williams* (US); Hackberry lake (Cherry county), *E. R. Walker & E. N. Anderson* (D, F, NEB); Lincoln, *Walker* (D, F, NEB), *G. H. Giles* (D).

Kansas: Lawrence, Yates Center and Bakers Bluff (Woodson county), *K. C. Fan 10669, 10702, 10703* (D, Fan).

Oklahoma: Norman and Broken Bow, *C. E. Taft 202, 209, 306* (D); Medicine Park, Lake Overholser (Oklahoma City), Platt Park, and Eagle mountain (Comanche county), *E. O. Hughes 561, 656, 697, 776* (D).

Texas: Marble Falls, Lake Austin (Travis county), Bastrop state park, Austin, and Eagle Lake, *F. A. Barkley et al. Alg8, 1463-7, 13683, 15576, 46402, 46409* (D, F, TEX); Mesquite, Las Norias, and Brownsville, *R. Runyon 3744, 3848, 3853a* (D, F, RUNYON); Lake Worth (Tarrant county) and Dallas, *W. Kiener 3947, 12381, 12383* (D, F, NEB-Kiener); Austin, *H. C. Bold* (PH); San Jacinto Battle Monument (Harris county), *H. K. Phinney 4T41/2* (D, F, Phinney); Normanzee lake (Leon county), *W. A. Daily* (PH, Daily).

Saskatchewan: Regina, *N. H. Cowdry* (FH); Echo, Katepwa, Pasqua, and Shady lakes, *U. T. Hammer* (D); Bigstone and Churchill lakes (Prince Albert district), *D. S. Rawson* (D, Daily); Wollaston lake and Meadow Creek reservoir, *F. Brooks* (D, Daily); Last Mountain lake, *P. E. Kuehne* (D, F).

Montana: Great Falls, *F. W. Anderson* (UC); Lo Lo, *D. Griffiths* (TYPE of *Dichothrix montana* Tild. in Tild., Amer. Alg. 572, MIN); Missoula, *F. A. Barkley A24* (D, F, MONTU), *M. Forbes 4326a* (D, F); Glacier national park, *B. Maguire F28, F29* (D, F); Lake McGregor west of Kalispell, *H. F. Buell 466, 468, 469* (D, F).

Wyoming: Centennial, *W. G. Solheim* (D, F); Jackson lake (Lincoln county), *A. S. Hazzard* (D, F); Yellowstone national park, *G. M. Smith* (D, EMC, F), *N.*

Prât (W), *W. H. Weed* (TYPE of *Calothrix thermalis* fa. *Tildenii* Pol., in Tild., Amer. Alg. 287, F) *10, 75—78, 82, 85* (D, UC), *W. A. Setchell 1900—1902, 1910—1913, 1933, 1945, 1979, 2025, 6136, 6145* (D, F, UC).

Colorado: Eldora, *R. Prettyman* (D, F); Cañon City, *H. B. Louderback* (D, F); Longs Peak (Boulder county), *W. Kiener 2346* (D, F, NEB-Kiener); Fort Collins, *L. W. Durrell* (PH).

New Mexico: Rain creek in Mogollon mountains (Grant county), Middle Fork of Gila river (Catron county), Faywood Hot Springs, Gila hot springs north of Silver City, and San Francisco hot springs near Glenwood, *H. Habeeb* (Habeeb, PH); lava beds near Grants and Malpais spring near Tularosa, *A. A. Lindsey* (D, F); Montezuma, *F. Drouet & D. Richards 2628* (D, F), *Drouet & I. P. Bjornsson 14024* (D); Rio Pueblo, Tres Ritos, and Montezuma, *Drouet & H. B. Louderback 13581, 13582, 13586, 13952* (D); Grants, *N. E. Gray 29* (D, F); Mimbres hot springs, *J. F. Macbride et al. 8192* (D, F).

Alberta: Banff, *E. Butler* (D, UC).

Utah: Centerville, *F. Drouet et al. 4105* (D, F); Beacon, *C. D. Marsh* (W. R. Taylor); Garfield Beach, *J. E. Tilden* (TYPE of *Dichothrix utahensis* Tild. in Tild., Amer. Alg. 288, MIN); mouth of Bear river, *A. K. Fisher* (UC); Fish lake (Sevier county), *V. Tanner 23-30* (D, F); Scofield reservoir (Carbon county) and Strawberry reservoir (Wasatch county), *A. S. Hazzard F10, F16* (D, F); Johnson reservoir (Sevier county), *S. Wright* (D, F); Hyrum reservoir (Cache county), *B. Maguire* (D, F); Zion canyon, *N. E. Gray 64* (D, F).

Arizona: Tucson, Indian hot springs near Eden, Madera and Sabino canyons (Pima county), and Leupp—Wupatki road (Coconino county), *F. Drouet et al. 2783, 14539, 14602, 14610, 14623* (D); Mary's lake at Flagstaff, *H. S. Colton* (W. R. Taylor); Marble, Havasupai, and Emery canyons (Coconino county), *E. U. Clover* (D, F, MICH); Mormon lake (Coconino county), *Klingenberg* (D, F).

Nevada: Darrough hot springs (Nye county), Monte Neva hot springs (White Pine county), and Warm springs (southwestern White Pine county), *I. La Rivers 2651, 2652, 2661, 2736* (D, RENO); Birch creek (Eureka county), Bowman and Frenchman creeks (Lander county), Troy creek and Locke's Ranch (Nye county), and Whiteman creek (White Pine county), *T. C. Frantz 2264, 2282, 2426, 2978, 3014, 4015, 4025* (D, RENO); south end of Walker lake, *W. A. Archer 6960* (D, F); Steamboat and Reno Hot Springs, *M. J. Groesbeck 429* (D, F).

Alaska: Iliuliuk (Unalaska), *W. A. Setchell & A. A. Lawson 4043* (D, F; UC); Anchorage, Big Lake, Cape Thompson, Fort Yukon, Miles 88 and 163 on Glenallen highway, Kenai peninsula, and Miles 70, 83, and 234 on Richardson highway, *D. Hilliard* (D).

British Columbia: Spillamacheen river (Purcell range) and Beavermouth, *W. R. Taylor* (Taylor); Minnesota Seaside Station (Vancouver island), *J. E.*

Tilden (as *Rivularia Biasolettiana* in Tild., Amer. Alg. 570, D, F); Vancouver, *M. Ashton 3* (D).

Washington: Bellingham, *N. L. Gardner 544* (TYPE of *Dichothrix minima* Setch. & Gardn., UC); Whidbey island, *Gardner 13* (D, F); Sportsmans lake (San Juan island) and North Yakima, *T. C. Frye* (D, F, UC); Lake Union (King county), *Tilden* (as *Calothrix Braunii* in Tild., Amer. Alg. 286b, D, F); Falls, Alkali, and Lenore lakes (Lower Grand Coulee), *R. W. Castenholz 2, 10, 16* (D); Othello and Richland, *C. C. Palmiter 34, 52* (D, F); Hall, Washington, and Green lakes (Seattle), *W. T. Edmondson* (D); Richland, *R. G. Genoway* (D, F); White Salmon, *L. E. Griffin* (D, F).

Oregon: Corvallis, *M. A. Nash* (PH); Malheur river above Ontario, *D. Griffiths* (BKL); Hunters hot springs (Lake county), *J. A. Peary 1* (PH); Empire, *N. L. Gardner 2754* (D, UC); big lava lake in Crook county, *L. K. Mann 37* (D, F); Kitson Springs and Willamette river above Dell creek (Lane county), *M. S. Doty 7324, 8900* (D, F).

California: Azusa, Alturas, Weaverville, Stanford University, Springville, Palm Springs, and San Fernando, *F. Drouet et al. 3455, 4159, 4257, 4316, 4480, 4745, 5756* (D, F); Pasadena, *A. J. McClatchie* (as *C. juliana* in Tild., Amer. Alg. 163, D, F); Berkeley, North Berkeley, San Diego, San Francisco, and San Leandro, *N. L. Gardner* (as *C. Braunii, C. fusca, C. Kawraiskyi, C. parietina, Rivularia natans*, and *Aphanizomenon flos-aquae* in Coll., Hold. & Setch., Phyc. Bor.-Amer. 1107, 1360a, 1360c, 1360d, 1719a, 2211, 2212, 2263a,b, D, F); Arroyo Mocho (Alameda county), Carmel, the Geysers (Sonoma county), Humboldt Bay, Oakland, Smith River, Tassajara Hot Springs, Tomales bay, and West Berkeley, *Gardner 3355, 3450, 4562, 6594, 7705, 7738, 7743, 7950, 8010* (D, UC); Mill Valley, Vernal falls in Yosemite national park, Arrowhead and Waterman hot springs (San Bernardino county), and Bodega bay, *W. A. Setchell 704, 1388, 1538, 1552, 3022* (D, UC); Death Valley, *M. J. Groesbeck 3* (TYPE of *Dichothrix inyoensis* Dr., F); Shoshone, Lone Pine, and Little Lake (Inyo county) and Bridgeport, *Groesbeck 20, 42, 79, 151a* (D, F); Richvale, *R. L. Chapman BU15III5* (D); La Verne, Manker flats on Mount Baldy, Pacific Grove, Pomona, and Redlands, *G. J. Hollenberg 1537, 1612, 2381, 3331, 3506* (D, Hollenberg, UC); Berkeley, *W. J. V. Osterhout & Gardner* (as *Calothrix parietina* in Coll., Hold. & Setch., Phyc. Bor.-Amer. 1360c, D, F); Carmel Bay, *C. P. Nott & Setchell* (as *Rivularia Biasolettiana* in Coll., Hold. & Setch., Phyc. Bor.-Amer. 358, D, NY); Los Angeles, *G. R. Johnstone 14* (D, F, UC); Death Valley, *J. & H. W. Grinnell 7622* (D, F, UC); La Canada, *B. W. Chambers* (D, UC); Big Sur, *I. F. Lewis* (D, F); Stonyford, *E. Lee 2610* (D, F, UC); Orinda, *C. G. Hyde 7729* (D, F, UC); Friant, *D. Brohasta 7764* (D, F, UC); Honey lake, *J. B. Davy 3149* (D, F, UC); Davenport, *J. F. Macbride 8057* (D, F).

Puerto Rico: Mayaguez, Maricao national forest, and Minillas, *L. R. Almodóvar 22, 206, 227, 403* (D, Almodóvar, FSU); Arecibo, Arroyo de los

Corchos, Jayuya, Río Piedras, Maricao, Sabana Grande, San Lorenzo, San-turce, and Utuado, *N. Wille 49a, 126, 501, 914, 1037b, 1276a, 1415a, 1554, 1696, 1776a, 1946* (TYPES of *R. flagelliformis* Gardn., *Anabaena unispora* Gardn., *Calothrix Braunii* var. *mollis* Gardn., *C. evanescens* Gardn., *C. juliana* var. *tenuior* Gardn., *C. linearis* Gardn., *C. parietina* var. *torulosa* Gardn., *C. simplex* Gardn., *C. simulans* Gardn., *Dichothrix Willei* Gardn., and *Nodularia epiphytica* Gardn., NY); Adjuntas, *E. G. Britton 1571* (TYPE of *Microchaete tenera* var. *tenuior* Gardn., NY).

Anguilla: Long Bay, *P. Wagenaar Hummelinck 545* (D, L).

Guadeloupe: Baillif and Vieux-Fort, *H. Maze & A. Schramm 4, 1076* (TYPES of *Mastichothrix longissima* Crouan and *Calothrix submarina* Crouan, PC).

Dominican Republic: San Pedro de Macoris, *J. G. Scarff* (D, FH, NY, W).

Jamaica: Castleton, *J. E. Humphrey* (FH); Gordon Town, *W. Joshua* (as *Lyngbya inundata* in Wittr. & Nordst., Alg. Exs. 776b, D, F, FH).

Netherlands Antilles: Bonaire: Punt Vierkant, *Wagenaar Hummelinck 381* (D, L). Curaçao: San Pedro spring, *M. Díaz-Piferrer 2001* (D, FPDB), *Wagenaar Hummelinck 80Aa* (F, L); Julianadorp, *J. G. de Jong 391* (D, L).

Mexico: Baja California: Vinorama east of La Paz, *A. Carter 2624* (D, UC); La Rumarosa, *J. Zedler* (PH). Nuevo Léon: Monterrey and Santa Catarina, *F. A. Barkley 1462, 14608b* (D, F, TEX). Oaxaca: *F. Müller* (TYPE of *Physactis mexicana* Kütz., L). Sonora: Jécori, Hermosillo, and Navajoa, *F. Drouet & D. Richards 3003, 3035, 3181* (D, F).

Guatemala: Lago de Amatitlán, Antigua, Cobán–San Pedro Carchá, Ju-tiapa, and Zacapa, *P. C. Standley 52528, 72060, 75366, 89469, 90008* (D, F); Agua Blanca–Amatillo, Lago de Atitlán, Jalapa–Volcán Jumay, San Felipe, La Fragua–Río Motagua, and La Laguna (Alta Verapaz), *J. A. Steyermark 29212, 30371, 32273, 39613, 46379, 47286* (D, F); Lake Amatitlán, *W. A. Kellerman 5061, 5068* (as *Lyngbya Lagerheimii* and *L. Martensiana* in Tild., Amer. Alg. *636, 637*, D, F), *S. E. Meek 49, 68* (D, F).

Honduras: Comayagua, El Zamorano, La Lima, Quebrada de Santa Clara (Morazán), and Siguatepeque, *Standley 200a, 1581, 5445, 6745a, 7150* (D, F); Santa Lucía, *F. A. Barkley & J. Sweeney 40124* (PH).

El Salvador: Ahuachapán, *Standley & E. Padilla V. 2451* (D, F); Lago de Coatepeque, *M. C. Carlson* (D, F).

Nicaragua: La Libertad, *Standley 8953* (D, F).

Costa Rica: Turrialba, *R. W. Holm & H. H. Iltis 1013* (D, MO).

Panama: Barro Colorado island, Summit, Chagres river, Río Pilon, and Gatun, *G. W. Prescott CZ27a, CZ80, CZ98, CZ147, CZ170* (D, EMC); Gigante bay, *C. W. Dodge* (D, MO); Juan Mina, *H. H. Bartlett CZ215* (D, EMC).

Brazil: Alagôas: Maceió, *F. Drouet 1261* (D, L, NY). Amazonas: Manaos, *J. W. H. Trail* (TYPE of *Inactis obscura* Dick., BM). Ceará: Urubú and Barro Vermelho, Fortaleza, *Drouet 1336, 1381, 1498* (D). Goias: Formoso and Peixe,

E. Y. Dawson 14845, 15160 (D, LAM). Guanabara: Rio de Janeiro, *Drouet 1259* (D). Mato Grosso: Cáceres, *F. C. Hoehne 52* (TYPE of *Calothrix scytonemicola* var. *brasiliensis* Borge, S). Paraíba: Campina Grande, *S. Wright 2013* (D). Pernambuco: Villa Bella and Jatobá, *Wright 2021, 2045* (D). São Paulo: Bom Fim, *A. Löfgren 206* (as *Gloeotrichia Pisum* in Wittr. & Nordst., Alg. Exs. 660c, D, F), *219* (NY); Embú (Itapecerica da Serra) and Perequê-Assú, *A. B. Joly 1099-1952, 15-1953* (D, SPF). Rio Grande do Sul: Viciri and Cachoeira, *G. A. Malme 34, 57* (D, F, S); Lagôa dos Quadros and Porto Alegre, *H. Kleerekoper 536a, 598* (D, F).

Uruguay: Montevideo, *J. Arechavaleta* (NY; as *G. natans* in Wittr. & Nordst., Alg. Exs. 753c, D, F; TYPE of *Calothrix stellaris* Born. & Flah. with *Aulosira implexa* in Wittr. & Nordst., Alg. Exs. 787, PC); prov. San José, *Arechavaleta* (as *Calothrix fusca* in Wittr., Nordst. & Lagerh., Alg. Exs. 1303, D, F); La Palma (dept. Florida), *W. G. Herter 99553* (D, F).

Argentina: numerous cultures from Buenos Aires, *D. Rabinovich de Halperín* (D, F, de Halperín); Belgrano (Buenos Aires), *R. Thaxter* (D, FH); Lago Guillermo (Río Negro) and Río Trafui (Neuquén), *S. A. Guarrera 2060, 2064* (D, F); Chascomus and Necochea (Buenos Aires) and Laguna Viejo (San Luis), *S. Wright 2093, 2094, 2102* (D, FH, L, NY, S, W).

Colombia: Guajira, Laguna del Pájaro, *P. Wagenaar Hummelinck 114* (D, L).

Ecuador: prope Baños (Tungurahua), *G. Lagerheim* (as *Calothrix thermalis, Rivularia haematites*, and *Scytonema myochrous* in Wittr., Nordst. & Lagerh., Alg. Exs. 1306, 1310, 1321, D, F); Chillogallo prope Quito, *Lagerheim* (as *Aulosira implexa* in Wittr., Nordst. & Lagerh., Alg. Exs. 1323, D, NY); South Seymour island, Galapagos, *W. Schmitt 140* (D, F).

Peru: Ancash: Yungay, *A. Aldave P. 1000* (Aldave, PH). Arequipa: Pozo del Negro, *A. Maldonado 11* (D, F). Cajamarca: entre Llacanora y Cajamarca, *Aldave 166* (Aldave, PH). Huánuco: Puerto Nuevo near Tingo María, *M. H. Hohn 36* (D, PH). Ica: Pisco, *C. Acleto A231* (PH, USM); Lagunas Saraja y Huacachina, *Maldonado 93, 300b* (D, F). La Libertad: Ascope and Laguna de Sausacocha, *M. Fernandez H. 191, 230* (PH); Trujillo, *Fernandez 189, 220* (PH); Santiago de Chuco and Ascope, *Aldave 238, 275* (Aldave, PH). Lima: Laguna Pachacamac and Cascadas de Barranco, *Acleto A143, A281* (PH, USM); Lago Chilca, *Maldonado 66* (D, F). Loreto: Isla de Iquitos, *Hohn 12* (D, PH).

Bolivia: Lake Titicaca, Molinobampa, *Percy Sladen Trust Expedition* (D, BM).

Chile: culture from La Portada (Antofagasta), *G. H. Schwabe* (TYPE of *Calothrix desertica* Schwabe, Schwabe; duplicate: D); "i Ch. Malpasos utlopp" (Patagonia), *O. Borge 397* (TYPE of *Leptochaete crustacea* Borzi, S).

Clipperton Island: lagoon, *W. Schmitt 10-38* (D, W. R. Taylor).

Hawaii: laboratory cultures, Honolulu, *M. S. Doty 18763, 20088* (Doty, PH); Oahu, *R. Hitchcock* (D, F); Koolauloa, *J. E. Tilden* (TYPE of *Calothrix scytonemicola* Tild. on *Scytonema cincinnatum* in Tild., Amer. Alg. 480, MIN);

Makaleha Valley and Waialua (Oahu), *O. & I. Degener 23813, 24878, 24911* (D, Degener); Hilo and Akaka falls (Hawaii), *W. A. Setchell 5209, 10037* (D, UC).

Tahiti: Tapahi, *Setchell & H. E. Parks 5045* (TYPE of *Microchaete tapahiensis* Setch., UC; duplicates: D, F).

Antarctica: Gap pond, Winter Harbour, *Discovery 23* (TYPES of *Calothrix gracilis* Fritsch and *C. intricata* Fritsch, BM); island, Budd Coast, *D. C. Nutt 68* (D, US); Sentinel range in Ellsworth mountains, Cape Evans on Ross island, and Walcott Glacier Valley, *J. S. Zaneveld et al. 64010063, 64010202, 64010204, 64010209, 64120017* (PH, Zaneveld).

New Zealand: Orakei Korako and Great Wairakei Geyser, *W. A. Setchell 5973, 5979, 5963* (D, UC); Taurangi Harbour, *S. Berggren 86b* (S); Korareka point (Bay of Islands), *I. B. & V. R. Warnock 360* (D, F); Frenchmans Creek (Bay of Islands), *V. J. Chapman* (D, F); Whakarewarewa, *A. Nash & L. A. Doore 564* (D, F); Waitata stream (Bay of Islands) and Rotorua, *L. M. Jones 506, 539* (D, F); Rotorua, *P. J. Philson et al. 529, 569, 570* (D, F); Scotts point (Mogunui) *V. W. Lindauer 2314* (D, F, Lindauer); Keri Keri and Russell, *Lindauer* (as *Rivularia Beccariana, R. Vieillardii,* and *Dichothrix gypsophila* in Lind., Alg. N.-Zel. Exs. 77, 102, 103, D, F); Waitakere stream and Karekare, *J. Trevarthen* (D); Whangamumu (Bay of Islands) and Auckland, *K. W. Loach* (D).

New Caledonia: *Vieillard 2008* (TYPE of *Schizosiphon Vieillardii* Kütz., L).

Japan: Kawachi, *R. Hitchcock* (D, F); Mera, Tateyama, Chiba-ken, *Y. K. Okada* (as *Gloeotrichia natans* in Okada, Alg. Aq. Japon. 1, D, F).

Philippines: Laguna, *O. A. Ramirez 26R* (D); Quezon City, *M. B. Valero 109* (D); Manila, Tanay and San Juan (Rizal), Balanga (Bataan) and Iwahig penal colony (Palawan), *G. T. Velasquez 109, 439, 500, 2421, 2869* (D, F, PUH); Buenavista (Guimaras island) and Iloilo City, *J. D. Soriano 1094, 1564* (D, F, PUH).

New Guinea: *H. J. Rogers 15, 24* (D, F).

China: Fukien: University of Amoy, *H. H. Chung A54* (TYPE of *Calothrix linearis* Gardn., FH); Fuling hot springs, *Chung A461* (D, FH); Kushan near Foochow, *Chung 329a* (TYPE of *Dichothrix Chungii* Gardn., FH), *A417* (D, FH). Hopei: Tientsin, *M. S. Clemens 6011* (D, UC). Kiangsu: Nanking and Tsishiasan, *C. C. Wang 360, 371* (D, F, UC). Kwangtung: Nodoa, Ha Kung Leng (Hainan), *F. A. McClure 1692b* (D, UC). Szechwan: Mount Omei, *H.-j. Chu 1366* (TYPES of *Rivularia Jaoi* Chu and *Calothrix omeiensis* Chu, D); Lemoka, *H. Handel-Mazzetti 1620* (TYPE of *C. aequalis* Skuja, W; duplicate: D), *1621* (D, W); Ningyüen, *Handel-Mazzetti 1794* (TYPE of *Sacconema homoiochlamys* Skuja, W; duplicate: D); Yungning, *Handel-Mazzetti 3096* (TYPE of *Calothrix micromeres* Skuja, W; duplicate: D); Tjiaodjio, *Handel-Mazzetti 1553b* (D, W). Yunnan: Tali lake, *S. C. Hsiao 4, 5, 7* (D, F, PH), *Handel-Mazzetti 8730* (D, W); Bödö (Peti), *Handel-Mazzetti 4471* (TYPE of

Dichothrix Handelii Skuja, W; duplicate: D); Likiang, *Handel-Mazzetti 4207, 4208* (D, W); Mudidjiu and Tschamutong, *Handel-Mazzetti 3197, 9786* (D, W).

Malaysia: Batu Berendam, *G. A. Prowse 344* (D).

Singapore: *O. Beccari* (FI).

Indonesia: Tempé (Celebes), *A. Weber-van Bosse 853, 862* (TYPE of *Calothrix minuscula* Web.-v. B., L); in monte Idjen (eastern Java), *C. E. B. Bremekamp* (TYPE of *Rivularia oolithica* Bremek., in Miner.-Geol. Inst., Univ. of Utrecht); Lake Singkarah (Sumatra), *Weber-van Bosse 592* (D, L).

Australia: Ipswich and Teviot Creek (Queensland), *A. B. Cribb 13-1*, M (BRIU, D); Yirrkala (Arnhem Land), *R. L. Specht A80* (TYPE of *Calothrix australiensis* Scott & Presc., EMC); Murray river, Blanchtown (South Australia), *R. L. Raschke 441* (PH).

Burma: Akyab, *S. Kurz 3215* (L); Elephant Point (Pegu), *Kurz 3267* (TYPE of *Scytonema Rhizophorae* Zell., BM); Mandalay and Monywa, *L. P. Khanna 486, 845, 853* (D, F); Rangoon, *Khanna 479, 507, 521* (TYPES of *Fortiea incerta* Skuja, *Calothrix geitonos* Skuja, and *C. dolichomeres* Skuja, D); Thingangyun, *Khanna 838* (D, F).

Kerguelen Island: *Eaton* (TYPES of *Mastichothrix minuta* Reinsch and *Schizosiphon kerguelenensis* Reinsch, BM).

Tibet: Gogra (Nurla), *G. Henderson 5* (TYPE of *S. Hendersonii* Dick., BM).

India: hot spring in the Himalayas, *M. R. Suxena 539* (D); Calcutta, *S. Kurz 2673* (L); Howrah (Bengal), *Kurz 1932* (TYPE of *Phormidium oryzetorum* Mart., LD; duplicate: BM), *2666* (TYPE of *Mastigonema granulatum* Mart., L; duplicate: BM), *1870* (LD); Nimeta waterworks, Baroda, *I. S. Jayangoudar B23, C3* (PH); Ranchi (Bihar), *J. P. Sinha* (PH); Igatpuri (Bombay), *A. Hansgirg* (TYPE of *Gloeotrichia indica* Schmidle, as *G. natans* in Mus. Vindob. Krypt. Exs. Alg. 221, W; duplicates: D, F, US); Kaylana prope Jodhpur, *S. K. Goyal Myx-J19* (TYPE of *G. Raciborskii* var. *kaylanaensis* Goy., IARI; duplicate: D); culture of Indian soils, *A. K. Mitra 56* (TYPE of *Calothrix anomala* Mitra, BM).

Pakistan: Lahore, *S. L. Ghose* (TYPES of *Campylonema lahorense* Ghose and *Aulosira fertilissima* Ghose, BM).

Kazakhstan: soil cultures, Bet-tak-dala, *N. V. Sdobnikova*, comm. H. S. Forest (PH).

Afghanistan: *W. Simpson* (BM).

Iraq: south of Sulaimaniya, *F. A. Barkley & S. Y. Haddad 7489a, 7514a* (BUA, PH).

Calothrix crustacea Schousboe & Thuret

Conferva Mucor Roth, Catal. Bot. 1: 191. 1797. *C. confervicola* Dillwyn, Brit. Conf., pl. 8. 1802. *Oscillatoria confervicola* Agardh [Disp. Alg. Suec., p. 37. 1812] ex Gomont, Ann. Sci. Nat. VII. Bot. 16: 242. 1892. *Elisa confervicola* S. F. Gray,

Nat. Arr. Brit. Pl. 1: 284. 1821. *Calothrix confervicola* Agardh [Syst. Algar., p. 70. 1824] ex Bornet & Flahault, Ann. Sci. Nat. VII. Bot. 3: 349. 1886. [*Oscillatoriella confervicola* Gaillon, Mém. Soc. d'Emulation Abbeville 1833: 472. 1833.] [*Desmarestella confervicola* Gaillon, ibid. 1833: 476. 1833.] [*Scytonema confervicola* Agardh ex Fries, Fl. Scan., p. 334. 1835.] [*S. Mucor* Agardh ex Fries, loc. cit. 1835.] *Leibleinia confervicola* Endlicher (as "Leiblinia"), [Mant. Bot. Altera Sist. Gen. Pl., Suppl. 3: 21. 1843] ex Bornet & Flahault, ibid. 3: 350. 1886. *L. chalybea* Kützing [Phyc. Gener., p. 221. 1843] ex Bornet & Flahault, loc. cit. 1886. *Lyngbya confervicola* Rabenhorst [Deutschl. Krypt.-Fl. 2(2): 83. 1847] ex Gomont, ibid. 16: 153. 1892. —Roth and Dillwyn based their new names upon a common TYPE, "Conferva marina parasitica, tenuissima et brevissima glauca," described by Dillenius, Hist. Musc., p. 552, Tab. 85, fig 21 (1741), from Wales: Aberystwyth, *L. Brown*, in herb. Dillenius (OXF). Fig. 74.

Conferva scopulorum Weber & Mohr, Naturh. Reise durch e. Th. Schwedens, p. 195. 1804. *Oscillatoria scopulorum* Agardh [Disp. Alg. Suec., p. 37. 1812] ex Gomont, Ann. Sci. Nat. VII. Bot. 16: 245. 1892. *Elisa scopulorum* S. F. Gray, Nat. Arr. Brit. Pl. 1: 284. 1821. *Calothrix scopulorum* Agardh [Syst. Algar., p. 70. 1824] ex Bornet & Flahault, Ann. Sci. Nat. VII. Bot. 3: 353. 1886. *C. fasciculata* Agardh [ibid., p. 71. 1824] ex Bornet & Flahault, ibid. 3: 361. 1886. [*Scytonema fasciculatum* Agardh ex Fries, Fl. Scan., p. 334. 1835.] *Schizosiphon scopulorum* Kützing [Phyc. Gener., p. 233. 1843] ex Bornet & Flahault, ibid. 3: 353. 1886. [*Lyngbya scopulorum* Zanardini, Sagg. Class. Nat. Ficee, p. 63, 1843.] *Schizosiphon fasciculatus* Kützing [Sp. Algar., p. 330. 1849] ex Bornet & Flahault, ibid. 3: 361. 1886. *S. Sowerbyanus* Crouan [Fl. Finistère, p. 116. 1867] ex Bornet & Flahault, ibid. 3: 353. 1886. *S. Mandonii* Martens (as "Mandoni") [ex Bornet & Thuret *pro synon.*, Not. Algol. 2: 159. 1880] ex Bornet & Flahault, loc. cit. 1886. *Conishymene tingitana* Schousboe ex Bornet *pro synon.*, Mém. Soc. Nat. Sci. Nat. & Math. Cherbourg 28: 187. 1892. —TYPE specimen from Sweden: Warberg, *Mohr*, 1803, in herb. Hooker (K in BM).

Rivularia atra Roth, Catal. Bot. 3: 340. 1806. *Chaetophora atra* Agardh, Disp. Alg. Suec., p. 43. 1812. *Linckia atra* Lyngbye [Tent. Hydroph. Dan., p. 195. 1819] ex Bornet & Flahault, Ann. Sci. Nat. VII. Bot. 4: 353. 1887. [*Linckiella atra* Gaillon, Mém. Soc. d'Emulation Abbeville 1833: 473. 1833.] *Euactis atra* Kützing [Phyc. Gener., p. 241. 1843] ex Bornet & Flahault, ibid. 4: 354. 1887. *Rivularia dura* c *atra* Rabenhorst, Deutschl. Krypt.-Fl. 2(2): 93. 1847. *Zonotrichia atra* Rabenhorst [Fl. Eur. Algar. 2: 219. 1865] ex Bornet & Flahault, loc. cit. 1887. *Chaetophora crustacea* Schousboe ex Bornet & Flahault *pro synon.*, loc. cit. 1887. *Scytomene rupestris* Schousboe ex Bornet *pro synon.*, Mém. Soc. Nat. Sci. Nat. & Math. Cherbourg 28: 189. 1892. *Gloeotrichia atra* Biswas, Hedwigia 74(1): 17. 1934. —TYPE specimen, here designated, in Jürgens, Alg. Aquat. 4: 4 (1817), from Oldenburg, Germany: Jever, *G. H. B. Jürgens* (PC). Fig. 75.

Thorea viridis Bory de Saint-Vincent, Ann. Mus. d'Hist. Nat. Paris 12: 134.

1808. —TYPE specimen, collected probably in North America by A. Michaux, labeled by Bory: "Thorea viridis. Type de l'erreur," in herb. Bornet—Thuret (PC).

Ulva bullata Poiret in Lamarck, Encycl. Méthod., Bot. 8: 175. 1808. *Alcyonidium bullatum* Lamouroux, Ann. Mus. d'Hist. Nat. Paris 20: 286. 1813. *Clavatella viridissima* Bory de Saint-Vincent, Dict. Class. d'Hist. Nat. 4: 197. 1823. *Nostoc bullatum* Duby [Bot. Gall., ed. 2, 2: 960. 1829] ex Bornet & Flahault, Ann. Sci. Nat. VII. Bot. 7: 222. 1888. *Rivularia bullata* Berkeley, Gleanings of Brit. Alg., p. 8. 1832. [*Scytochloria bullata* Trevisan, Nomencl. Algar. 1: 28. 1845.] *Physactis bullata* Kützing [Sp. Algar., p. 332. 1849] ex Bornet & Flahault, ibid. 4: 359. 1886. *Chaetophora chlorites* Schousboe ex Bornet & Flahault *pro synon.*, loc. cit. 1886. *Rivularia deformis* Schousboe ex Bornet & Flahault *pro synon.*, loc. cit. 1886. —TYPE specimen, from Brittany, labeled "Ulva bullata fl. fr. Nostoch bullatum bot. gall." in the bound collection, "Herbier de la Flore Française (Bot. Gall.) donné au Muséum par A. P. De Candolle, 1822" (PC). Fig. 83.

Rivularia nitida Agardh, Disp. Alg. Suec., p. 44. 1812. [*Scytochloria nitida* Harvey *pro synon.* in Hooker, Brit. Fl. 2 (Engl. Fl. 5): 393. 1833.] *Physactis nitida* Westendorp [in Westendorp & Wallays, Herb. Crypt. Belg. 27: 1347. 1859] ex Bornet & Flahault, Ann. Sci. Nat. VII. Bot. 4: 359. 1887. *Potarcus nitidus* Kuntze (as "Portacus"), Rev. Gen. Pl. 2: 912. 1891. —TYPE specimen, here designated in the absence of the original material, from Sweden: in mari ad Båstad, *C. A. Agardh*, Sept. 1810, in herb. Agardh (LD). Fig. 77.

Ceramium pulvinatum Mertens (name corrected in Index from "pluvinatum") in Jürgens, Alg. Aquat. Dec. 4: 5. 1817. *Calothrix pulvinata* Agardh. [Syst. Algar., p. 71. 1824] ex Bornet & Flahault, Ann. Sci. Nat. VII. Bot. 3: 356. 1886. *Symphyosiphon pulvinatus* Kützing [Phyc. Gener., p. 218. 1843] ex Forti, Syll. Myxoph., p. 540. 1907. *Scytonema pulvinatum* Rabenhorst [Deutschl. Krypt.-Fl. 2(2): 86. 1847] ex Forti, ibid., p. 537. 1907. *Schizosiphon pulvinatum* Rabenhorst [Fl. Eur. Algar. 2: 242. 1865] ex Bornet & Flahault, ibid. 3: 357. 1886. —TYPE specimen from Germany: ad littora maris in Dynastia Jeverensi prope Heppenserdrift & Mariensiehl, in herb. Mertens (BREM).

Oscillatoria zostericola Hornemann [Fl. Dan. 9(27): 8. 1818] ex Gomont, Ann. Sci. Nat. VII. Bot. 16: 246. 1892. *Elisa zostericola* S. F. Gray, Nat. Arr. Brit. Pl. 1: 284. 1821. [*Scytonema zostericola* Lyngbye ex Fries, Fl. Scan., p. 334. 1835.] *Leibleinia zostericola* Endlicher (as "Leiblinia") [Mant. Bot. Altera Sist. Gen. Pl., Suppl. 3: 21. 1843] ex Bornet & Flahault, Ann. Sci. Nat. VII. Bot. 3: 350. 1886. *Calothrix confervicola* var. *zostericola* Crouan [Alg. Mar. Finistère 3: 341. 1852] ex Bornet & Flahault, loc. cit. 1886. *Lyngbya zostericola* Endlicher [ex Rabenhorst, Fl. Eur. Algar. 2: 141. 1865] ex Gomont, ibid. 16: 155. 1892. —TYPE specimen from Fyn, Denmark: ad littus Hofmansgave, *Lyngbye*, 25 Nov. 1815 (C).

[*Linckia atra β viridis* Lyngbye, Tent. Hydrophyt. Dan., p. 195. 1819.]
—TYPE specimen from Sjaelland, Denmark: ad littus Torbaek, *Lyngbye*, 18
June 1815 (C).

[*Linckia ceramicola* Lyngbye, loc. cit. 1819.] *Rivularia ceramicola* Lyngbye ex
Kützing, Sp. Algar., p. 338. 1849. —TYPE specimen, here designated in the
absence of the original material, from Sweden: Kullen (Scania), marked with
this name by Lyngbye, 1835 (C).

Rivularia pellucida Agardh, Syst. Algar., p. 25. 1824. —TYPE specimen from
Sweden: Landscrona, in herb. Agardh (LD).

Chaetophora aeruginosa Agardh, ibid., p. 27. 1824. —TYPE specimen from
Sweden: Leufsta, marin, in herb. Agardh (LD).

Linckia atra var. *coadunata* Sommerfelt [Suppl. Fl. Lappon., p. 201. 1826] ex
Foslie, Tromsø Mus. Aarsh. 14: 56. 1891. *Rivularia atra β coadunata* Sommerfelt
ex Bornet, Bull. Soc. Bot. France 36: 149. 1889. *Rivularia coadunata* Foslie, loc.
cit. 1891. —TYPE specimen, "in terra litorali Nordlandiae, leg. et com. Sommerfelt" (UC).

Calothrix pannosa Agardh, [Flora 10(1: 40): 635. 1827] ex Bornet & Flahault,
Ann. Sci. Nat. VII. Bot. 3: 370. 1886. [*C. lamellata* Harvey ex Crouan *pro synon.*,
Alg. Mar. Finistère 3: 344. 1852.] *Symploca pannosa* Desmazières [Pl. Crypt.
France, ed. 3, 3: 130. 1854] ex Gomont, Ann. Sci. Nat. VII. Bot. 16: 118. 1892.
Scytonema pannosum Rabenhorst [Fl. Eur. Algar. 2: 260. 1865] ex Bornet &
Flahault, ibid. 3: 357. 1886. *Schizosiphon pannosus* Crouan (as "pannosum") [Fl.
Finistère, p. 116. 1867] ex Bornet & Flahault, ibid. 3: 356. 1886. —TYPE
specimen from Italy: Trieste, Juni 1827, in herb. Agardh (LD).

Calothrix hydnoides Harvey [in Hooker, Engl. Fl. 5(Brit. Fl. 2): 368. 1833] ex
Bornet & Flahault, Ann. Sci. Nat. VII. Bot. 3: 356. 1886. *Scytonema hydnoides*
Carmichael *pro synon.* [in Harvey, loc. cit. 1833] ex Bornet & Flahault, ibid. 5:
113. 1887. *Symploca hydnoides* Kützing [Sp. Algar., p. 272. 1849] ex Gomont,
Ann. Sci. Nat. VII. Bot. 16: 106. 1892. *Schizosiphon hydnoides* Crouan [Fl.
Finistère, p. 116. 1867] ex Bornet & Flahault, ibid. 3: 357. 1886. *Conferva
fasciculata* Schousboe *pro synon.* ex Bornet, Mém. Soc. Nat. Sci. Nat. & Math.
Cherbourg 28: 183. 1892. *Calothrix rigida* Harvey *pro synon.* ex Gomont, ibid. 16:
107. 1892. —TYPE specimen from Argyll, Scotland: Appin, *Carmichael*, in
herb. Agardh (LD).

Rivularia plicata Carmichael in Harvey in Hooker, Brit. Fl. 2 (Engl. Fl. 5):
392. 1833. *Physactis plicata* Kützing [Sp. Algar., p. 332. 1849] ex Bornet &
Flahault, Ann. Sci. Nat. VII. Bot. 4: 357. 1887. *Potarcus plicatus* Kuntze (as
"Portacus"), Rev. Gen. Pl. 2: 912. 1891. —TYPE specimen from England: Tor
Abbey rocks, *Mrs. Griffiths*, Sept. 2, 1832 (K in BM).

Rivularia Contarenii Zanardini, Bibl. Ital. 96: 134. 1839. *Mastigonema Contarenii*
Kützing (as "Mastichonema") [Bot. Zeit. 5(11): 179. 1847] ex Bornet &
Flahault, Ann. Sci. Nat. VII. Bot. 3: 355. 1886. *Homoeoactis Contarenii* Zanar-

dini [Notizie Intorno alle Cellulari Marine delle Lagune e de'Litorali di Venezia, p. 77. 1847] ex Forti, Syll. Myxoph., p. 610. 1907. *Calothrix Contarenii* Bornet & Flahault [Mém. Soc. Nat. Sci. Nat. & Math. Cherbourg 25: 201. 1885] Ann. Sci. Nat. VII. Bot. 3: 355. 1886. —TYPE specimen from Italy: Venetiis, ad saxa, *Zanardini* in herb. Bornet—Thuret (PC).

Rivularia fucicola Zanardini, Bibl. Ital. 96: 134. 1839. —TYPE specimen from Italy: Triest, ad *Fucus Sherardii* Stackh., in herb. Bornet—Thuret (PC).

Calothrix stellulata Zanardini [ibid. 96: 135. 1839] ex Bornet & Flahault, ibid. 3: 370. 1886. *Leibleinia stellulata* Zanardini [Notizie Intorno alle Cellulari Marine, p. 79. 1847] ex Forti, ibid., p. 294. 1907. —TYPE specimen from Italy: Venetiis, ad *Polysiphoniam opacam, Zanardini*, in herb. Bornet—Thuret (PC).

Rivularia medusae Meneghini, Lettera del Prof. G. Meneghini al Dott. I. Corinaldi a Pisa, p. [1]. 1840. *Mastigonema medusae* Kützing (as "Mastichonema") [Bot. Zeit. 5(11): 179. 1847] ex Forti, Syll. Myxoph., p. 610. 1907. *Schizosiphon medusae* Kützing [Sp. Algar., p. 328. 1849] ex Forti, loc. cit. 1907. —TYPE specimen from Italy: scopulis, Spezia, in herb. Meneghini (FI).

Rivularia cerebrina Montagne in Barker-Webb & Berthelot, Hist. Nat. d. Iles Canaries 3(2): 190. 1840. —TYPE specimen from the Canary islands, *Despréaux*, in herb. Bornet—Thuret (PC).

Rivularia parasitica Chauvin, Recherches sur l'Organisation etc. de Plusieurs Genres d'Algues, p. 41. 1842. *Schizosiphon parasiticus* Le Jolis [Liste d. Alg. Mar. Cherbourg, p. 30. 1863] ex Bornet & Flahault, Ann. Sci. Nat. VII. Bot. 3: 357. 1886. *Gloeotrichia parasitica* Rabenhorst [Fl. Eur. Algar. 2: 205. 1865] ex Bornet & Flahault, loc. cit. 1886. *Calothrix parasitica* Thuret [Ann. Sci. Nat. VI. Bot. 1: 381. 1875] ex Bornet & Flahault, loc. cit. 1886. *Chaetophora lumbricalis* Schousboe *pro synon.* ex Bornet, Mém. Soc. Nat. Sci. Nat. & Math. Cherbourg 28: 188. 1892. *Rivularia lumbricalis* var. *viridis* Schousboe *pro synon.* ex Bornet, loc. cit. 1892. *Potarcus parasiticus* Kuntze (as "Portacus"), Rev. Gen. Pl. 2: 912. 1891. —TYPE specimen, here designated in the absence of the original material, from France: Marseille, no. 339 in herb. Lenormand (CN). Fig. 72.

Zonotrichia hemispherica J. Agardh [Alg. Mar. Medit. & Adriat., p. 9. 1842] ex Bornet & Flahault, Ann. Sci. Nat. VII. Bot. 4: 355. 1887. *Euactis hemispherica* Kützing (as "hemisphaerica") [Phyc. Gener., p. 242. 1843] ex Bornet & Flahault, loc. cit. 1887. *Rivularia hemispherica* Kützing *pro synon.*, loc. cit. 1843. *R. atra* var. *hemispherica* Bornet & Flahault (as "hemisphaerica"), loc. cit. 1887. *R. atra* fa. *hemispherica* Kossinskaia (as "hemisphaerica"), Opred. Morsk. Sinez. Vodor., p. 166. 1948. —TYPE specimen from Istria, Yugoslavia: ad saxa, Pola, *Biasoletto*, in herb. Agardh (LD). Fig. 76.

Diplotrichia polyotis J. Agardh [Alg. Mar. Medit. & Adriat., p. 10. 1842] ex Bornet & Flahault, Ann. Sci. Nat. VII. Bot. 4: 360. 1887. *Rivularia polyotis* Hauck [Meeresalg., p. 495. 1885] and Bornet & Flahault [Mém. Soc. Nat. Sci. Nat. & Math. Cherbourg 25: 207. 1885] ex Bornet & Flahault, Ann. Sci. Nat.

VII. Bot. 4: 360. 1887. —TYPE specimen from Croatia, Yugoslavia: in Insula Pago, *Biasoletto*, in herb. Agardh (LD). Fig. 68.

Leibleinia aeruginea Kützing [Phyc. Gener., p. 221. 1843] ex Bornet & Flahault, Ann. Sci. Nat. VII. Bot. 3: 358. 1886. [*Lyngbya villosa* Rabenhorst, Deutschl. Krypt.-Fl. 2(2): 83. 1847.] *Calothrix aeruginea* Thuret [Ann. Sci. Nat. VI. Bot. 1: 381. 1875] ex Bornet & Flahault, loc. cit. 1886. *C. aeruginosa* Thuret ex Cotton, Journ. Linn. Soc. Bot. 43: 159. 1915. —TYPE specimen from Istria, Italy: an Hutchinsia, Zaule bei Triest, Apr. 1835, in herb. Kützing (L). Fig. 70.

Leibleinia purpurea Kützing [loc. cit. 1843] ex Bornet & Flahault, ibid. 3: 350. 1886. *Lyngbya purpurea* Kützing [ex Rabenhorst, loc. cit. 1847] ex Gomont, Ann. Sci. Nat. VII. Bot. 16: 155. 1892. —TYPE specimen from England: m. atlant., *D. Turner*, in herb. Kützing (L).

Schizosiphon lutescens Kützing [Phyc. Gener., p. 233. 1843] ex Bornet & Flahault, Ann. Sci. Nat. VII. Bot. 3: 359. 1886. —TYPE specimen, in which the plant described is mixed with *Schizothrix calcicola* (Ag.) Gom., from Germany: Helgoland, an Fischkästen, Jul. 1839, in herb. Kützing (L).

Schizosiphon rupicola Kützing [loc. cit. 1843] ex Bornet & Flahault, ibid. 3: 353. 1886. —TYPE specimen from Dalmatia, Yugoslavia: Spalato, ad rupes in mari, 29 März 1835, in herb. Kützing (L).

[*Schizosiphon scopulorum* β *crassus* Kützing (as "crassa"), loc. cit. 1843.] *S. crassus* Kützing [Sp. Algar., p. 329. 1849] ex Forti, Syll. Myxophyc., p. 329. 1849. —TYPE specimen, in which the plant described is mixed sparingly with *Microcoleus lyngbyaceus* (Kütz.) Crouan, from Italy: Civitavecchia, in herb. Kützing (L).

Schizosiphon chaetopus Kützing [Phyc. Gener., p. 234. 1843] ex Bornet & Flahault, Ann. Sci. Nat. VII. Bot. 3: 355. 1886. —TYPE specimen, in which the plant described is mixed with *Microcoleus lyngbyaceus*, from Denmark: Batterie, Trekroner Hafniae, *Hofman-Bang*, in herb. Kützing (L).

Schizosiphon consociatus Kützing [loc. cit. 1843] ex Bornet & Flahault, ibid. 3: 351. 1886. *Calothrix consociata* Bornet & Flahault [Mém. Soc. Nat. Sci. Nat. & Math. Cherbourg 25: 201. 1885] ex Bornet & Flahault, Ann. Sci. Nat. VII. Bot. 3: 351. 1886. —TYPE specimen from Italy: Terracina, in herb. Kützing (L).

Schizosiphon flagelliformis Kützing [loc. cit. 1843] ex Bornet & Flahault, loc. cit. 1886. —TYPE specimen from Istria, Italy: Zaule, Triest, in herb. Kützing (L).

[*Schizosiphon flagelliformis* β *radiatus* Kützing, loc. cit. 1843.] *S. radiatus* Kützing [Tab. Phyc. 2: 17. 1852] ex Bornet & Flahault, loc. cit. 1886. —TYPE specimen from Istria, Italy: Triest, an *Laurencia obtusa*, in herb. Kützing (L).

Heteractis mesenterica Kützing [Phyc. Gener., p. 236. 1843] ex Bornet & Flahault, Ann. Sci. Nat. VII. Bot. 4: 359. 1887. *Rivularia magna* Kützing *pro synon.*, loc. cit. 1843. *R. mesenterica* Thuret, Ann. Sci. Nat. VI. Bot. 1: 382. 1875. —TYPE specimen from Istria, Yugoslavia: Pola, ad saxa inundata, in herb. Kützing (L). Fig. 69.

Heteractis pruniformis Kützing [loc. cit. 1843] ex Bornet & Flahault, loc. cit. 1887. —TYPE specimen from Schleswig, Germany: in der Geltinger Bucht, *Suhr*, in herb. Kützing (L).

Physactis lobata Kützing [loc. cit. 1843] ex Bornet & Flahault, loc. cit. 1887. —TYPE specimen from Schleswig, Germany: am Graswürzel auf feuchter Erde, Flensburg, *Suhr*, in herb. Kützing (L).

Dasyactis minutula Kützing [ibid., p. 239. 1843] ex Bornet & Flahault, ibid. 4: 353. 1887. *Zonotrichia minutula* Rabenhorst [Fl. Eur. Algar. 2: 218. 1865] ex Bornet & Flahault, ibid. 4: 354. 1887. —TYPE specimen from Italy: an *Chondria* sp. im Golf von Neapel, Jul. 1835, in herb. Kützing (L).

Dasyactis Biasolettiana Kützing [Phyc. Gener., p. 240. 1843] ex Forti, Syll. Myxophyc., p. 667. 1907. *Zonotrichia Biasolettiana* Rabenhorst [Fl. Eur. Algar. 2: 218. 1865] ex Bornet & Flahault, Ann. Sci. Nat. VII. Bot. 4: 352. 1887. —TYPE specimen from Istria, Italy: Triest, riva lunga, in fossis, *Biasoletto*, in herb. Kützing (L).

Euactis amoena Kützing [loc. cit. 1843] ex Bornet & Flahault, ibid. 4: 355. 1887. *Zonotrichia amoena* Rabenhorst [ibid. 2: 221. 1865] ex Bornet & Flahault, loc. cit. 1887. —TYPE specimen from Dalmatia, Yugoslavia: Spalato, in herb. Kützing (L).

Euactis marina Kützing [loc. cit. 1843] ex Bornet & Flahault, loc. cit. 1887. *Rivularia marina* Rabenhorst, Deutschl. Krypt.-Fl. 2(2): 92. 1847. [*Zonotrichia atra* fa. *adriatica* Rabenhorst, Fl. Eur. Algar. 2: 219. 1865.] —TYPE specimen from Croatia, Yugoslavia: Cherso, ad rupes maritimas, Mar. 1835, in herb. Kützing (L).

Euactis hospita Kützing [Phyc. Gener., p. 241. 1843] ex Bornet & Flahault, Ann. Sci. Nat. VII. Bot. 4: 360. 1887. *Zonotrichia hospita* Rabenhorst [ibid. 2: 220. 1865] ex Bornet & Flahault, ibid. 4: 361. 1887. *Rivularia hospita* Thuret, Ann. Sci. Nat. VI. Bot. 1: 382. 1875. —TYPE specimen from Italy: Neapel, in herb. Kützing (L).

Euactis ligustica Kützing [loc. cit. 1843] ex Bornet & Flahault, ibid. 4: 354. 1887. *Zonotrichia ligustica* Rabenhorst [ibid. 2: 221. 1865] ex Forti, Syll. Myxophyc., p. 665. 1907. —TYPE specimen from Italy: Genua, in herb. Kützing (L).

Euactis prorumpens Kützing [loc. cit. 1843] ex Bornet & Flahault, ibid. 4: 360. 1887. *Rivularia mediterranea* Kützing *pro synon.*, loc. cit. 1843. *Zonotrichia prorumpens* Rabenhorst [ibid. 2: 219. 1865] ex Bornet & Flahault, ibid. 4: 361. 1887. —TYPE specimen from Italy: Mare mediterr., Livorno, in herb. Kützing (L).

[*Euactis Juergensii* Kützing (as "Jürgensii"), Phyc. Gener., p. 242. 1843.] *Rivularia Juergensii* Rabenhorst (as "Jürgensii"), Deutschl. Krypt.-Fl. 2(2): 92.

1847. *Zonotrichia Juergensii* Rabenhorst [Fl. Eur. Algar. 2: 221. 1865] ex Bornet & Flahault, Ann. Sci. Nat. VII. Bot. 4: 354. 1887. —TYPE specimen from the Frisian islands, Germany: Wangerogae, *Jürgens 10*, in herb. Kützing (L).
[*Euactis Lens* Kützing, loc. cit. 1843.] [*Zonotrichia Lens* Rabenhorst, loc. cit. 1865.] *Z. numulitoidea* Roumeguère *pro synon.* in Roumeguère, Dupray & Mougeot, Alg. de France no. 423. 1883—89. —TYPE specimen from Germany: Nordsee?, *Jürgens*, in herb. Kützing (L).

Physactis durissima Kützing [Phyc. Germ., p. 186. 1845] ex Bornet & Flahault, loc. cit. 1887. *Rivularia durissima* Rabenhorst, Fl. Eur. Algar. 2: 209. 1865. —TYPE specimen from Mecklenburg, Germany: an *Ceramium rubrum* in der Ostsee bei Dobberan, *Flotow*, in herb. Kützing (L).

Physactis aggregata Kützing [Bot. Zeit. 5(11): 178. 1847] ex Bornet & Flahault, loc. cit. 1887. *Rivularia aggregata* Rabenhorst, loc. cit. 1865. *Potarcus aggregatus* Kuntze (as "Portacus"), Rev. Gen. Pl. 2: 911. 1891. —TYPE specimen from Germany: Nordsee, *Jürgens*, in herb. Kützing (L).

Schizosiphon fucicola Kützing [loc. cit. 1847] ex Bornet & Flahault, ibid. 3: 379. 1886. *Dichothrix fucicola* Bornet & Flahault [Mém. Soc. Nat. Sci. Nat. & Math. Cherbourg 25: 204. 1885] ex Bornet & Flahault, Ann. Sci. Nat. VII. Bot. 3: 379. 1886. *Schizosiphon antillarum* Martens ex Bornet & Flahault *pro synon.*, loc. cit. 1886. *Calothrix fucicola* Kossinskaia, Opred. Morsk. Sinez. Vodor., p. 160. 1948. —TYPE specimen from Yucatan, Mexico: Campeche Bank, in herb. Kützing (L).

Schizosiphon villosus Kützing [Bot. Zeit. 5(11): 179. 1847] ex Forti, Syll. Myxoph., p. 634. 1907. —TYPE specimen from Italy: Livorno, in mari, 1835, in herb. Kützing (L).

Leibleinia coccinea Kützing [Bot. Zeit. 5: 193. 1847] ex Bornet & Flahault, ibid. 3: 350. 1886. *L. purpurea β coccinea* Kützing [Sp. Algar., p. 277. 1849] ex Gomont, Ann. Sci. Nat. VII. Bot. 16: 156. 1892. [*Lyngbya purpurea* b *coccinea* Rabenhorst, Fl. Eur. Algar. 2: 142. 1865.] —TYPE specimen from the Kattegat: in sinu Codano, ad *Cladophoram rupestrem*, in herb. Kützing (L).

Leibleinia flaccida Kützing [Bot. Zeit. 5: 193. 1847] ex Bornet & Flahault, loc. cit. 1886. [*Lyngbya Nemalionis* c *flaccida* Rabenhorst, ibid. 2: 143. 1865.] *L. Nemalionis* var. *flaccida* Mazé & Schramm ex Forti, ibid., p. 613. 1907. —TYPE specimen from Sweden: sinus Codani, *Areschoug*, in herb. Kützing (L).

Leibleinia virescens Kützing [loc. cit. 1847] ex Bornet & Flahault, loc. cit. 1886. [*Lyngbya Nemalionis* d *virescens* Rabenhorst, loc. cit. 1865.] —TYPE specimen from Schleswig, Germany: Flensburger Föhrde, in herb. Kützing (L).

Calothrix cyanea J. Agardh [Öfvers. K. Sv. Vet.-Akad. Förh. 4: 6. 1848] ex Bornet & Flahault, Ann. Sci. Nat. VII. Bot. 3: 370. 1886. —TYPE specimen from Yucatan, Mexico: in Sargassis ad Campeche Bank, *Liebmann*, in herb. Kützing (L).

Leibleinia rupestris Nägeli [in Kützing, Sp. Algar., p. 276. 1849] ex Gomont, Ann. Sci. Nat. VII. Bot. 16: 152. 1892. [*Lyngbya Nemalionis* b *rupestris* Rabenhorst, ibid. 2: 142. 1865.] *L. rupestris* Crouan ex Gomont, loc. cit. 1892. —TYPE specimen from Germany: Helgoland, in herb. Nägeli (ZT).

Leibleinia australis Kützing [ibid., p. 277. 1849] ex Bornet & Flahault, ibid. 3: 359. 1886. [*Lyngbya australis* Rabenhorst, ibid. 2: 148. 1865.] —TYPE specimen from Italy: Genua, in herb. Kützing (L).

Leibleinia chalybea β *bicolor* Kützing [loc. cit. 1849] ex Gomont, ibid. 16: 155. 1892. —TYPE specimen from the Netherlands: Zélande, *van den Bosch*, in herb. Kützing (L).

Leibleinia purpurea γ *amethystea* Kützing [loc. cit. 1849] ex Bornet & Flahault, ibid. 3: 350. 1886. *L. amethystea* Le Jolis [Liste Alg. Mar. Cherbourg, p. 30, 1863] ex Bornet & Flahault, loc. cit. 1886. [*Lyngbya purpurea* c *amethystea* Kützing ex Rabenhorst, loc. cit. 1865.] —TYPE specimen from France: Cherbourg, *Lenormand*, in herb. Kützing (L).

[*Symphyosiphon punctiformis* Kützing, Sp. Algar., p. 321. 1849.] —TYPE specimen from France: Iles Chausey, *Lenormand,* in herb. Kützing (L).

Symphyosiphon cespitosus Kützing [ibid., p. 322. 1849] ex Forti, Syll. Myxoph., p. 514. 1907.—TYPE specimen, here designated, the only material thus annotated by Kützing, from France: St.-Waast-la-Hougue, *Lenormand 79,* in herb. Kützing (L).

Symphyosiphon gallicus Kützing [loc. cit. 1849] ex Forti, ibid., p. 539. 1907. *Scytonema gallicum* Rabenhorst [Fl. Eur. Algar. 2: 262. 1865] ex Bornet & Flahault, Ann. Sci. Nat. VII. Bot. 3: 356. 1886. [*Symploca gallica* Crouan, Fl. Finistère, p. 115. 1867.] —TYPE specimen from France: Iles Chausey, *Lenormand*, in herb. Kützing (L).

Schizosiphon cuspidatus Kützing [ibid., p. 328. 1849] ex Bornet & Flahault, ibid. 3: 353. 1886. —TYPE specimen from Calvados, France: Longuer, *Lenormand 209*, in herb. Kützing (L).

Schizosiphon gregarius Kützing [ibid., p. 329. 1849] ex Bornet & Flahault, loc. cit. 1886. —TYPE specimen presumably from Germany: Mare balticum, in herb. Kützing (L).

[*Schizosiphon scopulorum* β *subsimplex* Kützing, loc. cit. 1849.]] [*S. subsimplex* Kützing, Tab. Phyc. 2: 17. 1852.] —TYPE specimen from England: Torquay, an Felsen, *Nägeli 99*, in herb. Kützing (L).

Schizosiphon dilatatus Kützing [Sp. Algar., p. 330. 1849] ex Forti, Syll. Myxoph., p. 634. 1907. —TYPE specimen from Italy: Genua, in herb. Kützing (L).

Schizosiphon Lenormandii Kützing (as "Lenormandi") [loc. cit. 1849] ex Bornet & Flahault, ibid. 3: 356. 1886. —TYPE specimen from Manche, France: Portbail, *Lenormand 80,* in herb. Kützing (L).

Physactis pilifera Kützing [ibid., p. 332. 1849] ex Bornet & Flahault, ibid. 4:

172

359. 1887. *Rivularia nitida* b *pilifera* Rabenhorst, Fl. Eur. Algar. 2: 208. 1865. —TYPE specimen from France: Cherbourg, *Lenormand*, in herb. Kützing (L). *Physactis pilifera β fuscescens* Kützing [loc. cit. 1849] ex Bornet & Flahault, Ann. Sci. Nat. VII. Bot. 4: 360. 1887. *P. fuscescens* Kützing [Tab. Phyc. 2: 19. 1852] ex Bornet & Flahault, ibid. 4: 361. 1887. —TYPE specimen from France: Marseille, *Giraudy*, 1846, in herb. Kützing (L).

Physactis spiralis Kützing [Sp. Algar., p. 332. 1849] ex Bornet & Flahault, ibid. 4: 359. 1887. *Rivularia nitida* c *spiralis* Rabenhorst, loc. cit. 1865. —TYPE specimen from England: Torquay, *Berkeley*, in herb. Kützing (L).

Euactis pachynema Kützing [ibid., p. 339. 1849] ex Bornet & Flahault, ibid. 4: 361. 1887. *Zonotrichia pachynema* Rabenhorst [ibid. 2: 220. 1865] ex Bornet & Flahault, loc. cit. 1887. —TYPE specimen from France: Marseille, *Lenormand 337*, in herb. Kützing (L).

[*Euactis atra β confluens* Kützing, ibid., p. 340. 1849.] [*Scytonema multifidum* Meneghini *pro synon.* in Kützing, loc. cit. 1849.] —TYPE specimen from Italy: ad rupes in sinu Neapolitano, *Meneghini*, in herb. Kützing (L).

Euactis Lenormandiana Kützing [loc. cit. 1849] ex Bornet & Flahault, Ann. Sci. Nat. VII. Bot. 4: 354. 1887. *Zonotrichia Lenormandiana* Rabenhorst [Fl. Eur. Algar. 2: 221. 1865] ex Bornet & Flahault, loc. cit. 1887. *Rivularia Lenormandiana* Crouan, Fl. Finistère, p. 117. 1867. —TYPE specimen from Manche, France: Granville, *Lenormand*, in herb. Kützing (L).

Euactis confluens Kützing [ibid., p. 341. 1849] ex Bornet & Flahault, loc. cit. 1887. *Rivularia confluens* Crouan, Alg. Mar. Finistère 3: 335. 1852. *Zonotrichia confluens* Kützing [ex Rabenhorst, ibid. 2: 220. 1865] ex Bornet & Flahault, loc. cit. 1887. *Rivularia atra* fa. *confluens* Lespinasse, Actes Soc. Linn. Bordeaux 36: 204. 1882. —TYPE specimen from France: ad rupes marinas, Brest, *Crouan*, in herb. Kützing (L).

Euactis plicata Nägeli [in Kützing, loc. cit. 1849] ex Forti, Syll. Myxoph., p. 660. 1907. *Zonotrichia plicata* Rabenhorst [ibid. 2: 221. 1865] ex Bornet & Flahault, ibid. 4: 357. 1887. —TYPE specimen from England: an Felsen, Torquay, *Nägeli 104*, in herb. Kützing (L).

Schizosiphon Warreniae Caspary [Ann. & Mag. of Nat. Hist., ser. 2, 6: 266. 1850] ex Bornet & Flahault, Ann. Sci. Nat. VII. Bot. 4: 352. 1887. *Rivularia Warreniae* Thuret, Ann. Sci. Nat. VI. Bot. 1: 382. 1875. —TYPE specimen from England: Falmouth, *Caspary*, in herb. Hooker (K in BM).

Calothrix Caulerpae Zanardini [Flora, N. R. 9(1: 3): 38. 1851] ex Bornet & Flahault, ibid. 3: 370. 1886. —TYPE specimen from the Red sea: ad surculos *Caulerpae Freycinetti, Portier*, in herb. Zanardini (Museo di Storia Naturale, Venezia).

Rivularia australis Harvey, Trans. R. Irish Acad. 22(1): 566. 1854. —TYPE specimen from Western Australia: Cape Riche, *Harvey 592* (BM). Fig. 81.

Symphyosiphon pulvinatus fa. *tenuior* Suringar [Obs. Phycol., index. 1857] ex

Forti, ibid., p. 611. 1907. *Calothrix pulvinata* fa. *tenuior* Suringar ex Forti, loc. cit. 1907. —TYPE specimen from the Netherlands: steenen zeeweringen, Lemmer, *W. F. R. Suringar DDD89*, Sept. 1854 (L).

Dichothrix penicillata Zanardini [Mem. R. Ist. Veneto Sci., Lett. & Arti 7(2): 297. 1858] ex Bornet & Flahault, ibid. 3: 373. 1886. *Calothrix penicillata* Kossinskaia, Opred. Morsk. Sinez. Vodor., p. 159. 1948. —TYPE specimen from the Red sea: on *Spyridia filamentosa, Figari Bey*, in herb. Bornet—Thuret (PC). Fig. 79.

Calothrix vivipara Harvey [Ner. Bor.-Amer. 3: 106. 1858] ex Bornet & Flahault, ibid. 3: 362. 1886. *C. scopulorum* var. *vivipara* Farlow [Mar. Alg. New England, p. 37. 1881] ex Farlow, ibid., p. 37. 1891. —TYPE specimen from Rhode Island: Seaconnet point, *Bailey* (TCD). Fig. 73.

Calothrix dura Harvey [ibid. 3: 107. 1858] ex Bornet & Flahault, ibid. 3: 363. 1886. *Tildenia dura* Poliansky, Izv. Glavn. Bot. Sada SSSR 27(3): 329. 1928. *Setchelliella dura* J. de Toni, Noter. di Nomencl. Algol. 8: [6]. 1936. —TYPE specimen, with that of *Microcoleus corymbosus* Harv., from Florida: mud flats, Key West, *W. H. Harvey* (TCD).

Microcoleus corymbosus Harvey [ibid. 3: 109. 1858] ex Gomont, Ann. Sci. Nat. VII. Bot. 15: 362. 1892. *Polythrix corymbosa* Grunow [in Bornet & Flahault, Mém. Soc. Nat. Sci. Nat. & Math. Cherbourg 25: 204. 1885] ex Bornet & Flahault, Ann. Sci. Nat. VII. Bot. 3: 380. 1886. *Gardnerula corymbosa* J. de Toni, ibid. 8: [5]. 1936. —TYPE specimen from Florida: mud flats, Key West, *W. H. Harvey* (TCD). Fig. 80.

[*Mastichothrix obscura* Kützing, Tab. Phyc. 9: 1. 1859.] —TYPE specimen from Victoria, Australia: on *Cladosiphon nigricans* Harv., Western Port, *W. H. Harvey 94* (TCD).

Rivularia opaca Harvey, Proc. Amer. Acad. Sci. 4: 334. 1859. —TYPE specimen from Japan: Loo Choo islands, *C. Wright* (US).

Calothrix fusco-violacea Crouan [Ann. Sci. Nat. IV. Bot. 12: 291. 1859] ex Bornet & Flahault, Ann. Sci. Nat. VII. Bot. 3: 352. 1886. —TYPE specimen from Finistère, France: dragué, rade de Camaret, 8 Jan. 1859, in herb. Crouan (Lab. Marit., Concarneau). Fig. 67.

Calothrix infestans Harvey [in Hooker, Fl. Tasmania 2:343. 1860] ex Forti, Syll. Myxoph., p. 628. 1907. —TYPE specimen from Tasmania: Port Arthur, *W. H. Harvey*, in herb. Hooker (K in BM).

[*Physactis gracilis* Crouan, Bull. Soc. Bot. France 7: 372. 1860.] *Rivularia plicata* fa. *gracilis* Crouan, Fl. Finistère, p. 117. 1867. —TYPE specimen from Finistère, France: rivière marine de Penfeld, Sept. 1854, in herb. Crouan (Lab. Marit., Concarneau).

Rivularia confinis Crouan, Bull. Soc. Bot. France 7: 836. 1860. —TYPE specimen from Finistère, France: à terre, Inez près Morlaix, 26 Jun. 1852, in herb. Crouan (Lab. Marit., Concarneau).

Rivularia investiens Crouan, loc. cit. 1860. —TYPE specimen from Finistère, France: dans les flaques de la pointe de Devin à l'ile Noirmoutier, *Crouan*, Aug. 1850, in Lloyd, Algues de l'Ouest de la France no. 301, in herb. Bornet—Thuret (PC).

Euactis Lenormandiana var. *Balanii* Le Jolis [Liste Alg. Mar. Cherbourg, p. 159. 1863] ex Bornet & Flahault, Ann. Sci. Nat. VII. Bot. 4: 354. 1887. —TYPE specimen from France: Cherbourg, in Le Jolis, Algues Marines de Cherbourg no. 189, in herb. Bornet—Thuret (PC).

Physactis atropurpurea Kützing [Ostern-Progr. Realsch. Nordhausen 1862—63: 9. 1863; in Le Jolis, Liste Alg. Mar. Cherbourg, p. 31. 1863] ex Bornet & Flahault, ibid. 4: 344. 1887. *Rivularia atropurpurea* Rabenhorst, Fl. Eur. Algar. 2: 209. 1865. *Potarcus atropurpureus* Kuntze (as "Portacus"), Rev. Gen. Pl. 2: 911. 1891. —TYPE specimen from France: Cherbourg, *A. Le Jolis 162*, in herb. Kützing (L).

Physactis obducens Kützing [Ostern-Progr. Realsch. Nordhausen 1862—63: 9. 1863] ex Bornet & Flahault, loc. cit. 1887. —TYPE specimen from France: Vendée, *Lloyd 23*, in herb. Kützing (L).

Euactis pulchra Cramer [Hedwigia 2(11): 61. 1863] ex Bornet & Flahault, ibid. 4: 361. 1887.—TYPE specimen from Sicily: in Acqua santa bei Palermo, *C. Cramer*, 22 Nov. 1856, in Rabenhorst, Algen Europas no. 1449 (PH).

Physactis pulchra Cramer [ibid. 2(11): 62. 1863] ex Bornet & Flahault, ibid. 4: 360. 1887.—TYPE specimen from Sicily: bei der Tonnara von Palermo, *C. Cramer*, 20 Nov. 1856, in Rabenhorst, Algen Europas no. 1450 (PH).

[*Amphithrix Sargassii* Crouan (as "Sargassi") in Schramm & Mazé, Essai Class. Alg. Guadeloupe, p. 30. 1865.] *Mastigonema Sargassii* Crouan (as "Mastichonema Sargassi") [in Mazé & Schramm, Essai Class. Alg. Guadeloupe, ed. 2, p. 32. 1870—77] ex Bornet & Flahault, ibid. 3: 379. 1886. —TYPE specimen from Guadeloupe: Moule, au fond du port, *Schramm 7*, Aug. 1862 (PC).

Schizosiphon juncicola Crouan [Fl. Finistère, p. 115. 1867] ex Bornet & Flahault, ibid. 3: 353. 1886. —TYPE specimen from Finistère, France: parois d'un vieux ponton dans la vase du Penfeld, *Crouan*, 15 Aug. 1864, in herb. Bornet—Thuret (PC).

Schizosiphon mucosus Crouan [loc. cit. 1867] ex Bornet & Flahault, loc. cit. 1886. —TYPE specimen from France: Brest, *Crouan*, in herb. Bornet-Thuret (PC).

Dasyactis fissurata Crouan [ibid., p. 116. 1867] ex Bornet & Flahault, Ann. Sci. Nat. VII. Bot. 4: 344. 1887. *Isactis plana* var. *fissurata* Bornet & Flahault, ibid. 4: 345. 1887. *I. plana* fa. *fissurata* Crouan ex Batters, Mar. Alg. Berwick-on-Tweed, p. 22. 1889. —TYPE specimen from Finistère, France, labeled *Dasyactis fissurata* Crouan (PC).

Dasyactis Saccorhizae Crouan [Fl. Finistère, p. 116. 1867] ex Bornet & Flahault, ibid. 4: 344. 1887. *Rivularia Saccorhizae* Crouan *pro synon.*, loc. cit. 1867. —TYPE specimen from Morbihan, France: sur *Saccorhiza bulbosa* à

Belle-Ile, Sept. 1850, in Lloyd, Algues de l'Ouest de la France no. 302, in herb. Bornet–Thuret (PC).

Schizosiphon trochicola Crouan [loc. cit. 1867] ex Bornet & Flahault, ibid. 3: 373. 1886. –TYPE specimen from France: sur *Trochus majus* dragué, rade de Brest, *Crouan*, 14 Dec. 1858 (Lab. Marit., Concarneau).

Calothrix trochicola Crouan [ibid., p. 118. 1867] ex Bornet & Flahault, ibid. 3: 348. 1886. –TYPE specimen with that of *Schizosiphon trochicola* Crouan above (Lab. Marit., Concarneau).

Thrichocladia nostocoides Zanardini [Phyc. Indic. Pugill., p. 25. 1872] ex Forti, Syll. Myxoph., p. 661. 1907.–TYPE specimen labeled: Singapoore, in herb. Beccari (FI).

Polythrix spongiosa Zanardini [ibid., p. 32. 1872] ex Bornet & Flahault, ibid. 3: 380. 1886.–TYPE specimen: Singapore, *O. Beccari*, 1865, in herb. Beccari (FI).

Calothrix crustacea Schousboe & Thuret [in Bornet & Thuret, Notes Algol. 1: 13. 1876] ex Bornet & Flahault, ibid. 3: 359. 1886. *Oscillatoria crustacea* Schousboe *pro synon.* [in Bornet & Thuret, loc. cit. 1876] ex Bornet, Mém. Soc. Nat. Sci. Nat. & Math. Cherbourg 28: 188. 1892. –TYPE specimen from Morocco: in saxis maritim., reg. Tingitana, *Schousboe*, Oct. 1827, in herb. Bornet–Thuret (PC). Fig. 78.

Limnactis neglecta Ardissone [in Ardissone & Strafforello, Enum. Alghe Liguria, p. 79. 1877] ex Bornet & Flahault, Ann. Sci. Nat. VII. Bot. 4: 361. 1887. –TYPE specimen from Italy: Genova, alla scogliera della cava, *Ardissone* (PAD).

Leibleinia moniliformis Reinsch [Journ. Linn. Soc. Bot. 16(92): 235. 1877] ex Forti, Syll. Myxoph., p. 294. 1907. –TYPE specimen from Cape province, South Africa: Salt River, *A. Eaton*, 1874 (BM).

Mastigonema velutinum Wolle [Bull. Torrey Bot. Club 6(49): 283. 1879] ex Bornet & Flahault, ibid. 3: 353. 1886. –TYPE specimen from New Jersey: ad Perth Amboy, *F. Wolle*, Jul. 1878, in Wittrock & Nordstedt, Algae Exs. no. 388 (PH).

Symploca violacea Hauck [Österr. Bot. Zeitschr. 29: 244. 1879] ex Forti, ibid., p. 311. 1907. –TYPE specimen from Italy: Triest, *Hauck 112* (L).

Microchaete grisea Thuret [in Bornet & Thuret, Notes Algol. 2: 127. 1880] ex Bornet & Flahault, ibid. 5: 85. 1887. *Fremyella grisea* J. de Toni, Noter. di Nomencl. Algol. 8: [4]. 1936. –TYPE specimen from Loire-Inférieure, France: coquille au Croisic, *G. Thuret*, 6 Aug. 1873, in herb. Bornet–Thuret (PC). Series of colorless attenuated cells are present on some trichomes of this material.

Rivularia atra var. *confluens* Farlow, Mar. Alg. New England, p. 38. 1881. –TYPE specimen from Massachusetts: Wood's Holl, *W. G. Farlow*, in Farlow, Anderson & Eaton, Algae Exs. Amer. Bor. no. 221, 2 (FH).

[*Calothrix Harveyi* Kjellman, K. Svenska Vet. Akad. Handl. 20(5): 322.

1883.] —TYPE specimen from Ireland: Malbay, *W. H. Harvey*, 1831 (TCD).

Calothrix prolifera Bornet & Flahault [Mém. Soc. Nat. Sci. Nat. & Math. Cherbourg 25: 202. 1885] ex Bornet & Flahault, Ann. Sci. Nat. VII. Bot. 3: 361. 1886. *C. crustacea* fa. *prolifera* Collins in Collins, Holden & Setchell, Phyc. Bor.-Amer. 24: 1168. 1904. —TYPE specimen from Pyrénées-Orientales, France: derrière le laboratoire de Banyuls, *C. Flahault 140*, 29 Dec. 1881, in herb. Bornet-Thuret (PC). Fig. 71.

Microchaete vitiensis Askenasy [in Bornet & Flahault, Mém. Soc. Nat. Sci. Nat. & Math. Cherbourg 25: 214. 1885] ex Bornet & Flahault, ibid. 5: 85. 1887. *Fremyella vitiensis* J. de Toni, Noter. di Nomencl. Algol. 8: [5]. 1936. —TYPE specimen from Fiji islands: Matuku, *Naumann 512-20*, in herb. Bornet—Thuret (PC). The trichomes terminate in conical cells.

Calothrix confervicola var. *purpurea* Bornet & Flahault, Ann. Sci. Nat. VII. Bot. 3: 350. 1886. —TYPE specimen from Dalmatia, Yugoslavia: Capocesto, Sept. 1833, comm. Grunow, in herb. Bornet-Thuret (PC).

Calothrix confervicola var. *mediterranea* Bornet & Flahault, ibid. 3: 351. 1886. *C. confervicola* fa. *mediterranea* Kossinskaia, Opred. Morsk. Sinez. Vodor., p. 149. 1948. —TYPE specimen from Pyrénées-Orientales, France: Banyuls, derrière le laboratoire, *C. Flahault,* 29 Dec. 1881, in herb. Bornet-Thuret (PC).

Calothrix Contarenii var. *spongiosa* Bornet & Flahault, ibid. 3: 356. 1886. —TYPE specimen from New Caledonia: sur les rochers à Touho, *Vieillard 2180*, in herb. Bornet—Thuret (PC).

Rivularia atra var. *confluens* Bornet, Mém. Soc. Nat. Sci. Nat. & Math. Cherbourg 28: 189. 1892. —TYPE specimen from Morocco: Aglah, Tingi, *P. K. A. Schousboe*, Dec. 1827, in herb. Bornet-Thuret (PC).

Calothrix scopulorum fa. *ramosa* J. Schmidt, Dansk Bot. Tidsskr. 22: 392. 1899. —TYPE specimen from Denmark: Frederikshavn, *Schmidt*, 4 Jun. 1897 (C).

Calothrix scopulorum fa. *simplex* J. Schmidt, loc. cit. 1899. —TYPE specimen from Denmark: Hellebaek, paa stene, *Schmidt*, 26 Jun. 1898 (C).

Calothrix scopulorum fa. *subsimplex* J. Schmidt, loc. cit. 1899. —TYPE specimen from Denmark: Bornholm, paa subm. klippe stykker, *Schmidt*, Jun. 1898 (C).

Calothrix fasciculata fa. *incrustans* Collins in Collins, Holden & Setchell, Phyc. Bor.-Amer. 12: 561. 1899. —TYPE specimen from Massachusetts: on rocks, Revere Beach, *F. S. Collins*, 29 Sept. 1895, in Phycotheca Boreali-Americana no. 561 (NY).

Isactis centrifuga Bornet in Collins, Rhodora 3: 136. 1901. —TYPE specimen from Rhode Island: on rocks, Ochre point, Newport, *Collins 3820*, 10 June 1900, in herb. Bornet-Thuret (PC).

Dichothrix rupicola Collins, ibid. 3: 290. 1901. *Calothrix rupicola* Kossinskaia, Opred. Morsk. Sinez. Vodor., p. 160. 1948. —TYPE specimen from Maine: on exposed rocks, Pemaquid Point, *Collins 4144*, 18 July 1901 (NY).

Brachytrichia maculans Gomont, Bot. Tidsskr. 24: 210. 1902. *Kyrtuthrix maculans* Umezaki, Mem. Coll. Agric. Kyoto Univ., Fisheries Ser. Spec. No., p. 64. 1958. —TYPE specimen from Cambodia: Lem Dan, Koh Chang, *J. Schmidt XIII*, in herb. Gomont (PC). Fig. 82.

Richelia intracellularis J. Schmidt, Vidensk. Medd. Naturh. Fören. Kjöbenhavn 1901: 146. 1902. —TYPE specimen from the Gulf of Siam: off Koh Kram, *Schmidt*, 21 Mar. 1900 (C). The trichomes here inhabit broken shells of *Rhizosolenia* sp.

Calothrix crustacea fa. *simulans* Collins in Collins, Holden & Setchell, Phyc. Bor.-Amer. 29: 1406. 1907. —TYPE specimen from Massachusetts: on Zostera, Mattapoisett, *F. S. Collins*, in Phycotheca Boreali-Americana no. 1406 (NY).

Scytonema fuliginosum Tilden, Amer. Alg. 7(1): 629. 1909. *Tildenia fuliginosa* Kossinskaia, Bot. Mat. Inst. Sporov. Rast. Glavn. Bot. Sada SSSR 4(5–6): 77. 1926. *Setchelliella fuliginosa* J. de Toni, Noter. di Nomencl. Algol. 8: [6]. 1936. *Scytonematopsis fuliginosa* Copeland, Ann. New York Acad. Sci. 36: 103. 1936. —TYPE specimen from Hawaii: tide pool, Pahala Plantation beach, Hawaii, *J. E. Tilden*, 4 July 1900, in Tilden, American Algae no. 629 (MIN).

Calothrix endophytica Cotton, Proc. R. Irish Academy 31(15): 104. 1912. *C. Cottonii* J. de Toni, ibid. 1: 5. 1934. *C. Chapmanii* Lami & Meslin, Bull. Lab. Marit. Dinard 43: 47. 1959. —TYPE specimen from Ireland: in salt marsh, Annagh island, Clew bay, Oct. 1910 (BM).

Microchaete nana Howe & Hoyt, Mem. New York Bot. Gard. 6: 105. 1916. *Fremyella nana* J. de Toni, ibid. 8: [4]. 1936. —TYPE specimen from North Carolina: reef 23 miles off Beaufort, dredged, *L. Radcliffe*, 11 Aug. 1914 (US).

Calothrix rectangularis Setchell & Gardner in Gardner, Univ. Calif. Publ. Bot. 6(17): 472. 1918. —TYPE specimen from Washington: East sound, Orcas island, *N. L. Gardner 2365*, Aug. 1910 (UC).

Calothrix robusta Setchell & Gardner in Gardner, ibid. 6(17): 473. 1918. —TYPE specimen from California: in tide pools, Cypress Point, Monterey county, *Gardner 3397*, May 1916 (UC).

Dichothrix seriata Setchell & Gardner in Gardner, loc. cit. 1918. *Calothrix seriata* Kossinskaia, Opred. Morsk. Sinez. Vodor., p. 153. 1948. —TYPE specimen from Washington: on rocks, Neah Bay, *Gardner 3832* (UC).

Rivularia mamillata Setchell & Gardner in Gardner, ibid. 6(17): 475. 1918. —TYPE specimen from Washington: Neah Bay, *Gardner 3829*, June 1917 (UC).

Calothrix nodulosa Setchell & Gardner, Proc. Calif. Acad. Sci., ser. 4, 12(29): 702. 1924. —TYPE specimen from Baja California Sur, Mexico: San Marcos island, *I. M. Johnston 9e*, June 1921 (CAS).

Calothrix nidulans Setchell & Gardner, ibid. 12(29): 703. 1924. —TYPE specimen from Gulf of California, Mexico: *I. M. Johnston 162*, 1921 (CAS).

Dichothrix Bornetiana Howe, Bull. Torrey Bot. Club 51: 357. 1924. *Calothrix Bornetiana* Kossinskaia, Opred. Morsk. Sinez. Vodor., p. 158. 1948. —TYPE specimen from the Bahamas: Bimini harbor, *M. A. Howe 3268*, 17 Apr. 1904 (NY).

Sirocoleum Jensenii Weber-van Bosse, Vidensk. Medd. Dansk Naturh. Fören. 81: 68. 1926. *Hydrocoleum Jensenii* Weber-van Bosse ex Frémy, Ann. Crypt. Exot. 1(1): 49. 1928. —TYPE specimen from Kei islands, Indonesia: Tual, sur des rochers, *Jensen 216*, 1922 (C).

Aulosira marina Weber-van Bosse, ibid. 81: 72. 1926. —TYPE specimen from Kei islands, Indonesia: sur des rochers, Tual, *Jensen 216*, 1922 (C).

Scytonema keiense Weber-van Bosse, ibid. 81: 73. 1926. —TYPE specimen from Kei islands, Indonesia: sur des rochers, Tual, *Jensen 216*, 1922 (C).

Calothrix javanica fa. *marina* Weber-van Bosse, ibid. 81: 75. 1926. *C. javanica* var. *marina* Weber-van Bosse ex J. de Toni, Gen. & Sp. Myxoph. 1(A–C): 100. 1949. —TYPE specimen from Kei islands, Indonesia: on *Bostrychia tenella*, *Jensen 181* (C).

Tildenia fuliginosa var. *symmetrica* Kossinskaia, Not. Syst. Inst. Crypt. Hort. Bot. Petropol. 4(5–6): 85. 1926. *T. fuliginosa* fa. *symmetrica* Kossinskaia, Opred. Morsk. Sinez. Vodor., p. 174. 1948. —TYPE specimen from Hawaii: tide pool, Pahala Plantation beach, Hawaii, *J. E. Tilden*, 4 July 1900, in Tilden, American Algae no. 629 (MIN).

Calothrix longifila W. R. Taylor, Papers Carnegie Inst. Washington 25: 51. 1928. *Fremyella longifila* Drouet, Field Mus. Bot. Ser. 20(6): 130. 1942. —TYPE specimen from the Dry Tortugas, Florida: in the moat, Garden key, *W. R. Taylor 134*, 22 June 1924 (herb. W. R. Taylor).

Calothrix rosea W. R. Taylor, ibid. 25: 52. 1928. —TYPE specimen from Dry Tortugas, Florida: dredged, White Shoal sta. 3, *W. R. Taylor 254*, 11 July 1924 (herb. W. R. Taylor).

Kyrtuthrix dalmatica Ercegović, Arch. f. Protistenk. 66(1): 173. 1929. *Brachytrichia dalmatica* Frémy, Mém. Soc. Nat. Sci. & Math. Cherbourg 41: 162. 1934. —TYPE specimen, designated by Umezaki, Mem. Coll. Agric. Univ. Kyoto, Fisheries Ser. Spec. No., p. 66 (1958), from Dalmatia, Yugoslavia: Split, *A. Ercegović*, 16 Apr. 1957 (herb. I. Umezaki).

Calothrix codicola Setchell & Gardner, Proc. Calif. Acad. Sci., ser. 4, 19: 124. 1930. —TYPE specimen from Mexico: Guadalupe island, *H. L. Mason 39*, Apr. 1925 (CAS).

Calothrix aeruginea var. *abbreviata* Setchell & Gardner, ibid. 19: 126. 1930. —TYPE specimen from Mexico: Guadalupe island, *Mason 113*, Apr. 1925 (CAS).

Calothrix clausa Setchell & Gardner, loc. cit. 1930. —TYPE specimen from Mexico: Guadalupe island, *Mason 59*, Apr. 1925 (CAS).

Calothrix Cystosirae Schiffner, Österr. Bot. Zeitschr. 82(4): 300. 1933. —TYPE specimen from Dalmatia, Yugoslavia: Halbinsel Sabioncello, Küste von Vigan, *F. Berger*, Sept. 1931, in Schiffner, Algae Marinae no. 992 (FH).

Calothrix Laurenciae Setchell & Gardner, ibid. 22(2): 72. 1937. —TYPE specimen from Revilla Gigedo islands, Mexico: on *Laurencia* sp., Sulphur bay, Clarion island, *J. T. Howell 231b*, 24 Mar. 1932 (CAS).

Rivularia firma Womersley, Trans. R. Soc. South Australia 70(2): 130. 1946. —TYPE specimen from Kangaroo island, South Australia: *H. B. S. Womersley* (UAD).

Dichothrix eylathensis Rayss & Dor, Bull. Israel Nat. Council for Res. & Devel., Sea Fisheries Res. Sta. Haifa 34: 15. 1963. —TYPE specimen from Eylath, Israel (in slide collection of Inka Dor).

Original specimens have not been available to me for the following names; their original descriptions must serve as Types until the specimens are found:

Tremella hemisphaerica Linnaeus [Sp. Pl. 2: 1158. 1753] ex Agardh, Syst. Algar., p. 25. 1824. *Fucus Tremella-hemisphaerica* Gmelin, Hist. Fuc., p. 225. 1768. *Rivularia hemisphaerica* Linnaeus ex Fries, Fl. Scan., p. 323. 1835.

Batrachospermum hemisphaericum Lamarck & De Candolle, Fl. Franç., ed. 3, 2: 591. 1805. [*Euactis hemisphaerica* Trevisan, Nomencl. Algar. 1: 56. 1845.]

Rivularia applanata Carmichael in Harvey in Hooker, Brit. Fl.2 (Engl. Fl. 5): 392. 1833. [*Zonotrichia applanata* Rabenhorst, Fl. Eur. Algar. 2: 221. 1865.]

Calothrix minutissima Moris & De Notaris [Mem. Accad. Sci. Torino, ser. 2, 2: 270. 1840] ex Bornet & Flahault, Ann. Sci. Nat. VII. Bot. 3: 370. 1886.

Rivularia monticulosa Montagne in Barker-Webb & Berthelot, Hist. Nat. des Iles Canaries 3(2): 191. 1840. *Euactis monticulosa* Kützing [Sp. Algar., p. 340. 1849] ex Forti, Syll. Myxoph., p. 665. 1907.

Rivularia Mesogloiae De Notaris, Mem. R. Accad. Sci. Torino, ser. 2, 4: 315. 1842.

[*Schizosiphon gypsophilus* fa. *juvenilis* Rabenhorst, ibid. 2: 237. 1865.]

Schizosiphon Lacosteanus Rabenhorst [ibid. 2:242. 1865] ex Forti, ibid., p. 635. 1907.

Schizosiphon capensis Rabenhorst [ibid. 2: 245. 1865] ex Forti, loc. cit. 1907.

Physactis Wichurae Martens [Preuss. Exped. n. Ost-Asien, Bot., Die Tange, p. 21. 1866] ex Forti, ibid., p. 674. 1907.

Schizosiphon obscurus Dickie (as "obscurum") [Journ. Linn. Soc. Bot. 11: 459. 1871] ex Forti, Syll. Myxoph., p. 633. 1907.

Mastigonema halos Wood [Smithsonian Contrib. Knowl. 241: 52. 1872] ex Wolle, Fresh-w. Alg. U. S., p. 242. 1887.

Microchaete aeruginea Batters, Journ. of Bot. 30: 86. 1892. *Fremyella aeruginea* J. de Toni, Noter. di Nomencl. Algol. 8: [3]. 1936.

Calothrix Rhizosoleniae Lemmermann, Abh. Naturw. Ver. Bremen 16: 355. 1899.

Calothrix Hansgirgii Schmidle, Allg. Bot. Zeitschr. 1900(3): 35. 1900. *Homoeothrix Hansgirgii* Lemmermann, Krypt.-Fl. Mark Brandenb. 3: 240. 1907.

Diplocolon Codii Batters, Journ. of Bot. 44: 1. 1906.

Calothrix fusca var. *marina* Ercegović, Rad Jugosl. Akad. Znan. & Umjetn. 244: 159. 1932.

Calothrix foveolarum Ercegović, Rad Jugosl. Akad., Razr. Mat.-Prirod. 75: 159. 1932.

Scytonema endolithicum Ercegović, loc. cit. 1932.

Microchaete adriatica Vouk, Rad Jugosl. Akad., Razr. Mat.-Prirod. 79: 6. 1936. *Fremyella adriatica* J. de Toni, Archivio Bot. 20: 1. 1946.

Calothrix litoralis Anand, Journ. of Bot. 75 (Suppl. 2): 43. 1937.

Calothrix pulvinata var. *prostrata* Anand, ibid. 75(suppl. 2): 44. 1937. *C. pulvinata* fa. *prostrata* Powell in Parke & Dixon, Journ. Mar. Biol. Assn. United Kingd. 44(2): 504. 1964.

Dichothrix Cavanillesii González Guerrero, Anal. Jard. Bot. Madrid 6: 251. 1946.

Calothrix aestuarii Gayral & Seizilles de Mazancourt, Bull. Soc. Bot. France 105: 346. 1958.

Trichomata aeruginea, luteo-viridia, olivacea, fusca, rosea, violacea, vel cinereo-viridia, cylindrica vel ad extrema abrupte attenuata vel ad unum extremum tumida, ad dissepimenta constricta vel non-constricta, diametro 3—22μ crassa, partim aut passim increscentia partim decrescentia, ambitu recta vel curvantia vel spiralia, longitudine indeterminata, per destructionem cellulae intercalaris aut per constrictionem ad dissepimentum frangentia, ad unum extremum vel ad utrinque extrema per paucas pluresve cellulas attenuatas decolorantes terminantia. Cellulae diametro trichomatis breviores vel longiores, 2—20μ longae, protoplasmate homogeneo vel granuloso raro pseudovacuolato, dissepimentis non granulatis; heterocystae vulgo ad solum extremum nonnumquam ad utrinque extrema terminales, vel intercalares, interdum seriatae, hemisphaericae vel quasi-sphaericae vel discoideae vel cylindricae, diametro 4—23μ crassae; sporae ignotae; cellulae vegetativae terminales primum hemisphaericae demum obtuse conicae. Materia vaginalis primum hyalina demum lutea vel fusca. Planta trichomata longa vel brevia vulgo in vaginis discretis laminosis cylindricis saepe ramosis aut in gelatina distributa comprehens. Fig. 67—83.

Trichomes blue-green, yellow-green, olive, brown, red, violet, or gray-green, cylindrical or abruptly attenuated at one end or at both ends or swollen at one end, constricted or not constricted at the cross walls, $3-22\mu$ in diameter, in part here and there increasing or decreasing in diameter, straight or curving or spiraled, indeterminate in length, breaking by the destruction of an intercalary cell or by constriction at a cross wall, often terminating at one end or at both ends in few or many attenuated colorless cells. Cells shorter or longer than the diameter of the trichome, $2-20\mu$ long, the protoplasm homogeneous or granulose, rarely pseudovacuolate, the cross walls not granulated; heterocysts terminal at one end or at both ends, or intercalary, sometimes seriate, hemispherical or almost spherical or discoid or cylindrical, $4-23\mu$ in diameter; spores unknown; terminal vegetative cells at first hemispherical, becoming obtuse-conical. Sheath material at first hyaline, becoming yellow or brown. Plant consisting of long or short trichomes commonly in discrete laminate cylindrical, often branching, sheaths, or distributed in gelatinous matrices. Figs. 67–83.

Calothrix crustacea appears to occupy marine and brackish-water habitats exclusively. The common ecophene covers rocks, wood, shells, metal, and larger plants and animals between and below tide limits throughout the world, often in association with *Schizothrix calcicola* (Ag.) Gom., *Microcoleus lyngbyaceus* (Kütz.) Crouan, and *Entophysalis deusta* (Menegh.) Dr. & Daily. Each cylindrical trichome commonly develops at one end a single heterocyst or seriate heterocysts and at the other an abruptly attenuated series of short cells, the terminal ones of which lengthen, lose their pigments, and become vacuolated. The trichome is often surrounded by more or less firm, cylindrical sheath material (Fig. 67). The colorless cells are frequently broken off by outside agents and are not regenerated, as in plants which Bornet & Flahault designated as *Calothrix pulvinata* (Mert.) Ag. (PC). Often heterocysts are produced from intercalary cells. As trichomes break and the fragments of them elongate within a sheath, the tips of the new elongating trichomes burst through the sides of the sheath material and cause the sheath to be further elaborated by single- or twin-branching. Many trichomes are found with attenuated colorless cells at both ends, or with heterocysts at both ends. On or in the tissues of larger plants, each trichome may become somewhat swollen at one end.

Where the colloidal material becomes gelatinous, usually in waters below low tide mark, or at times between tide marks, the sheath material becomes chiefly, although more or less obscurely, single-branched; and the trichomes grow radiately within hemispheres or spheres of gelatinous matter (such ecophenes have customarily been segregated as species of the historically irrelevant genus *Rivularia* Roth) or in upright-parallel gelatinous sheaths distributed in flat plates or discs (placed erroneously in the past in the genus *Isactis* Thur.). Among the latter ecophenes, the sheath material often becomes very firm or cartilaginous at the surface; and the trichomes elongate in U-fashion, with the free end of each trichome directed downward into the softer sheath material (Fig. 82) —this ecophene has been assigned during recent years to a species of *Kyrtuthrix* Erceg. (see Umezaki, 1958). Where the trichomes grow parallel with each other within a cylindrical, usually dendroid, often externally cartilaginous matrix, the plants have been placed in species of *Gardnerula* J. de Toni (Fig. 80).

Trichomes of large diameters are usually composed of short cells, whereas those of small diameters develop longer cells. In brackish and some marine habitats, many collections have trichomes with diameters of 4μ or less and are sometimes confused with similar plants of *Calothrix parietina*. These small trichomes of *C. crustacea* are commonly cylindrical, but the free ends are somewhat less abruptly attenuated than are those of large trichomes (Fig. 77).

My attempts to grow material of this species from Cape May, New Jersey, and from Key West and Marathon, Florida, in fresh water in the laboratory have been unsuccessful. Darley (1968) reported the cultivation of numerous ecophenes of this species in sea water.

Bornet (1873) says that this species is the algal component of the lichens *Lichina* spp. Fungi, bacteria, *Schizothrix calcicola*, *Spirulina subsalsa* Oerst., and some other small algae are found growing within the sheath material, which they degrade or further augment.

The following specimens, selected from among the some 3,500 studied, are representative of *Calothrix crustacea* as described above:

Norway: Nordlandia, *Sommerfelt* (TYPE of *Linckia atra* β *coadunata* Sommerf., UC; duplicate: D), *Schübeler*, 1846 (NY); lac. Borgepolden, *M. H. Foslie*, Jul. 1881 (as *Rivularia atra* in Wittr. & Nordst., Alg. Exs. 663c, D, F, PH); Svinör pr. Lindesnäs, *Foslie*, Aug. 1185 (as *Calothrix scopulorum* in Wittr. &

Nordst., Alg. Exs. 852, D, F, PH); Kjelmo, *Foslie* , Aug. 1887 (as *C. vivipara* in Wittr. & Nordst., Alg. Exs. 1307a, D, F, PH).

Sweden: Båstad, *Agardh,* Sep. 1810 (TYPE of *Rivularia nitida* Ag., LD); Varholmen pr. Göteborg, *S. Åkermark,* Aug., Dec. 1865 (as *R. atra* in Wittr. & Nordst., Alg. Exs. 663a,b, D, F, PH, W. R. Taylor); in mari Bahusiae, *J. E. Areschoug* (as *Calothrix scopulorum* in Aresch., Alg. Scand. Exs., ser. nov. 139, 235, D, F), *S. Åkermark* (as *Rivularia hemisphaerica* in Aresch., Alg. Scand. Exs., ser. nov. 188, D, F; as *Schizosiphon fasciculatus* and *Calothrix scopulorum* in Rabenh., Alg. Eur. 1869, 1780, D, F, PH), *A. Areschoug* (as *Leibleinia confervicola* in Aresch., Alg. Scand. Exs., ser. nov. 192, D, F), *F. R. Kjellman & V. Wittrock* (as *Calothrix scopulorum* in Wittr. & Nordst., Alg., Exs. 191, D, F, PH); Grankullavik Oelandiae, *O. Nordstedt,* Aug. 1883 (as *Rivularia atra* in Wittr. & Nordst., Alg. Exs. 663e, D, F, PH); Kullen, 1835 (TYPE of *Linckia ceramicola* Lyngb., C); Landscrona, *Agardh* (TYPE of *Rivularia pellucida* Ag., LD); Leufsta, *Agardh* (TYPE of *Chaetophora aeruginosa* Ag., LD); Malmö, *Nordstedt,* Sep. 1877, Nov. 1878 (as *Calothrix scopulorum* in Wittr. & Nordst., Alg. Exs. 484a, b, D, F, PH); in sinu Codano (TYPES of *Leibleinia coccinea* Kütz. and *L. flaccida* Kütz., L); Varberg, *D. E. Hylmö,* Oct. 1903 (as *Calothrix confervicola* in Schiffn., Alg. Mar. 991, MICH); Warberg, *Mohr,* 1830 (TYPE of *Conferva scopulorum* Web. & Mohr, K in BM).

Denmark: Bornholm, *J. Schmidt,* Jun. 1898 (TYPE of *Calothrix scopulorum* fa. *subsimplex* Schm., C); Frederikshavn, *Schmidt,* Jun. 1897 (TYPE of *C. scopulorum* fa. *ramosa* Schm., C); Hellebaek, *Schmidt,* Jun. 1898 (TYPE of *C. scopulorum* fa. *simplex* Schm., C); Hofmansgave (Fyn), *Lyngbye,* Nov. 1815 (TYPE of *Oscillatoria zostericola* Hornem., C); Strand zwischen Kopenhagen und Helsingör, *Koråb,* Aug. 1925 (as *Calothrix aeruginosa* in Schiffn., Alg. Mar. 749, MICH); Torbaek, *Lyngbye,* Jun. 1815 (TYPE of *Linckia atra β viridis* Lyngb., C); Trekroner Hafniae, *Hofmann-Bang* (TYPE of *Schizosiphon chaetopus* Kütz., L); Boserup in Roskilde-fjord, *L. K. Rosenvinge,* Oct. 1882 (as *Rivularia atra* in Wittr. & Nordst., Alg. Exs. 663d, D, F, PH).

Russia: Königsberg (Kaliningrad), *Baenitz* (NY).

Poland: Ostsee bei Adlershorst (Gdansk), *C. Lakowitz,* Sep. 1887 (as *R. plicata* in Hauck & Richt.,Phyk. Univ. 188, D, L).

Romania: Constanta, *E. C. Teodorescu* (as *Calothrix scopulorum* in Mus. Vindob. Krypt. Exs. Alg. 1007, D, F), *104,* Apr. 1897 (S), *1278,* Apr. 1903 (D, F, W).

Germany: Nordsee, *Jürgens* (TYPES of *Euactis Lens* Kütz. and *Physactis aggregata* Kütz., L); Mare balticum (TYPE of *Schizosiphon gregarius* Kütz., L); in der Geltinger Bucht, Schleswig, *Suhr* (TYPE of *Heteractis pruniformis* Kütz., L); Flensburg (TYPES of *Leibleinia virescens* Kütz. and *Physactis lobata* Kütz., L); Quellenthal (Schleswig), *C. Jessen* (D, F); Neustädter Bucht pr. Niendorf (Lübeck), *F. Kuckuck* (as *Rivularia atra* in Mus. Vindob. Krypt. Exs. Alg. 748b,

D, F, W); Heppenserdrift, Jever (TYPES of *R. atra* Roth, PC; and *Ceramium pulvinatum* Mert., BREM, in Jürgens, Alg. Aq. 4: 4 and 4: 5); Bülk, *Reinke,* Oct. 1886 (D, L); Helgoland, *Kützing,* Jul. 1839 (TYPE of *Schizosiphon lutescens* Kütz., L), *Nägeli 213* (TYPE of *Leibleinia rupestris* Näg., ZT); Dobberan (Mecklenburg), *Flotow* (TYPE of *Physactis durissima* Kütz., L); Wangeroogae, *Jürgens 10* (TYPE of *Euactis Juergensii* Kütz., L).

Yugoslavia: Rovigno, *A. Hansgirg,* Apr. 1889 (D, F, W); Abbazia, *Hansgirg* (D, F, W); Pola, *Biasoletto* (TYPE of *Zonotrichia hemispherica* J. Ag., LD), *Kützing* (TYPE of *Heteractis mesenterica* Kütz., L), *J. Baumgartner* (as *Rivularia atra* in Mus. Vindob. Krypt. Exs. Alg. 748, D, F, W); Pago (Croatia), *Biasoletto* (TYPE of *Diplotrichia polyotis* J. Ag., LD); Cherso, *Kützing,* Mar. 1835 (TYPE of *Euactis marina* Kütz., L); Zengg, *Hansgirg,* 1891 (D, F, W); Vigan, Sabioncello, *F. Berger,* Sep. 1931 (TYPE of *Calothrix Cystosirae* Schiffn., Alg. Mar. 992, FH; duplicates: D, MICH, MIN, LD); Spalato, *Kützing,* Mar. 1835 (TYPES of *Schizosiphon rupicola* Kütz. and *Euactis amoena* Kütz., L); Capocesto, Sep. 1883 (TYPE of *Calothrix confervicola* var. *purpurea* Born. & Flah., PC); Zara, *Hansgirg,* Aug. 1888 (D, F, W); Gravosa pr. Ragusa, *Hansgirg,* 1891 (D, F, W).

Greece: Mykonos, *K. H. Rechinger,* Apr. 1927 (as *C. scopulorum* in Schiffn., Alg. Mar. 994, LD).

Italy: Alghero, *Marcucci,* 1866 (Un. Itin. Crypt. LIV, D, F,); Cagliari, *Marcucci,* 1866 (Un. Itin. Crypt. XXVII, D, F); Civitavecchia (TYPE of *Schizosiphon scopulorum β .crassus* Kütz., L); Genova (TYPES of *S. dilatatus* Kütz., *Leibleinia australis* Kütz., and *Euactis ligustica* Kütz., L; TYPE of *Limnactis neglecta* Ard., PAD); Grado, *F. Hauck 1170* (PC); Livorno (TYPES of *Euactis prorumpens* Kütz. and *Schizosiphon villosus* Kütz.,L), *G. Arcangeli* (as *E. prorumpens* and *Zonotrichia atra* in Erb. Critt. Ital. ser. II, 867 and 1135, D, F); Miramar pr. Tergeste, *F. Krasser* (as *Rivularia atra* in Mus. Vindob. Krypt. Exs. Alg. 748, D, F, W); Monfalcone, *Hauck,* Sep. 1882 (as *R. Biasolettiana* in Wittr. & Nordst., Alg. Exs. 576, D, F, PH); Neapel (TYPES of *Dasyactis minutula* Kütz., *Euactis hospita* Kütz., and *E. atra β confluens* Kütz., L); Palermo, *C. Cramer,* Nov. 1856 (TYPES of *E. pulchra* Cram. and *Physactis pulchra* Cram. in Rabenh., Alg. Eur. 1449, 1450, PH); Spezia (TYPE of *Rivularia medusae* Menegh., FI; duplicate: PC), *J. Doria,* 1856 (as *Physactis bullata* in Rabenh., Alg. Sachs. 571, D, F, PH); Trieste (TYPE of *Calothrix pannosa* Ag., LD; TYPE of *Rivularia fucicola* Zanard. in Mus. Storia Nat., Venezia; TYPES of *Dasyactis Biasolettiana* Kütz. and *Schizosiphon flagelliformis β radiatus* Kütz., L; TYPE of *Symploca violacea* Hauck, L), *F. Hauck* (as *Isactis plana* and *Rivularia hospita* in Wittr. & Nordst., Alg. Exs. 585b and 661a, D, F, PH); Terracina (TYPE of *Schizosiphon consociatus* Kütz., L); Venezia (TYPES of *Rivularia Contarenii* Zanard. and *Calothrix stellulata* Zanard., PC); Zaule, Trieste (TYPES of *Schizosiphon flagelliformis* Kütz. and *Leibleinia aeruginea* Kütz., L).

Netherlands: Goes, *van den Bosch 411* (CN); Zélande, *van den Bosch* (TYPE of

L. chalybea B *bicolor* Kütz., L); Lemmer, *W. F. R. Suringar DDD89*, Sep. 1854 (TYPE of *Symphyosiphon pulvinatus* fa. *tenuior* Sur., L).

Scotland: Appin, Argyll, *Carmichael* (TYPE of *Calothrix hydnoides* Harv., LD); Bellochantuy, Kintyre island, *J. R. Lewis 5*, 1954 (D); Arran, *W. Arnott*, Aug. 1853 (CN); Wemyss Bay, *Arnott* (CN); Troon, Ayrshire, *G. W. Traill 33* (NY); Proudfoot Rocks, Wick harbour, Caithness, *Lewis 3*, 1954 (D); North Uist, Outer Hebrides, *Lewis 1*, Jul. 1956 (D); Loch Duich, Ross, *Lewis 10*, 1954 (D); culture from Portencross, Ayrshire, *W. D. P. Stewart & J. T. Koster,* Jul. 1961 (D, L).

England: (TYPES of *Conferva Mucor* Roth, OXF, and *Leibleinia purpurea* Kütz., L); Falmouth, *Caspary* (TYPE of *Schizosiphon Warreniae* Casp., K in BM); Berwick-on-Tweed, *E. Batters*, Oct. 1884 (BM); Swanage, *Batters*, Sep. 1894 (BM); Studland, *E. George*, Sep. 1893 (LD); Teignmouth, *R. Creswell*, Oct. 1880 (S); Tor Abbey rocks (TYPE of *Rivularia plicata* Carm., K in BM; as *R. nitida* in Wyatt, Alg. Danmon. 50, D, F, MICH); Torbay (as *Calothrix confervicola* in Wyatt, Alg. Danmon. 229, D, F); Torquay (TYPES of *Schizosiphon scopulorum* β *subsimplex* Kütz., *Physactis spiralis* Kütz., and *Euactis plicata* Näg., L), *W. H. Harvey*, Aug. 1844 (D, F); Weymouth, *Leipner*, Jul. 1866 (as *Lyngbya confervicola* in Rabenh., Alg. Eur. 1881, D, F), *E. M. Holmes*, Apr. 1890 (BM).

Wales: Anglesey, *J. R. Lewis*, Nov. 1950 (D, F); Aberystwyth, *Lewis 1, 3, 4*, Apr. 1953 (D, F).

Ireland: Annagh island, Oct. 1910 (TYPE of *Calothrix endophytica* Cott., BM); Malbay, *Harvey*, 1831 (TYPE of *C. Harveyi* Kjellm., TCD); Cultra, Belfast bay, *W. Thompson*, Sep. 1839 (PH); Dublin bay, *Thompson*, Sep. 1838 (PH).

France: (TYPE of *Ulva bullata* Poir., PC); (as *Symploca pannosa* in Desmazières, Pl. Crypt. France, éd. 2, 130, NY, UC); Ambleteuse pr. Boulogne-sur-Mer, *C. Flahault*, Aug. 1882 (as *Calothrix pannosa* and *Isactis plana* in Wittr. & Nordst., Alg. Exs. 575 and 585a, D, F, PH); Antibes, *G. Thuret*, Dec. 1871 (D, F, PC, YU), *C. Flahault* (as *Rivularia hospita* in Wittr. & Nordst., Alg. Exs. 661b,c, D, F, PH); Banyuls, *Flahault* (TYPES of *Calothrix prolifera* Born. & Flah. and *C. confervicola* var. *mediterranea* Born. & Flah., PC; as *C. parasitica, C. crustacea*, and *Rivularia mesenterica* in Wittr. & Nordst., Alg. Exs. 854, 855a, and 863, D, F, PH), *J. Feldmann 2717*, Sep. 1932 (L); falaises de Batz, *M. Gomont*, Oct. 1891 (D, F, PC); Belle-Ile (TYPE of *R. Saccorhizae* Crouan in Lloyd, Alg. de l'Ouest de la France 302, PC); Calvados, Arromanches (as *Symphyosiphon gallicus* in Hohenack., Meeresalg. 458b, UC), *Lenormand 209* (TYPE of *Schizosiphon cuspidatus* Kütz., L); Brest (TYPES of *S. mucosus* Crouan, *S. trochicola* Crouan, and *Calothrix trochicola* Crouan, PC and Lab. Biol. Mar. Concarneau; TYPE of *Euactis confluens* Kütz., L; as *Rivularia parasitica* and *Schizosiphon fasciculatus* in Desmaz., Pl. Crypt. France 142, 549, UC, W);

Cannes, *A. Raphélis* (as *Rivularia mesenterica* and *Calothrix scopulorum* in Hamel, Alg. de France 3, 11, 52, D, F, MICH); Iles Chausey, *P. Frémy* (as *C. pulvinata* in Hamel, Alg. de France 10, D, F, MICH), *Lenormand* (TYPES of *Symphyosiphon punctiformis* Kütz. and *S. gallicus* Kütz., L); Cherbourg (TYPES of *Physactis pilifera* Kütz., *P. atropurpurea* Kütz., and *Leibleinia purpurea* γ *amethystea* Kütz., L; TYPE of *Euactis Lenormandiana* var. *Balanii* Le Jol., PC; as *Physactis pilifera, Schizosiphon lasiopus, S. Lenormandii,* and *Zonotrichia Lenormandiana* in Rabenh., Alg. Eur. 1508, 2129, 2130, 2132, D, F, PH, UC); Deauville, *Frémy* (as *Calothrix crustacea* in Hamel, Alg. de France 155, D, F); Finistère (as *C. confervicola* var. in Crouan, Alg. Mar. Finist. 341, FH; TYPES of *C. fusco-violacea* Crouan, *Physactis gracilis* Crouan, and *Rivularia confinis* Crouan, Lab. Biol. Mar. Concarneau; TYPES of *R. investiens* Crouan, *Schizosiphon juncicola* Crouan, and *Dasyactis fissurata* Crouan, PC); Manche, *Lenormand* (TYPES of *Euactis Lenormandiana* Kütz., *Symphyosiphon cespitosus* Kütz., and *Schizosiphon Lenormandii* Kütz., L); Le Croisic, *E. Bornet & C. Flahault*, 1887 (as *Calothrix Contarenii, C. crustacea, Rivularia bullata,* and *Microchaete grisea* in Wittr. & Nordst., Alg. Exs. 853, 855b, 862, 871, D, F, NY, PH), *G. Thuret* (TYPE of *M. grisea* Thur., PC); Marseille, *L. Berner*, Aug. 1931 (as *Calothrix confervicola* in Schiffn., Alg. Mar. 990, MICH); Marseille, herb. Lenormand (TYPES of *Euactis pachynema* Kütz. and *Physactis pilifera* β *fuscescens* Kütz., L; TYPE of *Rivularia parasitica* Chauv., CN); Saint-Jean-de-Luz, *Bory de Saint-Vincent*, Oct. 1813 (PC); Roscoff, *Flahault*, Sep. 1881 (S); Saint-Malo, *A. Hamel-J.* (as *R. bullata* in Hamel, Alg. de France 103, D, F, MICH); Saint-Servan, *Hamel-J.* (as *Calothrix confervicola* in Hamel, Alg. de France 102, D, F, MICH); Vendée, *Lloyd 23* (TYPE of *Physactis obducens* Kütz., L).

Spain: Pontevedra, Ria de Arosa, *C. van den Hoek 64/82*, Jul. 1964 (L).

Portugal: Setubal, *F. Welwitsch 325*, 1842–50 (BM, D).

Tanzania: Mafia island, Chole bay, *W. E. Isaac 3760* (D, L), *3765a* (L), Sep. 1967; 35 miles south of Tanga-Kekombe, *W. A. Lawson A3220*, Oct. 1962 (D, GC).

South Africa: Durban, *A. A. Weber-van Bosse*, 1894 (D, F. L); Salt River, Cape, *A. Eaton* (TYPE of *Leibleinia moniliformis* Reinsch, BM).

Cameroons: Victoria, *Lawson A737*, Dec. 1953 (D, GC).

Dahomey: Cotonou, *D. M. John & Lawson 6714*, Mar. 1971 (GC, PH).

Togo: Lomé, *John & Lawson 6645,* Mar. 1971 (GC, PH).

Ghana: Princes Town, *Lawson A1022*, Dec. 1955 (D, GC); Komenda, *Lawson A1289*, Jan. 1957 (D, GC).

Morocco: Tangier, *P. K. A. Schousboe* (TYPES of *Calothrix crustacea* Schousb. & Thur. and *Rivularia atra* var. *confluens* Born., PC).

Sierra Leone: Goderich, Freetown, *G. W. Lawson A628*, Jul. 1953 (D, GC).

Azores: Feraria (San Miguel) and Costa Flores, *W. Trelease 1451, 1462,* Aug. 1894 (MO).

Madeira: *Liebetrut* (L), *Mandon* (CN), *Lowe* (BM), ex herb. Lenormand (D, F).

Canary Islands: *Despréaux* (TYPE of *Rivularia cerebrina* Mont., PC); Puerto de la Cruz, Tenerife, *G. W. Lawson A2000*, Jun. 1962 (GC, PH).

Tristan da Cunha: saltpot, *G. Broekhuysen M49*, Feb. 1948 (BM, D).

Gough Island: *N. M. Wace 1527, 1544*, 1956 (BM, D).

Bermuda: Mangrove bay, south shore, Gravelly bay, Fairyland, north shore, *A. B. Hervey, F. S. Collins* (as *Polythrix corymbosa, Rivularia polyotis, Calothrix fusco-violacea, Brachytrichia maculans*, and *Dichothrix Baueriana* in Coll., Hold. & Setch., Phyc. Bor.-Amer. 1903, 1904, 2060, 2159, 2160, D, F); Whalebone bay and Castle island, *W. R. Taylor 49-15, 56-839* (D, W. R. Taylor); Cobbler's island and Smith's island, *T. A. & A. Stephenson BRC1, BRS1*, 1952 (D, F); Tuckers Town bay and south Paget island, *A. J. Bernatowicz 49-1904, 51-734* (D, F, MICH).

Newfoundland: Bay of Islands, *M. A. Howe 870*, 1901 (D, NY); Big Barachois, Placentia bay, *H. J. Humm 1*, 1948 (D, F, Humm).

Saint Pierre and Miquelon: Pointe Blanche, *P. Le Gallo 51*, 1945 (D, F, W. R. Taylor).

Nova Scotia: Halifax, *Howe 1013, 1033, 1047* (D, NY), *R. A. Lewin* (D, F); Pictou, *Howe 591* (D, UC); Chester, *Lewin* (D, F); Inner Sambro island, *R. D. Wood & C. MacFarlane 53-7-14-13* (D, F); Peggy's Cove and Caribou island, *T. A. & A. Stephenson NSP1, NSC5* (D, F).

Prince Edward Island: East Point and Cavendish, *Stephenson PES7, PEC23, PEC24* (D, F).

Maine: Pemaquid Point, *F. S. Collins 4144*, 1901 (TYPE of *Dichothrix rupicola* Coll., NY; duplicate: PC); Cape Rosier, Mount Desert, Harpswell, and Pemaquid Point, *Collins* (as *Microchaete grisea, Rivularia atra, Calothrix aeruginea, C. fasciculata, C. pulvinata, C. confervicola* var. *purpurea*, and *Dichothrix rupicola* in Coll., Hold. & Setch., Phyc. Bor.-Amer. 158, 261, 357c, 804, 957b, 958, 1456, D, F. NY); Northport and Cutler, *Collins 1017, 4350* (D, NY, UC); Eastport, *W. G. Farlow* (D, FH); Appledore, *Farlow* (D, FH), *R. D. Wood 52-7-29-1* (D, F); Seguin Island, *M. A. Howe 172, 262* (D, UC); Ogunquit, *H. Croasdale*, 1937 (D, FH, NY, S, W).

New Hampshire: Fort Stark, Newcastle, *M. S. Doty & J. Newhouse 7823*, 1951 (D, Doty, F).

Massachusetts: Nobska point (Falmouth), Gansett Estate (Falmouth), Eel pond (Woods Hole), Penikese island, Madaket (Nantucket), Gay Head (Marthas Vineyard), and Sconticut point (Fairhaven), *F. Drouet 2081, 2121, 2132, 2139, 2153, 2166, 2241* (D, NY); Woods Hole, *W. A. Setchell* (as *Calothrix vivipara* in Wittr., Nordst. & Lagerh., Alg. Exs. 1307b, D, F, PH), *W. G. Farlow* (TYPE of *Rivularia atra* var. *confluens* Farl., FH; duplicate: PC; as *R. atra, Isactis plana, Calothrix pulvinata, C. parasitica*, and *C. confervicola* in Farl., Anders. & Eat.,

Alg. Exs. Amer. Bor. 50, 221-2, 222, 224, 225, D, F, FH); Gloucester, *Farlow* (as *Rivularia atra* in Farl., Anders. & Eat., Alg. Exs. Amer. Bor. 221-1, D, F, FH); Magnolia and Nahant, *Farlow* (FH); Revere Beach and Salem, *F. S. Collins 1010, 4654* (D, F); Cuttyhunk, *W. R. Maxon* (as *Calothrix parasitica* in Mus. Vindob. Krypt. Exs. Alg. 2550, D, F, NY, US); Woods Hole, *W. A. Setchell* (as *C. confervicola* in Coll., Hold. & Setch., Phyc. Bor.-Amer. 9, D, F), *C. P. Nott* (as *C. parasitica* and *C. fusco-violacea* in Coll., Hold. & Setch., Phyc. Bor.-Amer. 111, 217, D, F); Cuttyhunk, *Setchell* (as *Isactis plana* in Coll., Hold. & Setch., Phyc. Bor.-Amer. 156a, D, F); Medford, *Collins* (as *Rivularia nitida* in Coll., Hold. & Setch., Phyc. Bor.-Amer. 260, D, F); Quamquisset Harbor, Falmouth, *Setchell* (as *R. nitida* in Coll., Hold. & Setch., Phyc. Bor.-Amer. 260b, D, F); Revere Beach, *Collins* (TYPE of *Calothrix fasciculata* fa. *incrustans* Coll. in Coll., Hold. & Setch., Phyc. Bor.-Amer. 561, NY; duplicates: D, F); Marblehead, *Collins* (as *C. vivipara, C. scopulorum*, and *Rivularia atra* in Coll., Hold. & Setch., Phyc. Bor.-Amer. 357a, 560, 805, D, F); Mattapoisett, *Collins* (TYPE of *Calothrix crustacea* fa. *simulans* Coll. in Coll., Hold. & Setch., Phyc. Bor.-Amer. 1406, NY; duplicates: D, F) (as *Rivularia nitida* in Coll., Hold. & Setch., Phyc. Bor.-Amer. 1562, D, F).

Rhode Island: Grove point, Block Island, *F. Drouet 3655, 3658—3660* (D, F); Beaver Tail, Conanicut island, *Collins* (D, UC); Little Compton, *W. C. Simmons* (D, UC); Napatree point, *D. C. Eaton* (D, F, YU); Ochre point, Newport, *Collins 3820* (TYPE of *Isactis centrifuga* Born., PC; duplicates in Coll., Hold. & Setch., Phyc. Bor.-Amer. 757, D, F, NY, W. R. Taylor), *Collins* (as *I. centrifuga* in Coll., Hold. & Setch., Phyc. Bor.-Amer. 2213, D, NY); Sakonnet point, *Bailey* (TYPE of *Calothrix vivipara* Harv., TCD); Scarborough Beach, *R. D. Wood 66-5-20-1* (PH).

Connecticut: Bridgeport, *I. Holden 765, 967* (D, UC); East Bridgeport, *L. N. Johnson 142* (D, F); Groton, *W. A. Setchell* (D, UC); Charles island, Milford, *Holden* (as *Rivularia atra* in Coll., Hold. & Setch., Phyc. Bor.-Amer. 357b, D, F, NY); Noank harbor, *D. C. Eaton* (D, F, YU); Stratford, *Johnson 187* (D, F); Thimble islands, *Eaton* (D, F, YU); Woodmont, *Johnson 823* (D, F), *Holden* (as *Calothrix crustacea* in Coll., Hold. & Setch., Phyc. Bor.-Amer. 10, D, F).

Quebec: Ile du Phare, St. Mary's islands, *J. Brunel 719, 722* (D, F, MT); Kegashka river, *Brunel 735, 736, 738* (D, F, MT).

New York: Cold Spring Harbor, *D. S. Johnson* (D, F), *L. N. Johnson* (as *Isactis plana* in Coll., Hold. & Setch., Phyc. Bor.-Amer. 156b, D, F); College Point, *H. Averill* (D, F, YU); Gardiners island, *E. T. Moul 7600* (D, RUT); Laurelton, *L. N. Johnson 1008a* (D, F); Orient, *R. Latham* (D, F, NY); Shinnecock Hills, *I. C. G. Cooper* (D, F).

New Jersey: Cape May, *F. Drouet & H. B. Louderback 14893* (PH); Perth Amboy, *F. Wolle* (TYPE of *Mastigonema velutinum* Wolle in Wittr. & Nordst.,

Alg. Exs. 388, PH; duplicates: D, F); Atlantic City, *S. R. Morse* (PH), *F. S. Collins 2092* (D, F), *J. E. Peters* (D, F, PH).

Delaware: Rehoboth Beach, *Drouet & Louderback 8562* (D); Rehoboth Bay, *J. S. Zaneveld et al.* D87 (PH, Zaneveld); Refuge Bay, *Zaneveld et al. D79b* (PH, Zaneveld).

Maryland: Fairmount and Sandy Hill beach (Tyaskin), *Drouet & P. W. Wolle 2253, 3642a, 3643* (D); Indian River Inlet and Snow Hill, *Zaneveld et al. 46b, 47* (PH, Zaneveld); Abington Beach, *C. F. Reed 72080* (PH, Reed); Chesapeake bay, Calvert county, *C. W. Reimer 5a* (PH); Shelltown, *P. W. Wolle* (D).

Virginia: York river at Croaker and Queens Creek, *J. C. Strickland 48, 1224* (D, Strickland); York river 15 miles downstream from York River Bridge, *J. S. Zaneveld* (PH, Zaneveld); Gloucester Point, *H. J. Humm* (PH), *B. L. Wulff* (PH), *Zaneveld 600181* (PH, Zaneveld).

North Carolina: Beaufort, *Drouet & Humm 14747* (PH), *L. Radcliffe* (TYPE of *Microchaete nana* Howe & Hoyt, US; duplicate: NY), *L. G. Williams 3* (D), *H. L. Blomquist 13700* (D, F), *T. A. & A. Stephenson BS14* (D, F), *Humm* (D, F), *C. S. Nielsen 1734* (D, F, FSU); New River Inlet and Cape Lookout, *Williams* (D, F); Wilmington and Wrightsville Beach, *Humm* (D, F).

South Carolina: Charleston harbor, *D. R. Wiseman* (PH).

Georgia: Sapelo, St. Simons, and Cabretta islands, *R. L. Chapman C103, C150, C161* (PH, Chapman).

Florida: Adams Key, *A. H. Curtiss* (D, F); Alligator Harbor, *Drouet, C. S. Nielsen & H. B. Louderback 12263* (D); Carrabelle, *Drouet & Nielsen 10963* (D, F, FSU); Cutler, *Humm* (D, F); Dania Beach, *Drouet & Louderback 10276* (D, F); Marineland, Daytona Beach, *T. A. & A. Stephenson MIB* (D, F); Dry Tortugas, *W. R. Taylor 134* (TYPE of *Calothrix longifila* W. R. Taylor, Taylor), *254* (TYPE of *C. rosea* W. R. Taylor, Taylor); Dunedin, *S. Earle Ericson 216* (D); Elliotts Key, *M. A. Howe 2975* (D, NY); Punta Gorda Beach near Englewood, *H. J. Humm* (D); 10 miles south of Fort Myers, *P. C. Standley 73186* (D, F); Jacksonville Beach, *Humm 2* (D, F); Old Mallory square, Key West, *Drouet & Louderback 14940* (PH); Key West, *W. H. Harvey* (TYPES of *C. dura* Harv. and *Microcoleus corymbosus* Harv., TCD), *Howe* (as *Rivularia polyotis* and *Polythrix corymbosa* in Coll., Hold. & Setch., Phyc. Bor.-Amer. 1015, 1016, D, NY); Pelican point, Marathon, *Drouet & Louderback 15030* (PH); Mullet Key, Pinellas county, *R. C. Phillips 360* (D); Palm Beach, *Drouet & Louderback 10205* (D, F); Panama City, *Drouet & Nielsen 11609* (D, F, FSU); Port Leon, *Drouet & E. M. Atwood 11461* (D, F, FSU); Rock islands, Taylor county, *J. Morrill* (D); east end of Santa Rosa island, *Drouet et al. 10634* (D, F, FSU); Piney point, Tampa bay, *Humm* (D, F).

Alabama: Point Clear, *Drouet & Louderback 10162* (D, F).

Mississippi: Bay St. Louis, *Drouet 9860* (D, F); near Waveland, *Drouet 9815* (D, F); Biloxi, *Drouet 10002* (D, F); Fort Massachusetts, Ship island, *L. H. Flint 2* (D); Ocean Springs, *Drouet & R. L. Caylor 9894* (D, F).

Keewatin: Point Fisher, Winter island, *P. A. Dutilly* (D, F).

Manitoba: Churchill, *O. & I. Degener 25390* (D, Degener).

Louisiana: Manchac pass, *Drouet & Flint 9668* (D, F); Lake Pontchartrain at Mandeville, *Drouet & Flint 9541, 9542, 9545* (D, F); Lake Pontchartrain at R. S. Maestri bridge, *Drouet & P. Viosca Jr. 9323* (D, F).

Texas: Port Aransas, *H. J. Humm* (D); Padre island, *E. Whitehouse 24530* (D, F, SMU).

British Columbia: Ucluelet arm, Barkley sound, Vancouver island, *J. Macoun 99* (D, UC); Brandon island near Nanaimo, *T. A. & A. Stephenson NB32* (D, F); Minnesota Seaside Station, Vancouver island, *C. K. Leavitt & C. M. Crosby* (as *Rivularia nitida* in Tilden, Amer. Alg. 571, D, F).

Washington: Cape Flattery, *N. L. Gardner 3829* (TYPE of *R. mamillata* Setch. & Gardn., UC; duplicates: D, F), *3832* (TYPE of *Dichothrix seriata* Setch. & Gardn., UC; duplicates: D, F), *3853* (D, F, US); Orcas island, *Gardner 2365* (TYPE of *Calothrix rectangularis* Setch. & Gardn., UC; duplicates: D, F); San Juan island, *Gardner 6094* (D, F, UC); Whidbey island, *Gardner 303* (D, F, UC; as *C. pulvinata* in Coll., Hold. & Setch., Phyc. Bor.-Amer. 957a, D, F).

Oregon: Coos Bay, *V. J. Chapman* (PH).

California: cultures of uncertain origin, *M. B. Allen 1* (D), *G. J. Hollenberg 2647* (D, F, Hollenberg); Alameda, *Gardner* (as *C. crustacea* in Coll., Hold. & Setch., Phyc. Bor.-Amer. 1168, 1212, D, F); Balboa, *Hollenberg 1552* (D, F, UC, Hollenberg); Capitola, *M. Doty 7325* (Doty, D, F); Carmel Bay, *Gardner* (as *Rivularia atra* var. *confluens* in Coll., Hold. & Setch., Phyc. Bor.-Amer. 2107, D, F); Corona del Mar, *Hollenberg 3041.5* (D, F, Hollenberg); Cypress point, Monterey county, *Gardner 3397* (TYPE of *Calothrix robusta* Setch. & Gardn., UC; duplicate: D); Fort Ross, *W. A. Setchell 1713* (D, F, UC); culture from La Jolla, *M. B. Allen*, comm. T. E. Jensen (PH); near Point Carmel, *Setchell* (as *C. pilosa* in Coll., Hold. & Setch., Phyc. Bor.-Amer. 859, D, F, MO, NY); San Pedro, *N. L. Gardner* (as *C. scopulorum, C. crustacea, C. vivipara,* and *Rivularia atra* var. *hemisphaerica* in Coll., Hold. & Setch., Phyc. Bor.-Amer. 1561, 1720, 2104, 2105, D, F); Santa Catalina island, *E. Y. Dawson 5661* (D, F).

Bahama Islands: Castle island near Acklin island, *M. A. Howe 5735* (D, NY); Andros, *C. Monty SB2* (PH); Frozen cay, Berry islands, *Howe 3535* (D, NY); Bimini, *Howe 3268* (TYPE of *Dichothrix Bornetiana* Howe, NY; duplicate: D), *H. J. Humm 21* (D, F), *E. T. Moul 10282* (D, RUT); Smiths point, Great Bahama, *Howe 3787* (D, UC); Gun cay, *Howe 3129* (D, NY); Hog island, Nassau, *Howe 3367* (D, NY); Abraham cay, Mariguana, *Howe 5412* (D, NY); Ship Channel cay, *Howe 3931* (D, NY); Shrouds cay, Exuma chain, *Howe*

3949a (D, F); Silver cay, *Howe 3033* (D, UC); Watling island, *Howe 5124* (D, NY).

Puerto Rico: *Howe* (as *D. Baueriana* in Coll., Hold. & Setch., Phyc. Bor.-Amer. 1169, D, F); Boca de Congrejos, *H. L. Blomquist 12509a* (D, F); Culebra island, *Howe 4277* (D, NY); Guanica Bay, *Blomquist 13013* (D, F); La Parguera, Lajas, *L. R. Almodóvar 3454, 3509* (D, Almodóvar); Cataño, San Juan, *Howe 2255* (D, NY); Santurce, *Howe* (as *Calothrix Contarenii* in Coll., Hold. & Setch., Phyc. Bor.-Amer. 1113, D, F).

Virgin Islands: Rendezvous bay, St. John, *H. J. Humm* (D).

Antigua: Port Royal bay and Hodge point, *W. R. Taylor 66-403, 66-487* (Taylor).

Guadeloupe: Moule, *Mazé & Schramm* (TYPE of *Amphithrix Sargassii* Crouan, PC), *G. Hamel & A. Hamel-Joukov* (as *Dichothrix fucicola* in Hamel, Alg. Antill. Franç. 101, W. R. Taylor); Rivière des Pères, *Hamel & Hamel-J.* (MICH); Les Saintes—Terre-de-haut, *J. Feldmann 3295* (Taylor); Anse au S. de Bouillante et Pointe de Chateaux, *Feldman 3261, 3775* (Taylor).

Martinique: Pointe debout, Fort de France, *E. Jardin* (PC).

St. Lucia: La Brellotte bay, Gros Islet Qtr., *W. R. Taylor 68-304B* (Taylor).

Barbados: St. Andrew parish, *W. R. Taylor 66-223* (Taylor); Bathsheba, *A. Vickers* (as *Calothrix aeruginea* in Vick., Alg. Barbad. 10, LD); Hastings Rocks, *Vickers* (D, F).

Sargasso Sea: at 34° 14′ N, 64° 30′ W, *J. H. Sharp* (PH).

Haiti: Petit Grove, *C. H. Arndt 287* (W. R. Taylor); Port Salut, *C. R. Orcutt 9651* (D, US); south of Dame Marie, *H. H. Bartlett 17841* (D, F, MICH); Tortuga island, *W. L. Schmitt & G. R. Lunz 3* (D, F).

Cuba: Gibara, Banes, Arroyos de Mantua, Guanabo, and Batabanó, *M. Díaz-Piferrer 5817, 5878, 6097, 6366, 6666* (FPDB, PH); Guantánamo, *R. Jervis 549* (D, W. R. Taylor).

Jamaica: Port Maria, *J. E. Humphrey* (as *Dichothrix penicillata* in Coll., Hold. & Setch., Phyc. Bor.-Amer. 62, D, F); Ocho Rios and Gun cay off Port Royal, *V. J. Chapman* (D, IJ); Northeast cay, Morant cays, *C. B. Lewis A 983* (D, IJ); Fort Clarence near Kingston and Montego Bay, *M. A. Howe 4654, 4772* (D, NY); Priory (St. Ann), Soldier's Bay (Portland), and Lucea (Hanover), *W. R. Taylor 56-40, 56-497, 56-634* (D, MICH).

Cayman Islands: between West Indian Beach Club and Pageant Beach Hotel, Grand Cayman, *W. R. Taylor 67-96, 67-97* (Taylor).

Netherlands Antilles: Aruba: Punta Basora, *W. R. Taylor 39-339, 39-340* (D, MICH). Curaçao: Piscadera Baai, Koraal Tabak (Yoca), Boca Djegoe, Awa di Oostpunt, Playa Boca Canoa, and Kust van Hato, *M. Díaz-Piferrer 14, 232, 770, 930, 1134, 1193* (D, FPDB).

Mexico: Baja California: Ensenada, *N. L. Gardner 4986, 4961* (D, UC); Guadalupe island, *H. L. Mason 39* (TYPE of *Calothrix codicola* Setch. & Gardn.,

CAS), *59* (TYPE of *C. clausa* Setch. & Gardn., CAS), *113* (TYPE of *C. aeruginea* var. *abbreviata* Setch. & Gardn., CAS), *E. Y. Dawson 8201, 8589, 8622* (D). Baja California Sur: San Marcos island, *I. M. Johnston 9e* (TYPE of *C. nodulosa* Setch. & Gardn., CAS); Punta Palmilla near San José del Cabo, Cabeza Ballena near Cabo San Lucas, and Punta Santa Rosalía, *Dawson 3297, 3313, 1479* (D). Gulf of California: *Johnston 162* (TYPE of *C. nidulans* Setch. & Gardn., CAS). Colima: Clarion island, Revilla Gigedo islands, *J. T. Howell 231b* (TYPE of *C. Laurenciae* Setch. & Gardn., CAS), *W. R. Taylor 39-9* (D, MICH). Jalisco: Bahía Chamela, *Dawson 21289* (D). Nayarit: María Magdalena island, Tres Marías, *Taylor 39-673* (D, MICH). Sonora: Bahía Kino, *F. Drouet & D. Richards 2903* (D, F); Puerto Lobos, Bahía Tepoca, *Dawson 814* (D); Guaymas, *Drouet & Richards 3332, 3371* (D, F). Yucatán: Campeche Bank, *Liebmann* (TYPE of *C. cyanea* J. Ag., LD; duplicate: PC; TYPE of *Schizosiphon fucicola* Kütz., L).

British Honduras: *D. R. Stoddart HMC5* (W. R. Taylor).

Costa Rica: Río Sándoval—Río Tigre and Río Sandalo, *C. W. Dodge 4537, 10208* (D, MO); Bahía Ballena and Punta Gallardo, *E. Y. Dawson 16734, 16735, 16872* (D); Puerto Culebra, *W. R. Taylor 34-538a* (D, MICH); Dominical, *R. A. Lewin 6* (D).

Panama: Islas Secas near Puerto Nuevo, *W. R. Taylor 39-122* (D, MICH); Isla Brincanco bay (Islas Contreras) and Isla del Rey (Bahía San Telmo), *Dawson 21050, 21175* (D).

Venezuela: Anzoátegui: Playa de Juan Pedro, Guanta, *M. Díaz-Piferrer 20285* (FPDB, PH). Nueva Esparta: Isla de Coche and Isla Margarita, *Díaz-Piferrer 20654, 20893* (FPDB, PH). Sucre: Ensenada El Tigrillo, Isla Venado, Ensenada de las Mujeres, and Playa de Punta Arenas, *Díaz-Piferrer 20142, 20152, 20314, 20456* (FPDB, PH).

Brazil: Paraná: Ilha de Farol (Caiobá), Guaratuba, and Ilha do Sai, *A. B. Joly 490, 523, 580* (D, SPF). Rio de Janeiro: Emboacyca, Angra dos Reis, *Joly 252* (D, SPF). Santa Catarina: Laguna, *Joly 1054* (D, SPF). São Paulo: Ilhabela and Itanhaen, *Joly 270, 453* (D, SPF).

Argentina: Chubut: Punta Castro and Faro Golfo Nuevo La Farola, *I. K. Paternoster 71, 72* (D. R. de Halperín, PH); Playa Santa Isabel, *de Halperín 125* (de Halperín, PH); Punta Delgada, *de Halperín 52* (de Halperín, PH); Deseado, *de Halperín 20, 26, 30, 50, 72* (de Halperín, PH); Playa Magagna, Punta Cuevas, and Punta Pardelas, *de Halperin 104, 105, 130B, 131A, 133A, 135B* (de Halperín, PH).

Chile: Puerto Montt, *G. H. Schwabe 8* (D, F).

Clipperton Island: *E. Y. Dawson 19968* (D, UC).

Hawaii: Hawaii: Pahala plantation beach, *J. E. Tilden* (TYPES of *Scytonema fuliginosa* Tild. and *Tildenia fuliginosa* var. *symmetrica* Kossinsk. in Tild., Amer. Alg. 629, MIN; duplicates: D, F, MO); Laupahoehoe and Punaluu Landing,

W. A. Setchell 5284, 10705 (D, F, UC); Hookena, South Kona, M. S. Doty & A. J. Bernatowicz 13422 (D, Doty). Kauai: Koloa, M. Reed 359 (D, UC); Poipu beach, W. J. Newhouse 53-101 (D, Newhouse). Oahu: Kawela, Moku Manu, Rabbit island, and Kahuku, Doty 8063, 8450, 8838, 8921 (D, Doty); Coconut island (Kaneohe bay) and Hanauma bay, Newhouse 1953-21, 1953-62 (D, F, Newhouse); Kaena point, O. & I. Degener 25113 (D, Degener); Kakaako and Waikiki, Honolulu, Setchell 5545, 5578, 5617 (D, UC).

Tuamotu Islands: Raroia, with cultures, Doty & Newhouse 11876, 11877, 11320, 11372 (D, Doty, Newhouse).

Society Islands: Tiamaha, M. L. Grant 4798 (D, UC); Papeete, Tahiti, W. A. & C. B. Setchell & H. E. Parks 5060a (D, UC); Tahiti, C. Copeland 7230 (D, UC).

Marshall Islands: Arno atoll, L. Horwitz 9308, 9324, 9632 (D, F); Bikar atoll, F. R. Fosberg 34562 (D, F); Bikini atoll, R. F. Palumbo Y8a, Y29 (D, F), J. Tracy (D, F); Eniwetok atoll, E. Y. Dawson 13621, 13662, 13663, 13687, 13868, 14034 (D); Jaluit atoll, Dawson 13151a, 13160 (D, F); Kwajalein atoll, Dawson 12564b, 12594, 12623, 12641, 12645 (D); Likiep atoll, Fosberg 33836 (D, F); Majuro atoll, Dawson 12703 (D); Rongelap atoll, W. R. Taylor 46-499 (D, F, MICH); Rongerik atoll, Taylor 46-535 (D, F, MICH).

Gilbert Islands: Betio, Tarawa, J. Cooper 18807A (D, Doty).

Caroline Islands: Metalanim, Ponape, S. F. Glassman 2301 (D, F).

Kapingamarangi: W. J. Newhouse 1189, 1262, 1345, 1395, 1484, 1652 (D, Newhouse).

Fiji Islands: Matuku, Naumann 512-20 (TYPE of Microchaete vitiensis Asken., PC).

Kermadec Islands: Raoul island, V. J. Chapman 5 (D).

New Zealand: Stanmore Bay, New Brighton (Otago), Hokianga river, Anawhatta, Stanley bay (Waitemata), and Piha, V. J. Chapman (D); Auckland (as Calothrix scopulorum in Lindauer, Alg. N.-Zel. Exs. 301, D, F), Deepwater Cove, Bay of Islands (as Isactis plana in Lindauer, Alg. N.-Zel. Exs. 52, D, F), Campbell's Beach (Pihama), and Stewart island, V.W. Lindauer (D, F); Styx river near Christchurch, M. Ratnasabapathy 22-6-1961 (D); Parua bay near Whangarei, Island bay near Wellington, and Devonport opposite Auckland, W. A. Setchell 6070, 6161, 6175 (D, UC); Stewart island, J. E. Tilden (as Rivularia atra in Tild., So. Pac. Pl. 15, MO); L. M. & M. H. Jones, A. & V. K. Nash (Tild., So. Pac. Pl., 2nd Ser., 613, 618, 729, 770, 793, D).

Solomon Islands: Kokomtambu island (Florida islands) and Kopiu (Guadalcanal), A. Bailey 30, 659 (BM, PH).

New Caledonia: Touho, Vieillard 2180 (TYPE of Calothrix Contarenii var. spongiosa Born. & Flah., PC); Anse Vata near Nouméa, Mrs. R. Catala 41 (D, F, NSW).

Japan: Hokkaido: Reibun and Rishiri islands, I. Umezaki 1315, 1336 (D, Umezaki). Honshu: Fukui pref.: Obama bay, Umezaki 3, 8, 55, 57, 127, 241,

194 REVISION OF THE NOSTOCACEAE

242, 425, 941 (D, Umezaki), Yashiro bay, *Umezaki 122* (D, Umezaki), Wakasa-Takehama, *Umezaki 105* (D, Umezaki); Hyogo pref.: near Kinosaki, *Umezaki 2121a* (D, Umezaki); Kyoto pref.: Maizuru bay, *Umezaki 37, 438* (D, Umezaki); Miyagi pref.: Matsushima bay, *Umezaki 1163* (D, Umezaki). Honjima island, Seto Inland Sea: *Umezaki 1957* (D, Umezaki). Kyushu: Fukuoka pref.: Tsuyasaki, *Umezaki 1983a* (D, Umezaki); Miyazaki pref.: Ao-shima, *Umezaki 2069, 2070a, 2073a* (D, Umezaki); Kagoshima pref.: Ushinohama, *Umezaki 2046, 2047* (D, Umezaki). Ryukyu Islands: *C. Wright* (TYPE of *Rivularia opaca* Harv., US; duplicate: D); Amami-oshima, *Umezaki 1759, 1796, 2198, 2251, 2443, 2447, 2467* (D, Umezaki); Kikaijima, *Umezaki 2384* (D, Umezaki).

Philippines: Babuyan islands: Calayan island, *R. C. McGregor 23, 30* (D, UC). Luzon: Narvacan beach, Ilocos Sur, *M. M. Kalaw 4* (D, PUH). Mindoro: Puerto Galera, *G. T. Velasquez 795, 1043, 1076, 1930, 1931* (D, F, PUH). Palawan: Taytay, *E. D. Merrill* (D). Sulu Archipelago: Bungao, Tawi-Tawi, *Velasquez 3213* (D, F, PUH).

China: Liu-wu-tien, Amoy, *C. Y. Chiao A30* (D, UC).

South Viet Nam: Bays of Nhatrang (Cauda and Ile de Tré) and Binhcang, *E. Y. Dawson 11085, 11254, 11394* (D, F).

Cambodia: Koh-Chang, *J. Schmidt XIII* (TYPE of *Brachytrichia maculans* Gom., PC).

Thailand: off Koh Kram, *Schmidt* (TYPE of *Richelia intracellularis* Schmidt, C).

Singapore: *O. Beccari*, 1865 (TYPES of *Polythrix spongiosa* Zanard. and *Thrichocladia nostocoides* Zanard., FI; duplicate: L).

Indonesia: Kai islands: Tual, *Jensen* (TYPES of *Aulosira marina* Web.-v. B., *Scytonema keiense* Web.-v. B., and *Sirocoleum Jensenii* Web.-v. B., C), Aru islands, *Jensen 181* (TYPE of *Calothrix javanica* fa. *marina* Web.-v. B., C); Irian: Seget-Selee-straat, *A. Weber-van Bosse* (L); Gisser, *Weber-van Bosse* (D, L).

Queensland: Port Denison, *A. Dietrich* (L); Miami, Moreton Bay, and Noosa, *A. B. Cribb* (BRIU, D, F).

New South Wales: Collaroy, *C. M. Crosby & L. A. Doore 325* (D); Twofold bay, Eden, *E. Pope 2031* (D, NSW); Corrimal headland, *V. May 648* (D, NSW); Kiama, *W. H. Harvey 593* (PH); Cabbage Tree Creek, *A. B. Cribb 139.9* (BRIU, D).

Victoria (Australia): Western Port, *W. H. Harvey 94* (TYPE of *Mastichothrix obscura* Kütz., TCD), *J. B. Wilson* (BM, D); Victoria lake, Paynesville, *Cribb 79-4* (BRIU, D); Portland, *C. Beauglehole 691B* (AD-U, D).

Tasmania: Low Head, *Cribb 73-22* (BRIU, D); Port Arthur, *W.H. Harvey* (TYPE of *Calothrix infestans* Harv., K in BM), *Cribb 30-1, 33-4, 38-2, 116-8, 116-26, 149-18, 152-12* (BRIU, D), *M. H. & L. M. Jones 260* (D), *T. T. Earle & P. J. Philson 268* (D); Pelican island, Southport, *Cribb 44-4* (BRIU, D).

South Australia: Kangaroo island, *H. B. S. Womersley* (TYPE of *Rivularia firma* Womersl., AD-U; duplicate: D); Vivonne bay, D'Estree's bay, Pennington bay, Middle River, Cape de Couedie, and Cape Willoughby, Kangaroo island, *Womersley A2206d, A4146i, A7041e, A9465g, A12894, A12943* (AD-U, D); Pt. Willunga reef, Marino, Granite island in Victor Harbour, Cape Carnot, Point Westall, Mangrove point in Spencer gulf, and Myponga, *Womersley* (AD-U, D).

Western Australia: Cape Riche, *W. H. Harvey 592* (TYPE of *R. australis* Harv., BM; duplicates: LD, PC); Rottnest island, *M. H. Jones 63* (D), *C. M. Crosby 85, 86* (D).

Maldive Islands: Male, Addu, Ari, and Fadiffolu atolls, *H. E. Hackett 8, 11, 23, 54a, 59, 90* (PH).

Chagos Archipelago: Diego Garcia, *C. Rhyne 356, 395, 515, 552, 698* (PH, US).

St. Paul Island (in the Indian Ocean): *Exped. Novara* (D, W).

Seychelles: Beau Vallon, Mahé, *W. E. Isaac 4018* (PH).

Mauritius: *N. Pike* (BKL, D).

Red Sea: *Portier* (TYPE of *Calothrix Caulerpae* Zanard. in Mus. di Storia Nat., Venezia); *Figari* (TYPE of *Dichothrix penicillata* Zanard., PC).

Israel: Eylath, *T. Rayss & I. Dor* (TYPE of *D. eylathensis* Rayss & Dor, in the collection of I. Dor); Caesarea and Tantura, *Rayss 8, 10* (D).

Sinai: Taba, *I. Dor* (Dor).

Raphidiopsis Fritsch & Rich

Raphidiopsis Fritsch & Rich, Trans. R. Soc. So. Africa 18(1): 91. 1929.
—Type species: *R. curvata* Fritsch & Rich.

Trichomata cylindrica vel acicularia, extrema gradatim attenuantia, septata, ambitu recta vel curvantia vel spiralia, cellulis terminalibus primum
hemisphaericis demum acute conicis, heterocystis terminalibus acute conicis,
heterocystis intercalaribus ovoideo-cylindricis, sporis ovoideo-cylindricis.
Materia vaginalis dispersa aut mucosa. Planta trichomata nuda vel in muco
distributa comprehens.

Trichomes cylindrical or acicular, gradually attenuated toward the ends,
septate, straight or curving or spiraled, terminal cells at first hemispherical
then acute-conical; terminal heterocysts acute-conical, intercalary heterocysts
ovoid-cylindrical; spores ovoid-cylindrical. Sheath material dispersed or
mucous. Plant consisting of naked trichomes or trichomes distributed in
mucus.

One species:

Raphidiopsis curvata Fritsch & Rich

Raphidiopsis curvata Fritsch & Rich, Trans. R. Soc. So. Africa 18(1): 91. 1929.
R. spiralis Fritsch & Rich ex Skuja, N. Acta R. Soc. Sci. Upsal., ser. 4, 14(5): 29.
1949. —TYPE specimen from Newlands, Griqualand West, South Africa, in
the slide collection (BM). Fig. 65.
Anabaenopsis philippinensis W. R. Taylor, Amer. Journ. Bot. 19: 458, 462.

1932. —TYPE specimen from Laguna province, Philippines: plankton of Sampaloc lake, San Pablo, *E. Quisumbing,* 1929 (W. R. Taylor).

Aphanizomenon americanum Reinhard, Bull. Torrey Bot. Club 68: 328. 1941. —TYPE specimen from Minnesota: plankton, Minnesota river at Mendota, *E. G. Reinhard,* Jul. 1928 (D). Fig. 63.

Anabaenopsis seriata Prescott in Prescott & Andrews, Hydrobiologia 7(1—2): 61. 1955. —TYPE specimen from Emporia, Kansas: Wooster lake, *T. F. Andrews,* Sept. 1951 (EMC).

Anabaenopsis wustericum Obuchova in Obuchova & Kosenko, Bot. Mat. Herb. Inst. Bot. Akad. Nauk Kazakhst. SSR 2: 84. 1964. —TYPE specimen mixed with that of *A. seriata* Presc. from Kansas (EMC).

Original specimens have not been available to me for the following names; the original descriptions must serve as Types until the specimens are found:

Aphanizomenon Kaufmannii Schmidle (as "Kaufmanni") in Kaufmann, Rev. d'Egypte 4: 113. 1897. *Cylindrospermum Kaufmannii* Huber-Pestalozzi, Phytopl. d. Süssw. 1: 191. 1938.

Anabaena Raciborskii Woloszyńska, Bull. Int. Acad. Sci. Cracovie, Cl. Sci. Math. & Nat., ser. B, 1912: 684. 1913. *Anabaenopsis Raciborskii* Woloszyńska ex Geitler in Pascher, Süsswasserfl. 12: 331. 1925.

Cylindrospermum doryophorum Brühl & Biswas, Journ. & Proc. Asiatic Soc. Bengal, N. S. 18(10): 577. 1923.

Anabaenopsis Raciborskii var. *lyngbyoides* Geitler & Ruttner, Arch. f. Hydrobiol. Suppl.-Bd. 14(6): 455. 1935.

Raphidiopsis mediterranea Skuja, Hedwigia 77: 23. 1937. *R. subrecta* Frémy ex Skuja, N. Acta R. Soc. Sci. Upsal., ser. 4, 14(5): 27. 1949.

Raphidiopsis indica Singh, Journ. Indian Bot. Soc. 21(3—4): 239. 1942.

Aphanizomenon Elenkinii Kisselev, Bot. Mat. Otd. Sporov. Rast. Inst. Bot. Akad. Nauk SSSR 7: 65. 1951.

Raphidiopsis mediterranea fa. *major* Yoneda, Mem. Coll. Agric. Kyoto Univ. 66: 53. 1953.

Aphanizomenon Elenkinii var. *gracile* Kaschtanova, Bot. Mat. Otd. Sporov. Rast. Bot. Inst. Komarova Akad. Nauk SSSR 10: 22. 1955.

Aphanizomenon Elenkinii fa. *maeoticum* Pitzyk, Bot. Mat. Otd. Sporov. Rast. Bot. Inst. Komarova Akad. Nauk SSSR 11: 27. 1956.

Anabaenopsis Raciborskii fa. *minor* Behre, Arch. f. Hydrobiol. Suppl. 23(1): 27. 1956.

Anabaenopsis Raciborskii fa. *major* Behre (as "maior"), loc. cit. 1956.

Aphanizomenon Issatschenkoi Proschkina-Lavrenko, Bot. Mat. Otd. Sporov. Rast. Bot. Inst. Komarova Akad. Nauk SSSR 15: 30. 1962. *Anabaena Issatschenkoi* Usachev *pro synon.* ex Proschkina-Lavrenko, loc. cit. 1962.

Anabaenopsis gangetica Nair, Hydrobiologia 30(1): 147. 1967.

Raphidiopsis turcomanica Kogan, Nov. Sist. Nizsh. Rast. Bot. Inst. Komarova Akad. Nauk SSSR 1967: 9. 1967; Bot. Zhurn. 52(7): 955. 1967.

Raphidiopsis mediterranea var. *grandis* Hill, Phycologia 9(1): 76. 1970.

Raphidiopsis Brookii Hill, Phycologia 11(2): 214. 1972.

Trichomata aeruginea, cylindrica vel per totam longitudinem aut ad utrinque extrema longe et sensim attenuata, ad apices acute conica, ad dissepimenta constricta vel non-constricta, diametro 1.5—5μ crassa, ambitu recta vel curvata vel spiralia, longitudine indeterminata, per destructionem cellulae intercalaris vel per constrictionem ad dissepimentum frangentia. Cellulae diametro trichomatis longiores vel breviores, 1—6μ longae, protoplasmate saepe pseudovacuolato, dissepimentis non granulatis; heterocystae terminales nonnumquam intercalares interdum seriatae, acute conicae vel ovoideo-cylindricae, diametro 1.5—6μ crassae; sporae cylindricae vel ovoideae raro sphaericae, solitariae vel seriatae, muris hyalinis; cellulae vegetativae terminales primum hemisphaericae demum acutissime conicae. Materia vaginalis dispersa vel mucosa. Planta trichomata longa vel brevia nuda vel in muco saepe in flore aquae composita. Fig. 61—66.

Trichomes blue-green, cylindrical or long and noticeably attenuated through their entire lengths or at both ends, acute-conical at the tips, constricted or not constricted at the cross walls, 1.5—5μ in diameter, straight or curved or spiraled, indeterminate in length, breaking by the destruction of an intercalary cell or by constriction at a cross wall. Cells longer or shorter than the diameter, 1—6μ long, protoplasm often pseudovacuolate, cross walls not granulated; heterocysts terminal or intercalary, sometimes seriate, acute-conical or ovoid-cylindrical, 1.5—6μ in diameter; spores cylindrical or ovoid or rarely spherical, solitary or seriate, the walls hyaline; terminal vegetative cells at first hemispherical becoming very acute-conical. Sheath material dispersed or mucous. Plant composed of long or short naked trichomes, or trichomes in mucus, often in water blooms. Figs. 61—66.

Raphidiopsis curvata is found occasionally in pools of fresh water of long standing and in the plankton of lakes and rivers, in tropical regions as well as in temperate zones where the water becomes very warm during the summer. It sometimes develops as water blooms, coloring the water verdigris-blue-green. Occasionally only vegetative cells are found; heterocysts and spores usually develop later when conditions become favorable to their formation. Frequently pseudovacuoles are lacking, especially where the trichomes are found in shallow water. This species, from Florida, has been grown in almost pure culture by Dr. Don E. Henley. I have had this culture under observation for many months. The acute-conical terminal cells often mature into heterocysts (Figs. 63, 64), which die or are lost from the trichome when spores are matured adjacent to them. Or, here and there, terminal cells are lost, so that the trichomatal tips appear blunt-ended or truncate; or a terminal cell may become long-attenuate and rounded at the apex.

The following specimens are representative of *Raphidiopsis curvata* as described above:

Czechoslovakia: pond Dolejší near Lnáre (Bohemia), *B. Fott* (D, Daily, F).

South Africa: Newlands (Griqualand West) (TYPE of *Raphidiopsis curvata* Fritsch & Rich, BM).

Connecticut: Linsley pond (North Branford), *G. E. Hutchinson* (D, F).

North Carolina: Raleigh, *L. A. Whitford 1, W1003* (PH); Westbrook pond near Garner (Wake county), *R. C. Phillips* (D).

Florida: culture from Lake Warren (Orange county), *D. E. Henley ASM1* (D); Mulberry (Polk county), *G. K. Reid 18* (PH); Gainesville, *M. A. Brannon 83* (D, F, PC); St. Johns river at Welaka, *E. L. Pierce* (D, F).

Ohio: Inwood park, Cincinnati, *W. A. Daily 139* (D, Daily, F); Beaver creek near Celina (Mercer county), *J. B. Lackey* (D, F).

Minnesota: Minnesota river at Mendota, *E. G. Reinhard* (TYPE of *Aphanizomenon americanum* Reinh., D); Galpin lake and Chaska, *H. Hill* (D, Hill).

Missouri: Ashland wild life area (Boone county), *V. W. Proctor* (D, F); lakes at Lake Hill and Webster Groves, *F. Holtzwart* (D).

Nebraska: lakes at Grand Island and Kearney, *W. Kiener 15178, 15185a* (D, F, NEB).

Kansas: Wooster lake, Emporia, *T. F. Andrews G506* (TYPES of *Anabaenopsis seriata* Presc. and *A. wustericum* Obuch. & Kos., EMC).

Brazil: Ceará: Siqueira near Porangaba, *S. Wright* (BM, D, FH, MICH, NY, RB). Paraíba: Campina Grande, *Wright 1573* (D, NY); Açude Simão near Campina Grande, *Wright 1569* (BM, D, NY, RB), *1572* (D, NY), *1582* (D, NY), *1603* (D, NY, W), *1967* (D, NY), *1999* (D, NY); Açude Lapa near Campina Grande, *Wright 1559* (D, NY); Açude Mendonça near Mogeiro

Baixo, *Wright 2022* (BM, D, MICH, NY, RB); Açude Linda Flôr near Mogeiro Baixo, *Wright 2028* (D, NY).

Peru: Laguna Huancaco, Trujillo, *M. Fernández H. 185* (HUT, PH); Laguna Toledo, Trujillo, *A. Aldave P. 55* (HUT, PH).

Philippines: Sampaloc lake, San Pablo (Laguna), *E. Quisumbing* (TYPE of *Anabaenopsis philippinensis* W. R. Taylor, Taylor), *G. T. Velasquez 678* (D, F, PUH).

China: West lake, Hangchow, *J. E. Nielsen 68* (D, F).

Malaysia: Batu Berendam (Malacca), *G. A. Prowse 336* (D, F).

Burma: Rangoon, *L. P. Khanna 692* (D, F).

India: Hyderabad (Deccan), *Rao* (D, G, W).

Nomina Excludenda

The following names of taxa were published on the bases of objects which are not members of the Nostocaceae with cylindrical trichomes as circumscribed here. These taxa were described originally as members of this group or were later transferred into it. Most taxa are herein interpreted by study of the original (Type) specimens. Where Type specimens were unavailable to me, the original descriptions, including illustrations, are designated as provisional Types, to serve until appropriate Type specimens are found. Names dependent upon single Type specimens or Type descriptions are presented below in paragraphs arranged in alphabetical order of the originally published species or infraspecific names (basionyms). Paragraphs beginning with an asterisk (*) list only basionyms and nomenclature directly pertinent to this revision; additional synonymy has been listed by Drouet (1968).

Ammatoidea murmanica Petrov (as "Hammatoidea"), Bot. Mat. Otd. Sporov. Rast. Bot. Inst. Komarova Akad. Nauk SSSR 14: 109. 1961. —In the absence of the original material, the original description is here designated as the Type = *Schizothrix calcicola* (Ag.) Gom.

*Amphithrix Kützing [Phyc. Gener., p. 220. 1843] ex Gomont, Ann. Sci. Nat. VII. Bot. 16: 156. 1892. *Kuetzingina* Kuntze, Rev. Gen. Pl. 3(2): 411. 1898. —Type species: *Amphithrix amoena* Kützing [loc. cit. 1843] ex Gomont, ibid. 16: 178. 1892. *Kuetzingina amoena* Kuntze, loc. cit. 1898. —Type specimen from Sachsen, Germany: Nordhausen, in herb. Kützing (L) = *Microcoleus vaginatus* (Vauch.) Gom.

*Amphithrix barbata Kützing [Bot. Zeit. 5: 194. 1847] ex Forti, Syll. Myxoph., p. 603. 1907. —Type specimen from Calvados, France: Falaise, in herb. Kützing (L) = *Schizothrix calcicola* (Ag.) Gom.

*Amphithrix hospita Kützing [Sp. Algar., p. 275. 1849] ex Forti, loc. cit. 1907. [Dasyactis hospita Crouan, Fl. Finistère, p. 117. 1867.] —Type specimen from Oldenburg, Germany: Jade Busen, Jürgens, in herb. Kützing (L) = Schizothrix calcicola (Ag.) Gom.

Amphithrix janthina fa. fuscescens Frémy, Bull. Soc. Linn. Normandie, sér. 7, 9: 123. 1927. —Type specimen from Manche, France: Saint-Lô, P. Frémy, Jun. 1922 (NY) = Schizothrix calcicola (Ag.) Gom.

Amphithrix Laminariae Kuckuck, Wiss. Meeresunters. Komm. z. Wiss. Unters. d. Deutshe. Meere in Kiel & d. Biol. Anst. auf Helgoland, N. F. 1: 263 1896. —In the absence of the original material, the original description is here designated as the Type = Schizothrix calcicola (Ag.) Gom.

Amphithrix Leprieurii Montagne [Ann. Sci. Nat. IV. Bot. 14: 168. 1860] ex Forti, Syll. Myxoph., p. 603. 1907. —Type specimen, with that of Symphyothrix cartilaginea Mont., from French Guyana: Cayenne, Leprieur 1287bis (PC) = Schizothrix calcicola (Ag.) Gom.

Amphithrix Meneghiniana Kützing [Phyc. Gener., p. 220. 1843] ex Forti, loc. cit. 1907. Rivularia scytonemoidea Meneghini [Mem. R. Accad. Sci. Torino, ser. 2, 5: 142. 1843; pro synon. in Kützing, ibid., p. 221. 1843] ex Bornet & Flahault, Ann. Sci. Nat. VII. Bot. 4: 364. 1887. Amphithrix scytonemoidea Trevisan [Nomencl. Algar. 1: 36. 1845] ex Forti, loc. cit. 1907. Kuetzingina scytonemoidea Kuntze (as "scytonemodea"), Rev. Gen. Pl. 3(2): 411. 1898. —Type specimen from Italy: in Horto Patavino, Meneghini, in herb. Kützing (L) = Schizothrix calcicola (Ag.) Gom.

Amphithrix thermalis Kützing [Sp. Algar., p. 275. 1849] ex Hansgirg, Prodr. Algenfl. Böhm. 2: 53. 1892. Arthrotilum thermale Rabenhorst [Fl. Eur. Algar. 2: 231. 1865] ex Hansgirg, loc. cit. 1892. —Type specimen from Venezia Euganea, Italy: Abano, in herb. Kützing (L) = Hapalosiphon sp.

Amphithrix villosa Kützing [Phyc. Germ., p. 178. 1845; in Römer, Alg. Deutschl., p. 36. 1845] ex Forti, Syll. Myxoph., p. 603. 1907. Kuetzingina villosa Kuntze, loc. cit. 1898. —Type specimen from Sachsen, Germany: Clausthal, Römer 109, in herb. Kützing (L) = Schizothrix calcicola (Ag.) Gom.

Amphithrix violacea fa. aeruginea Frémy, Rev. Algol. 12: 146. 1941. —Type specimen, from the original locality on Bonaire, Netherlands Antilles: Goto, P. Wagenaar Hummelinck 1106A, Sept. 1949 (L) = Schizothrix calcicola (Ag.) Gom.

Amphithrix violacea fa. lutescens Printz, Norske Vid.-Akad. Oslo Skr., Mat.-nat. Kl. 1(5): 45. 1926. —In the absence of original material, the original description is here designated as the Type = Schizothrix calcicola (Ag.) Gom.

Anhaltia flabellum Heufler, Verh. Zool.-bot. Ver. Wien 2(1852; Abh.: "Drei neue Algen"): 7. 1853. —Type specimen from Hungary: e thermis Budae, de Heufler (W) = Stigeoclonium sp.

Anhaltia Schwabe, Linnaea 9: 127. 1835. —Type species: A. Fridericae Schwabe, loc. cit. 1835. —Type specimen from Anhalt, Germany: Dessau, Herzogin von Dessau, Dec. 1833 (W) = Fungus hyphae in an animal gel.

Aphanizomenon Manguinii Bourrelly in Bourrelly & Manguin, Alg. d'Eau Douce Guadeloupe, p. 155. 1952. —In the absence of the original material, the original description is here designated as the Type = Nostocacea with torulose trichomes.

Aphanizomenon ovalisporum Forti, Atti Accad. Agric., Sci., Lett., Arti & Comm. Verona, ser. 4, 12: 125. 1911.—In the absence of the original material, the original description is here designated as the Type = Nostocacea with torulose trichomes.

Aphanizomenon ovalisporum fa. *brevicellum* Ulomsky, Bot. Mat. Otd. Sporov. Rast. Bot. Inst. Komarova Akad. Nauk SSSR 11: 31. 1956. —In the absence of the original material, the original description is here designated as the Type = Nostocacea with torulose trichomes.

Aphanizomenon ovalisporum fa. *oblongum* Ulomsky, loc. cit. 1956. In the absence of the original material, the original description is here designated as the Type = Nostocacea with torulose trichomes.

Aphanizomenon sphaericum Kisselev, Tr. Zool. Inst. Akad. Nauk SSSR 16: 573. 1954; Bot. Mat. Otd. Sporov. Rast. Bot. Inst. Komarova Akad. Nauk SSSR 10: 37. 1955. —In the absence of the original material, the original description is here designated as the Type = Nostocacea with torulose trichomes.

Aulosira aenigmatica Frémy, Blumea Suppl. 2: 37. 1942. —In the absence of the original material, the original description is here designated as the Type = Nostocacea with torulose trichomes.

Aulosira laxa var. *microspora* Lagerheim [Öfvers. K. Vet.-Akad. Förh. 1883(2): 49. 1883] ex Forti, Syll. Myxoph., p. 481. 1907. *A. laxa* fa. *microspora* Elenkin in Hollerbach, Kossinskaia & Poliansky, Opred. Presn. Vodor. SSSR 2: 299. 1953. —In the absence of the original material, the original description is here designated as the Type = Nostocacea with torulose trichomes.

Aulosira minor Wille, Deutsch. Südpolar-Exped. 8(Bot.): 420. 1928. —Type specimen from Kerguelen island, Indian ocean: *35*, in the slide collection of N. Wille (O) = *Nostoc* sp.

Aulosira planctonica Elenkin, Not. Syst. Inst. Crypt. Hort. Bot. Petropol. 1(8): 128. 1922.—In the absence of the original material, the original description is here designated as the Type = Nostocacea with torulose trichomes.

Aulosira planctonica var. *cylindrica* Aptekar, Dnevn. Vses. S'ezda Bot. v Leningrade 1928: 140. 1928; (as "zylindrica"), Zapisk. Dnepropetrovsk. Inst. Narodn. Osviti 2: 206, 210. 1928. *A. planctonica* fa. *cylindrica* Elenkin, Monogr. Alg. Cyanoph., Pars Spec. 1: 871. 1938. —In the absence of the original material, the original descriptions are here designated as the Type = Nostocacea with torulose trichomes.

Aulosira Schauinslandii Lemmermann, Bot. Jahrb. 34: 622. 1905. —In the absence of the original material, the original description is here designated as the Type = Nostocacea with torulose trichomes.

Aulosira thermalis G. S. West, Journ. of Bot. 40: 244. 1902. —In the absence of the original material, the original description is here designated as the Type = *Hapalosiphon* sp.

*Bangia tenuis Kützing, Alg. Aq. Dulc. Dec. 9: 86, footnote. 1834. *Hypheothrix fontana* Kützing [ex Rabenhorst, Fl. Eur. Algar. 2: 75. 1865] ex Hansgirg, Prodr. Algenfl. Böhm. 2: 85. 1892. —Type specimen from Sachsen, Germany: ad saxa in fonte, Kölme bei Halle, in herb. Kützing (L) = *Schizothrix calcicola* (Ag.) Gom.

Batrachospermum myurus Lamarck & De Candolle, Fl. Franç., ed. 3, 2: 591. 1805. *Conferva myurus* Brousson in Lamarck & De Candolle *pro synon.*, loc. cit. 1805. *Rivularia myurus* Mougeot & Nestler, Stirp. Crypt. Vogeso-rhen. 5: 500. 1815. —In the absence of the original material, the specimen in Mougeot & Nestler, Stirpes Cryptogamae Vogeso-rhenanae no. 500, is here designated as the Type = *Palmella myosurus* (Ducluz.) Lyngb.

*Beggiatoa hinnulea Wolle, Bull. Torrey Bot. Club 6: 182. 1877. *Hypheothrix hinnulea* Forti, Syll. Myxoph., p. 336. 1907. —Type specimen from Pennsylvania: trenches, warm water, *F. Wolle* (D) = Bacteria.

*Blennothrix vermicularis Kützing [Phyc. Gener., p. 226. 1843] ex Gomont, Ann. Sci. Nat. VII. Bot. 15: 339. 1892. *Calothrix vermicularis* Kützing *pro synon.* [loc. cit. 1843] ex Gomont, loc. cit. 1892. —Type specimen from Italy: Neapel in mari, in herb. Kützing (L) = *Microcoleus lyngbyaceus* (Kütz.) Crouan.

Brachytrichia affinis Setchell & Gardner, Univ. Calif. Publ. Bot. 6(17): 475. 1918. —Type specimen from California: Laguna Beach (UC) = *B. Quoyi* (Ag.) Born. & Flah.

Brachytrichia Balanii fa. *purpurea* Frémy, Comptes Rend. Acad. Sci. Paris 195: 1413. 1932. —In the absence of the original material, the original description is here designated as the Type = Stigonematacea.

Brachytrichia Codii Setchell, Univ. Calif. Publ. Bot. 12(5): 66. 1926. —Type specimen from Tahiti (UC) = *B. Quoyi* (Ag.) Born. & Flah.

Brachytrichia Zanardini [Phyc. Indic. Pugill., p. 24. 1872] ex Bornet & Flahault, Ann. Sci. Nat. VII. Bot. 3: 341. 1886. —Type species: *B. rivulariiformis* Zanardini (as "rivulariaeformis") [ibid., p. 25. 1872] ex Forti, Syll. Myxoph., p. 680. 1907. *B. rivularioides* Zanardini ex Bornet & Flahault, ibid. 4: 373. 1887. —Type specimen from Sarawak, Malaysia: Tangion Datu, *O. Beccari 4*, 1867, in herb. Beccari (FI) = *B. Quoyi* (Ag.) Born. & Flah.

*Limnanthe Kützing [Linnaea 17: 86. 1843] ex Forti, ibid., p. 468. 1907. *Limnochlide* Kützing [Phyc. Gener., p. 203. 1843] ex Hansgirg, Prodr. Algenfl. Böhm. 2: 73. 1892. —Type species: *Byssus flos·aquae* Linnaeus, Sp. Pl., p. 1168. 1753. *Limnanthe Linnaei* Kützing [Linnaea 17: 86. 1843] ex Forti, ibid., p. 469. 1907. *Limnochlide flos-aquae* Kützing [Phyc. Gener., p. 203. 1843] ex Hansgirg, loc. cit. 1892. *Trichormus flos-aquae* Ralfs [Ann. & Mag. of Nat. Hist., ser. 2, 5: 327. 1850] ex Bornet & Flahault, ibid. 7: 229. 1888. *Aphanizomenon flos-aquae* Ralfs [ibid. 5: 340. 1850] ex Bornet & Flahault, ibid. 7: 241. 1888. —Type

specimen: "Byssus flos-aquae" (LINN) = *Microcoleus vaginatus* (Vauch.) Gom.

*Calothrix Agardhii Crouan [Fl. Finistère, p. 118. 1867] ex Bornet & Flahault, ibid. 3: 370. 1886. —Type specimen from Finistère, France: Banc de St.-Marc, Aug. 1847 (Lab. Biol. Mar., Concarneau) = *Schizothrix mexicana* Gom.

Calothrix ambigua Meneghini in Zanardini [Bibl. Ital. 99: 200. 1840] ex Bornet & Flahault, Ann. Sci. Nat. VII. Bot. 3: 370. 1886. —Type specimen from Italy: in aquis subsalsis Venetiis, ex Zanardini in herb. Montagne (PC) = *Microcoleus lyngbyaceus* (Kütz.) Crouan.

Calothrix Braunii var. *contorta* Gardner, Mem. New York Bot. Gard. 7: 67. 1927. *C. Braunii* fa. *contorta* Poliansky in Elenkin, Monogr. Alg. Cyanoph., Pars Spec. 2: 1068. 1949. —Type specimen from Puerto Rico: wall of a house, Maricao, *N. Wille 1049*, Feb. 1915 (NY) = *Fischerella* sp.

Calothrix Brebissonii Kützing [Bot. Zeit. 5(11): 180. 1847] ex Bornet & Flahault, loc. cit. 1886. *Dillwynella Brebissonii* Kuntze, Rev. Gen. Pl. 2: 892. 1891. —Type specimen from Calvados, France: Falaise, *190*, in herb. Kützing (L) = *Porphyrosiphon Notarisii* (Menegh.) Kütz.

*Calothrix caespitula Harvey [in Hooker, Engl. Fl. 5(Brit. Fl. 2): 369. 1833] ex Bornet & Flahault, loc. cit. 1886. *Leibleinia caespitula* Kützing [Sp. Algar., p. 278. 1849] ex Gomont, Ann. Sci. Nat. VII. Bot. 16: 136. 1892. *Calothrix obscura* Crouan [Alg. Mar. Finistère 3: 339. 1852] ex Bornet & Flahault, ibid. 3: 371. 1886. —Type specimen from Ireland: Malbay, 1831, in herb. Agardh (LD) = *Microcoleus lyngbyaceus* (Kütz.) Crouan.

*Calothrix comoides Harvey [Phyc. Austral., Synopt. Catal., p. lxii. 1863] ex Bornet & Flahault, Ann. Sci. Nat. VII. Bot. 3: 371. 1886. —Type specimen from Victoria, Australia: Port Phillip Heads, *W. H. Harvey 598*, in Australian Algae (BM) = *Microcoleus lyngbyaceus* (Kütz.) Crouan.

*Calothrix Cresswellii Harvey [Phyc. Brit. 1: pl. LXXVI. 1846] ex Bornet & Flahault, loc. cit. 1886. —Type specimen from England: Sidmouth (BM) = *Schizothrix calcicola* (Ag.) Gom.

Calothrix De-Notaris Fiorini Mazzanti [Atti Accad. Pontif. N. Lincei Roma 14(4): 239. 1861] ex Gomont, ibid. 16: 141. 1892. —Type specimen from near Roma, Italy: in aquis mineralibus, *Fiorini Mazzanti*, 1861 (PC) = *Oscillatoria lutea* Ag.

*Calothrix janthiphora Fiorini Mazzanti (as "jantiphora") [Sovra Due Nuove Alghe delle Acque Albule, p. 2. 1857] ex Gomont, ibid. 16: 233. 1892. [*C. iantifera* Fiorini ex Viale, Atti Accad. Pontif. N. Lincei Roma 8–9 (1855–56): 38. 1874.] —Type specimen from near Roma, Italy: in Aquis Albulis, in herb. Montagne (PC) = *Porphyrosiphon Notarisii* (Menegh.) Kütz.

[*Calothrix janthiphora* β *ecolor* Fiorini Mazzanti, Atti Accad. Pontif. N. Lincei Roma 17(3): 103. 1864.] —In the absence of the original material, the original description is here designated as the Type = *Porphyrosiphon Notarisii* (Menegh.) Kütz.

Calothrix limbata Harvey [Phyc. Austral. 5: 792. 1863] ex Bornet & Flahault, Ann. Sci. Nat. VII. Bot. 3: 371. 1886. —Type specimen from Australia: *W. H. Harvey 596* (BM) = *Microcoleus lyngbyaceus* (Kütz.) Crouan.

Calothrix luteofusca Agardh [Flora 10(1: 40): 635. 1827] ex Bornet & Flahault, loc. cit. 1886. *Leibleinia luteofusca* Kützing [Phyc. Gener., p. 221. 1843] ex Gomont, Ann. Sci. Nat. VII. Bot. 16: 136. 1892. *L. sordida* Kützing [Phyc. Germ., p. 179. 1845] ex Gomont, ibid. 16: 138. 1892. —Type specimen from Italy: Trieste, in herb. Agardh (LD) = *Microcoleus lyngbyaceus* (Kütz.) Crouan.

Calothrix luteola Carmichael [in Greville, Scott. Crypt. Fl. 5: 299. 1827] ex Bornet & Flahault, loc. cit. 1886. *Leibleinia luteola* Kützing [Sp. Algar., p. 276. 1849] ex Gomont, ibid. 16: 152. 1892. *Calothrix melaleuca* Carmichael *pro synon.* [ex Harvey, Phyc. Brit. 3: pl. CCCXLII. 1851] ex Bornet & Flahault, loc. cit. 1886. —Type specimen from Argyll, Scotland: Appin, Carmichael (K in BM) = *Schizothrix calcicola* (Ag.) Gom.

Calothrix maxima Martens [Proc. Asiatic Soc. Bengal 1870: 183. 1870] ex Forti, Syll. Myxoph., p. 636. 1907. —Type specimen from Indonesia: Java, in herb. Agardh (LD) = *Oscillatoria princeps* Vauch.

Calothrix mutabilis Zanardini [ex Frauenfeld, Alg. Dalmat. Küste, p. 5. 1855] ex Bornet & Flahault, Ann. Sci. Nat. VII. Bot. 3: 370. 1886. —Type specimen from Dalmatia, Yugoslavia: Capocesto, *Vidovich* (W) = *Schizothrix mexicana* Gom.

Calothrix purpurea Meneghini [ex Frauenfeld, loc. cit. 1855] ex Bornet & Flahault, loc. cit. 1886. —Type specimen presumably from Dalmatia, Yugoslavia: "C. purpurea," in herb. Meneghini (FI) = *Microcoleus lyngbyaceus* (Kütz.) Crouan.

Calothrix radiosa fa. *aeruginea* Rabenhorst [Fl. Eur. Algar. 2: 272. 1865] ex Gomont, Ann. Sci. Nat. VII. Bot. 16: 101. 1892. [*C. aeruginea* A. Braun *pro synon.* in Rabenhorst, loc. cit. 1865.] *C. radiosa* b *aeruginea* Rabenhorst ex Bornet & Flahault, ibid. 3: 372. 1886. *Plectonema radiosum* Gomont, ibid. 16: 100. 1892. —Type specimen from Upper Austria: Milchdorf, *Schiedermayr*, Dec. 1861, in Rabenhorst, Algen Europas no. 1305 (PC) = *Porphyrosiphon Notarisii* (Menegh.) Kütz.

[*Calothrix radiosa* fa. *fuscescens* Rabenhorst, Fl. Eur. Algar. 2: 272. 1865.] —Type specimen from Sardinia, Italy: Tacco di Seui, Barbagia, sui rupelli, *Marcucci*, 1866, in Un. Itin. Crypt. (F) = *Porphyrosiphon Notarisii* (Menegh.) Kütz.

Calothrix rhizomatoidea Reinsch [in Rabenhorst, Hedwigia 5(10): 153. 1866] ex Bornet & Flahault, loc. cit. 1886. *Tolypothrix rhizomatoidea* Reinsch [Algenfl. Mittl. Th. Franken, p. 52. 1867] ex Bornet & Flahault, Ann. Sci. Nat. VII. Bot. 5: 126. 1887. *Hapalosiphon pumilus* var. *rhizomatoidea* Hansgirg, Prodr. Algenfl. Böhm. 2: 26. 1892. —Type specimen from Franken, Germany: in stagnis pr. Beiersdorf, *P. Reinsch*, Oct. 1864, in Rabenhorst, Algen Europas no. 1904 (FH) = *Hapalosiphon* sp.

Calothrix rufescens Carmichael [in Harvey in Hooker, Engl. Fl. 5(Brit. Fl. 2): 368. 1833] ex Bornet & Flahault, ibid. 3: 370. 1886. *Tolypothrix rufescens* Hassall [Hist. Brit. Freshw. Alg. 1: 242. 1845] ex Bornet & Flahault, ibid. 5: 122. 1887. *Schizosiphon rufescens* Kützing [Bot. Zeit. 5(11): 179. 1847] ex Forti, Syll. Myxoph., p. 638. 1907. —Type specimen from Argyll, Scotland: Appin, *Carmichael* (TCD) = *Schizothrix calcicola* (Ag.) Gom.

Calothrix Sandriana Frauenfeld [Alg. Dalmat. Küste, p. 5. 1855] ex Bornet & Flahault, ibid. 3: 372. 1886. —Type specimen from Yugoslavia: Dalmatie, *P. Titius*, in herb. Lenormand (CN) = *Microcoleus lyngbyaceus* (Kütz.) Crouan.

Calothrix semiplena Agardh [Flora 10(1: 40): 634. 1827] ex Bornet & Flahault, loc. cit. 1886. *Leibleinia semiplena* Kützing [Phyc. Gener., p. 221. 1843] ex Gomont, Ann. Sci. Nat. VII. Bot. 16: 138. 1892. —Type specimen from Italy: Trieste, Jun. 1827, in herb. Agardh (LD) = *Microcoleus lyngbyaceus* (Kütz.) Crouan.

Calothrix tenella Gardner, Mem. New York Bot. Gard. 7: 67. 1927. —Type specimen from Puerto Rico: near Maricao, *N. Wille 1036a*, Feb. 1915 (NY) = *Fischerella* sp.

Calothrix tinctoria Agardh [Syst. Algar., p. 72. 1824] ex Bornet & Flahault, Ann. Sci. Nat. VII. Bot. 3: 372. 1886. *Hypheothrix tinctoria* Rabenhorst [Fl. Eur. Algar. 2: 75. 1865] ex Gomont, ibid. 15: 303. 1892. —Type specimen from Vosges, France: circa Bruyerum, in herb. Agardh (LD) = *Schizothrix calcicola* (Ag.) Gom.

Calothrix Tomasinii Kützing & Biasoletto (as "Tomasini") [in Kützing, Alg. Aq. Dulc. Dec. 13: 130. 1836] ex Bornet & Flahault, loc. cit. 1886. *C. Tomasiniana* Kützing [Phyc. Gener., p. 229. 1843] ex Bornet & Flahault, loc. cit. 1886. *Plectonema Tomasinianum* Gomont (as "Tomasiniana"), Journ. de Bot. 4(20): 353. 1890. —Type specimen from Venezia Giulia, Italy: aus der Timavo in Montfalcone, in herb. Kützing (L) = *Porphyrosiphon Notarisii* (Menegh.) Kütz.

Calothrix variegata Zanardini [Bibl. Ital. 96: 135. 1839] ex Drouet, Monogr. Acad. Nat. Sci. Philadelphia 15: 87. 1968. *Leibleinia variegata* Zanardini [Not. Intorno alle Cellul. Mar. d. Lag. & Lit. Venezia, p. 79. 1847] ex Drouet, loc. cit. 1968. *L. polychroa β confervicola* Kützing [Sp. Algar., p. 279. 1849] ex Drouet, loc. cit. 1968. —Type specimen from Italy: Venetiis, in herb. Montagne (PC) = *Schizothrix mexicana* Gom.

Camptothrix W. & G. S. West, Journ. of Bot. 35: 269. 1897. —Type species: *C. repens* W. & G. S. West, loc. cit. 1897. —Type specimen, with that of *Schizothrix natans* W. & G. S. West, from Angola: prope Humpata etc., *F. Welwitsch 15*, May 1860 (BM) = *Stigonema* sp.

Chamaenema Kützing, Linnaea 8: 364. 1833. —Type species: *C. carneum* Kützing, loc. cit. 1833. *Plectonema carneum* Lemmermann, Krypt.-Fl. Mark Brandenb. 3: 206. 1910. —Type specimen from Germany: an Fensterscheiben, with *Cryptococcus mollis* Kütz. in Kützing, Algarum Dec. 3, no. 28 (L) = Fungi.

Chroolepus janthinus Montagne in Barker-Webb & Berthelot, Hist. Nat. d. Iles Canaries 3(2:4): 188. 1840. *Hypheothrix janthina* Kützing [ex Rabenhorst, Fl. Eur. Algar. 2: 76. 1865] ex Gomont, Ann. Sci. Nat. VII. Bot. 15: 330. 1892. *Amphithrix janthina* Bornet & Flahault [Mém. Soc. Nat. Sci. Nat. & Math. Cherbourg 25: 200. 1885] ex Bornet & Flahault, Ann. Sci. Nat. VII. Bot. 3: 344. 1886. *Homoeothrix janthina* Starmach, Acta Hydrobiol. 1(3—4): 149. 1959. —Type specimen from the Canary islands, in herb. Montagne (PC) = *Schizothrix calcicola* (Ag.) Gom.

[*Coenocoleus* Berkeley & Thwaites in Smith & Sowerby, Suppl. Engl. Bot. 4: 2940. 1849.] —Type species: [*C. Smithii* Berkeley & Thwaites, loc. cit. 1849.] [*Calothrix Smithii* Cooke, Brit. Fresh-w. Alg., p. 277. 1884.] —Type specimen from England: Wareham, Sept. 1847 (BM) = *Schizothrix calcicola* (Ag.) Gom.

Conferva aegagropila Linnaeus, Sp. Pl., p. 1167. 1753. *Calothrix aegagropila* Kützing [Alg. Aq. Dulc. Dec. 1: 7. 1833] ex Bornet & Flahault, ibid. 3: 370. 1886. *Tolypothrix aegagropila* Kützing [Phyc. Gener., p. 228. 1843] ex Bornet & Flahault, ibid. 5: 121. 1887. *T. lanata* var. *aegagropila* Hansgirg, Prodr. Algenfl. Böhm. 2: 38. 1892. *T. tenuis* fa. *aegagropila* Kossinskaia in Elenkin, Monogr. Alg. Cyanoph., Pars Spec. 1: 954. 1938. *T. lanata-aegagropila* Kossinskaia *pro synon.*, loc. cit. 1938. —Type specimen, selected by van den Hoek in Rev. Eur. Sp. Cladophora, p. 51 (1963): no. 1277.49 in herb. Linnaeus (LINN) = *Cladophora* sp.

Conferva atro-virens Dillwyn, Brit. Conf., pl. 25. 1803. *Scytonema atro-virens* Agardh (as "atrovirens") [Disp. Alg. Suec., p. 39. 1812] ex Bornet & Flahault, Ann. Sci. Nat. VII. Bot. 5: 112. 1887. —Type specimen from England, in Dillwyn's bound collection (LINN in BM) = parasitized *Stigonema* sp.

Conferva chthonoplastes Mertens in Hornemann, Fl. Dan. 9(25): 6. 1817. [*Scytonema chthonoplastes* Sprengel, Linn. Syst. Veget., ed. 16, 4(1): 363. 1827.] —Type specimen from Denmark: in vadis sinus Othoniensis, Jan. 1816, in herb. Lyngbye (C) = *Schizothrix arenaria* (Berk.) Gom.

Conferva comoides Dillw., Brit. Conf., pl. 27. 1803. *Scytonema comoides* Agardh [Syn. Alg. Scand., p. 112. 1817] ex Bornet & Flahault, ibid. 5: 113. 1887. —Type specimen from England, in Dillwyn's bound collection (LINN in BM) = Diatoms.

Conferva foetida Villars, Hist. Pl. Dauphiné 3: 1010. 1789. *Rivularia foetida* Lamarck & De Candolle, Fl. Franç., ed. 3, 2: 5. 1805. —Type specimen, from near the original locality, here designated: Bex, Switzerland, *Schleicher*, 1825, in herb. Bory (PC) = *Palmella myosurus* (Ducluz.) Lyngb.

Conferva fontinalis Linnaeus, Sp. Pl., p. 1164. 1753. *Calothrix fontinalis* Agardh [Syst. Algar., p. 73. 1824] ex Bornet & Flahault, Ann. Sci. Nat. VII. Bot. 5: 127. 1887. —Type specimen from England: "Conferva minima, Byssi facie" in herb. Dillenius (OXF) = *Vaucheria* sp.

Conferva Hofmannii Agardh, ibid., p. 100. 1824. *Leibleinia Hofmannii* Kützing

[Phyc. Gener., p. 222. 1843] ex Gomont, Ann. Sci. Nat. VII. Bot. 16: 141. 1892. —Type specimen from Denmark: ad munimentum Trecroner p. Hafniam, 1822, in herb. Agardh (LD) = *Oscillatoria lutea* Ag.

*Conferva limosa Dillwyn, Brit. Conf., pl. 20. 1802. [*Calothrix tenuis* Ainé, Pl. Crypt.-Cell. Dept. Saône-et-Loire, p. 257. 1863.] *C. radiosa* var. *aeruginea* Durairatnam & Amaratunga, Bull. Fish. Res. Sta. Ceylon 20(2): 154. 1969. —Type specimen from England, in Dillwyn's bound collection (LINN in BM) = *Microcoleus lyngbyaceus* (Kütz.) Crouan.

*Conferva nivea Dillwyn, ibid., p. 54, pl. C. 1809. *Calothrix nivea* Agardh [Syst. Algar., p. 70. 1824] ex Bornet & Flahault, ibid. 3: 370. 1886. *Tolypothrix nivea* Hassall [Hist. Brit. Freshw. Alg. 1: 241. 1845] ex Bornet & Flahault, ibid. 5: 125. 1887. —Type specimen from England: "sp. orig. Dillwynii," *Backhouse*, in herb. Agardh (LD) = Bacteria.

Hassallia Berkeley [in Hassall, ibid. 1: 231. 1845] ex Bornet & Flahault, Ann. Sci. Nat. VII. Bot. 5: 115. 1887. *Tolypothrix* Sect. *Hassallia* Kirchner in Engler & Prantl, Natürl. Pflanzenfam. 1(1A): 79. 1900. —Type species: *Conferva ocellata* Dillwyn, Brit. Conf., p. 60. 1809. [*Scytonema myochrous* β *ocellatum* Dillwyn (as "ocellata") ex Agardh, Disp. Alg. Suec., p. 38. 1812.] *Conferva myochrous* β *ocellata* Dillwyn *pro synon.* ex Agardh, loc. cit. 1812. *Scytonema myochrous* var. *ocellatum* Agardh (as "ocellata") [Syn. Alg. Scand., p. 114. 1817] ex Bornet & Flahault, ibid. 5: 70. 1887. *S. ocellatum* Lyngbye [Tent. Hydroph. Dan., p. 97. 1819] ex Bornet & Flahault, ibid. 5: 95. 1887. [*S. compactum* var. *ocellatum* Agardh ex Biasoletto, Viaggio di S. M. Federico Augusto, Re di Sassonia, per l'Istria, Dalmazia e Montenegro, p. 257. 1841.] *Hassallia ocellata* Hassall [Hist. Brit. Freshw, Alg. 1: 231. 1845] ex Bornet & Flahault, ibid. 5: 117. 1887. *Scytonema minutum* var. *ocellatum* Agardh ex Forti, Syll. Myxoph., p. 537. 1907. —Type specimen from England: "Scyt. ocellatum from Dillwyn," in herb. Agardh (LD) = *Stigonema* sp.

Conferva radicans Dillwyn, Brit. Conf., Introd., p. 57. 1809. [*Scytonema radicans* S. F. Gray, Nat. Arr. Brit. Pl., p. 286. 1821.] —Type specimen from England: Bantry, *Miss Hutchins*, in the Dillwyn collection (NMW) = *Sphacelaria* sp.

Conferva seriata Wahlenberg, Fl. Lappon., p. 514. 1812. [*Scytonema seriatum* S. F. Gray, loc. cit. 1821.] —Type specimen from Lapland: ad Arjeploug, in herb. Agardh (LD) = *Stigonema* sp.

Conferva vellea Roth, Catal. Bot. 3: 300. 1806. *Scytonema velleum* Agardh [Syst. Algar., p. 38. 1824] ex Bornet & Flahault, Ann. Sci. Nat. VII. Bot. 5: 112. 1887. —Type specimen, "Conferva vellea dedit Roth," in herb. Agardh (LD) = Fungi.

Cornicularia hispidula Acharius, Lichenogr. Univ., p. 617. 1810. [*Scytonema atro-virens* β *proliferum* Agardh (as "prolifera"), Disp. Alg. Suec., p. 39. 1812.] —Type specimen from Sweden: Lasjöfers, in herb. Agardh (LD) = parasitized *Stigonema* sp.

Cylindrospermum polyspermum Kützing [Phyc. Gener., p. 212. 1843] ex Bornet & Flahault, ibid. 7: 234. 1888. *Aulosira polysperma* Lagerheim [Öfvers. K. Vet.-Akad. Förh. 1883(2): 48. 1883] ex Bornet & Flahault, ibid. 7: 257. 1888. —Type specimen from Germany: Halle, Sprengel comm., in herb. Kützing (L) = Nostocacea with torulose trichomes.

Dictyonema membranaceum Agardh, Syst. Algar., p. 85. 1824. *Scytonema Dictyonema* Rabenhorst [Fl. Eur. Algar. 2: 264. 1865] ex Bornet & Flahault, ibid. 5: 113. 1887. —Type specimen from Oceania: Iles Marianas, *Gaudichaud 40*, 1822, in herb. Kützing (L) = Lichen containing parasitized *Scytonema Hofmannii* Ag.

*Dictyothrix lateritia Kützing [Phyc. Gener., p. 202. 1843] ex Forti, Syll. Myxoph., p. 308. 1907. *Hypheothrix Dictyothrix* Rabenhorst [ibid. 2: 86. 1865] ex Gomont, Ann. Sci. Nat. VII. Bot. 16: 115. 1892. —Type specimen from Venezia Euganea, Italy: Abano, in herb. Kützing (L) = *Schizothrix calcicola* (Ag.) Gom.

Ernstiella Chodat, Bull. Soc. Bot. Génève, sér. 2, 3: 126. 1911. —Type species: *E. rufa* Chodat, loc. cit. 1911. —Type specimen, here designated, from the original locality in Switzerland: in a small fountain, Parc de Mon Repos, Geneva, *F. Drouet & H. B. Louderback 15023*, Sept. 1970 (D) = *Schizothrix calcicola* (Ag.) Gom. mixed with *Entophysalis rivularis* (Kütz.) Dr.

[*Erythronema* Berkeley in Hooker, Himalayan Journ. 2: 376. 1854.] —Type species: [*E. Hookerianum* Berkeley, Introd. Crypt. Bot., p. 143. 1857.] —In the absence of the original material, the original descriptions are here designated as the Type = *Porphyrosiphon Notarisii* (Menegh.) Kütz.

Fischera muscicola Thuret [in Bornet & Thuret, Not. Algol. 2: 155. 1880] ex Bornet & Flahault, Ann. Sci. Nat. VII. Bot. 5: 67. 1887. *Scytonema muscicola* Borzi *pro synon.* [in Bornet & Thuret, loc. cit. 1880] ex Borzi, Studi sulle Mixof., p. 130. 1917. —Type specimen from France: Antibes, sables humides du Golfe Jouan, *G. Thuret*, Dec. 1874, in herb. Bornet—Thuret (PC) = *Fischerella* sp.

Fucus Opuntia Goodenough & Woodward, Trans. Linn. Soc. London 3: 219. 1797. *Rivularia Opuntia* Smith & Sowerby, Engl. Bot. 34: pl. 1868. 1808. —In the absence of the original material, the original description is here designated as the Type = Rhodophycea.

Glaucothrix gracillima Zopf, Sitzungsber. Bot. Ver. Brandenb. 1882: 44. 1882. *Plectonema gracillimum* Hansgirg, Physiol. & Algol. Stud., p. 108. 1887. —Type specimen from Germany: Anger prope Leipzig, *P. Richter*, in Wittrock & Nordstedt, Algae Exsiccatae no. 593a (PH) = *Schizothrix calcicola* (Ag.) Gom.

Glaucothrix Kirchner, Krypt.-Fl. Schles. 2(1): 229. 1878. *Plectonema* Sect. *Glaucothrix* Hansgirg, loc. cit. 1887. —Type species: *Glaucothrix putealis* Kirchner, loc. cit. 1878. *Plectonema puteale* Hansgirg, loc. cit. 1887. —In the absence of the original material, the original description is here designated as the Type = Bacteria. See Drouet, Monogr. Acad. Nat. Sci. Philadelphia 15: 313 (1968).

Handeliella sparsa Jao, Sinensia 15(1—6): 82. 1944. —In the absence of the

original material, the original description is here designated as the Type =
Stigonema sp.

Homoeothrix Batrachospermorum Skuja, N. Acta R. Soc. Sci. Upsal., ser. 4,
18(3): 68. 1964. —Type specimen from Swedish Lapland: Abisko (H. Skuja)
= *Schizothrix calcicola* (Ag.) Gom.

Homoeothrix brevis Kufferath, Ann. de Biol. Lac. 7: 278. 1914. *Camptothrix
brevis* Geitler in Pascher, Süsswasserfl. 12: 340. 1925. —In the absence of the
original material, the original description is here designated as the Type =
Schizothrix calcicola (Ag.) Gom.

Homoeothrix cerknicensis Lazar, Razprave Slovenska Akad. Znan. in Umetn.,
Razr. Prirod. in Medic. Vede, Odd. Prirod. Vede 14(2): 26. 1971. —In the
absence of the original material, the original description is here designated as
the Type = *Schizothrix calcicola* (Ag.) Gom.

Homoeothrix crustacea Voronikhin, Not. Syst. Inst. Crypt. Hort. Bot. Petropol.
2(8): 115. 1923. *H. Woronichinii* Margalef, Collect. Bot. 3(3): 231. 1953. —In the
absence of the original material, the original description is here designated as
the Type = *Schizothrix calcicola* (Ag.) Gom.

Homoeothrix Desikacharyensis Vasishta, Journ. Indian Bot. Soc. 42: 575. 1963.
—In the absence of the original material, the original description is here
designated as the Type = *Schizothrix calcicola* (Ag.) Gom.

Homoeothrix fusca Starmach, Acta Soc. Bot. Polon. 11(3): 288. 1934. —In the
absence of the original material, the original description is here designated as
the Type = *Schizothrix calcicola* (Ag.) Gom.

Homoeothrix fusca fa. *britannica* Godward, Journ. of Ecol. 25(2): 564. 1937.
—Type specimen, here designated, from the original locality in England: rocks
in Lake Windermere at Watbarrow Point, *J. W. G. Lund*, 1971 (D) =
Schizothrix calcicola (Ag.) Gom.

Homoeothrix fusca fa. *elongata* Starmach, ibid. 11(3): 290. 1934. —In the
absence of the original material, the original description is here designated as
the Type = *Schizothrix calcicola* (Ag.) Gom.

Homoeothrix fusca fa. *longissima* Starmach, Acta Hydrobiol. 10(1—2): 158.
1968. —In the absence of the original material, the original description is here
designated as the Type = *Schizothrix calcicola* (Ag.) Gom.

Homoeothrix fusca fa. *minor* Starmach, Acta Soc. Bot. Polon. 11(3): 291. 1934.
—In the absence of the original material, the original description is here
designated as the Type = *Schizothrix calcicola* (Ag.) Gom.

Homoeothrix globulus Voronikhin, Journ. Bot. de l'URSS 17(3): 309.
1932.—In the absence of the original material, the original description is here
designated as the Type = *Schizothrix calcicola* (Ag.) Gom.

Homoeothrix minuta Seckt, Bol. Acad. Nac. Cienc. Córdoba (Argentina) 25:
425. 1921. —In the absence of the original material, the original description is
here designated as the Type = *Schizothrix calcicola* (Ag.) Gom.

Homoeothrix moniliformis Vasishta, Journ. Indian Bot. Soc. 42: 577. 1963. —In the absence of the original material, the original description is here designated as the Type = *Schizothrix calcicola* (Ag.) Gom.

Homoeothrix Poljanskyi Muzafarov, Not. Syst. Inst. Crypt. Bot. Inst. Acad. Sci. URSS 8: 84. 1952. —In the absence of the original material, the original description is here designated as the Type = *Schizothrix calcicola* (Ag.) Gom.

Homoeothrix schizothrichoides Muzaffarov, Bot. Mat. Otd. Sporov. Rast. Bot. Inst. Komarova Akad. Nauk SSSR 12(1): 32. 1959. —In the absence of the original material, the original description is here designated as the Type = *Schizothrix calcicola* (Ag.) Gom.

Homoeothrix simplex Voronikhin, Journ. Bot. de l'URSS 17(3): 309. 1932. —In the absence of the original material, the original description is here designated as the Type = *Schizothrix calciola* (Ag.) Gom.

Homoeothrix simplex fa. *elegans* Elenkin, Monogr. Alg. Cyanoph., Pars Spec. 2: 1832. 1949. —In the absence of the original material, the original description is here designated as the Type = *Schizothrix calcicola* (Ag.) Gom.

Homoeothrix sinensis Jao, Sinensia 10(1—6): 221. 1939. —In the absence of the original material, the original description is here designated as the Type = *Schizothrix calcicola* (Ag.) Gom.

Homoeothrix subtilis Skuja, N. Acta R. Soc. Sci. Upsal., ser. 4, 18(3): 69. 1964. —Type specimen from Swedish Lapland: near Abisko, *H. Skuja* (Skuja) = *Schizothrix calcicola* (Ag.) Gom.

Homoeothrix varians Geitler, Biol. Gener. 3: 801. 1927; Arch. f. Protistenk. 60(2): 445. 1928. —In the absence of the original material, the original description is here designated as the Type = *Schizothrix calcicola* (Ag.) Gom.

Homoeothrix varians var. *major* Geitler, Arch. f. Hydrobiol. Suppl.-Bd. 12: 630. 1933. —In the absence of the original material, the original description is here designated as the Type = *Schizothrix calcicola* (Ag.) Gom.

Hydrocoleum phormidioides Rabenhorst [Fl. Eur. Algar. 2: 151. 1865] ex Gomont, Ann. Sci. Nat. VII. Bot. 15: 344. 1892. *Scytonema phormidioides* Bulnheim & Rabenhorst *pro synon.* [in Rabenhorst, loc. cit. 1865] ex Bornet & Flahault, Ann. Sci. Nat. VII. Bot. 5: 114. 1887. —Type specimen from Upper Austria: bei Aussee, *O. Bulnheim*, in Rabenhorst, Algen Sachsens no. 532 (W) = *Microcoleus vaginatus* (Vauch.) Gom.

Hygrocrocis olivacea Agardh, Flora 10(1:40): 631. 1827. *Hypheothrix lutescens* Rabenhorst [Fl. Eur. Algar. 2: 76. 1865] ex Gomont, ibid. 15: 328. 1892. —Type specimen from Czechoslovakia: Carlsbad, *Agardh*, in herb. Hofmann-Bang (C) = Bacteria.

Hygrocrocis rigidula Kützing, Phyc. Gener., p. 152. 1843. [*Hypheothrix rigidula* Grunow, Reise d. Novara, Bot. 1: 29. 1870.] —Type specimen from Germany: Tennstädt, in herb. Kützing (L) = *Schizothrix calcicola* (Ag.) Gom.

Hypheothrix acutissima Gardner, Mem. New York Bot. Gard. 7: 49. 1927. —Type specimen from Puerto Rico: Maricao, *N. Wille 1271a*, Feb. 1915 (NY) = *Porphyrosiphon Notarisii* (Menegh.) Kütz.

Hypheothrix aeruginea Kützing [Sp. Algar., p. 269. 1849] ex Gomont, Ann. Sci. Nat. VII. Bot. 15: 327. 1892. —Type specimen, with that of *Euactis Lenormandiana* Kütz., from Manche, France: Granville, ex Lenormand, in herb. Kützing (L) = *Oscillatoria submembranacea* Ard. & Straff.

Hypheothrix aeruginea var. *subtorulosa* Zeller [Journ. Asiatic Soc. Bengal 42(2): 176. 1873] ex Desikachary, Cyanoph., p. 334. 1959. —Type specimen from Burma: Pegu, Kenbatee, *S. Kurz 3131* (BM) = *Cylindrospermum* sp.

Hypheothrix aeruginea fa. *thermalis* Rabenhorst [Fl. Eur. Algar. 2: 78. 1865] ex Gomont, loc. cit. 1892. —Type specimen from Lombardia, Italy: Bormio, *E. Levier*, Aug. 1871, in Erbario Crittogamico Italiano, ser. 2, no. 667 (D) = *Schizothrix calcicola* (Ag.) Gom.

Hypheothrix aikenensis Wolle [Bull. Torrey Bot. Club 6: 182. 1877] ex Forti, Syll. Myxoph., p. 329. 1907. —Type specimen from South Carolina: Aiken, *H. W. Ravenel 243* (D) = *Schizothrix Friesii* (Ag.) Gom.

Hypheothrix anguina Suringar [Alg. Japon., p. 19. 1870] ex Hansgirg, Prodr. Algenfl. Böhm. 2: 110. 1892. —In the absence of the original material, the original description is here designated as the Type = Bacteria or Fungi.

Hypheothrix Braunii β *continua* Kützing [Sp. Algar., p. 266. 1849] ex Gomont, Ann. Sci. Nat. VII. Bot. 15: 327. 1892. *H. thermalis* c *continua* Rabenhorst [Fl. Eur. Algar. 2: 82. 1865] ex Gomont, ibid. 15: 329. 1892. *H. thermalis* var. *continua* Forti, Syll. Myxoph., p. 330. 1907. —Type specimen from Germany: Baden, in herb. Kützing (L) = *Schizothrix calcicola* (Ag.) Gom.

Hypheothrix Bremiana Nägeli [in Kützing, Sp. Algar., p. 267. 1849] ex Gomont, ibid. 15: 327. 1892. *H. rufescens* c *Bremiana* Rabenhorst [ibid. 2: 84. 1865] ex Gomont, ibid. 15: 328. 1892. *H. rufescens* fa. *Bremiana* Forti, ibid., p. 338. 1907. —Type specimen from Switzerland: Zürich, *Nägeli 285*, in herb. Kützing (L) = *Schizothrix calcicola* (Ag.) Gom.

Hypheothrix bullosa Wolle [Bull. Torrey Bot. Club 6: 182. 1877] ex Wolle, Fresh-w. Alg. U. S., p. 321. 1887. —Type specimen from Pennsylvania: "on Lyngbya Wollei" (D) = *Schizothrix calcicola* (Ag.) Gom.

Hypheothrix calcarea Nägeli [in Kützing, ibid., p. 268. 1849] ex Gomont, ibid. 15: 327. 1892. *H. lateritia* f *calcarea* Rabenhorst [ibid. 2: 85. 1865] ex Gomont, ibid. 15: 328. 1892. *H. lateritia* var. *calcarea* Rabenhorst ex Forti, ibid., p. 335. 1907. —Type specimen from Switzerland: Zürich im Katzensee, *Nägeli 279*, in herb. Kützing (L) = *Schizothrix calcicola* (Ag.) Gom.

Hypheothrix calcicola fa. *dilute-aerugineo-viridis* Rabenhorst [Fl. Eur. Algar. 2: 78. 1865] ex Gomont, Ann. Sci. Nat. VII. Bot. 15: 307. 1892. —Type specimen from Austria: Wien, *L. v. Heufler*, in Rabenhorst, Algen Europas no. 1391 (W) = *Schizothrix calcicola* (Ag.) Gom.

Hypheothrix calcicola fa. *glabra* Stockmayer, Ann. K. K. Nat. Hofmus. Wien 22: 106. 1907. —Type specimen from Lower Austria: Rehberg prope Krems, *K. Rechinger*, in Musei Vindobonensis Kryptogamae Exsiccatae, Alg. no. 1520a (W) = *Schizothrix calcicola* (Ag.) Gom.

Hypheothrix calcicola fa. *lacuno-spongiosa* Stockmayer, ibid. 22: 107. 1907. —Type specimen from Lower Austria: Rehberg, *Rechinger*, in Mus. Vindob. Krypt. Exs. no. 1520b (W) = *Schizothrix calcicola* (Ag.) Gom.

Hypheothrix calcicola fa. *symplocoidea* Stockmayer, loc. cit. 1907. —Type specimen from Lower Austria: Rehberg, *Rechinger*, in Mus. Vindob. Krypt. Exs. no. 1520 (W) = *Schizothrix calcicola* (Ag.) Gom.

*Hypheothrix cataractarum Nägeli [in Kützing, Sp. Algar., p. 269. 1849] ex Gomont, Ann. Sci. Nat. VII. Bot. 16: 170. 1892. —Type specimen from Switzerland: Zürich, *Nägeli 87*, in herb. Kützing (L) = *Microcoleus vaginatus* (Vauch.) Gom.

Hypheothrix cataractarum γ *terebriformis* Kützing [loc. cit. 1849] ex Gomont, ibid. 15: 327. 1892. *H. terebriformis* Nägeli *pro synon.* [in Kützing, loc. cit. 1849] ex Hansgirg, Prodr. Algenfl. Böhm. 2: 90. 1892. —Type specimen from Switzerland: Zürich, *Nägeli 275*, in herb. Kützing (L) = *Microcoleus vaginatus* (Vauch.) Gom.

*Hypheothrix compacta b *symplociformis* Grunow [in Rabenhorst, Fl. Eur. Algar. 2: 79. 1865] ex Gomont, ibid. 15: 329. 1892. *H. compacta* var. *symplociformis* Rabenhorst ex Forti, Syll. Myxoph., p. 307. 1907. *H. laminosa* fa. *symplociformis* Grunow ex Elenkin, Monogr. Alg. Cyanoph., Pars Spec. 2: 1475. 1949. —Type specimen from Austria: Dampfkessel hinter dem Raaber Bahnhofe bei Wien, *Heufler*, 1860, in Rabenhorst, Algen Europas no. 1308 (W) = *Schizothrix calcicola* (Ag.) Gom.

*Hypheothrix Confervae Kützing [Phyc. Gener., p. 230. 1843] ex Gomont, ibid. 15: 327. 1892. —Type specimen from Sachsen, Germany: in der Salzke bei Langenbogen, in herb. Kützing (L) = *Schizothrix calcicola* (Ag.) Gom.

*Hypheothrix cyanea Nägeli [in Kützing, Sp. Algar., p. 269. 1849] ex Gomont, ibid. 15: 328. 1892. —Type specimen from Switzerland: Zürich, *Nägeli 270*, in herb. Kützing (L) = *Schizothrix calcicola* (Ag.) Gom.

Hypheothrix delicatula Kützing [loc. cit. 1849] ex Gomont, loc. cit. 1892. —Type specimen from France: *De Brébisson 582*, in herb. Kützing (L) = *Schizothrix calcicola* (Ag.) Gom.

*Hypheothrix fasciculata Nägeli [in Kützing, loc. cit. 1849] ex Gomont, Ann. Sci. Nat. VII. Bot. 15: 298. 1892. —Type specimen from Switzerland: Zürich, *Nägeli 282*, in herb. Kützing (L) = *Schizothrix calcicola* (Ag.) Gom.

Hypheothrix fenestralis Kützing [ibid., p. 268. 1849] ex Gomont, ibid. 15: 328. 1892. —Type specimen, with those of *Palmella testacea* A. Br. and *P. brunnea* A. Br., from Baden, Germany: Freiburg, *A. Braun*, Oct. 1847, in herb. Kützing (L) = *Schizothrix calcicola* (Ag.) Gom.

*Hypheothrix fonticola Nägeli [in Kützing, ibid., p. 893. 1849] ex Gomont, loc. cit. 1892. —Type specimen from Switzerland: Zürich, *Nägeli 587*, in herb. Kützing (L) = *Schizothrix calcicola* (Ag.) Gom.

Hypheothrix fucoidea Piccone [N. Giorn. Bot. Ital. 16: 291. 1884] ex Gomont, ibid. 15: 330. 1892. —Type specimen from Eritrea, Ethiopia: Baja di Assab, *G. Caramagna*, Nov. 1881 (PAD) = Fungi.

Hypheothrix fusco-violacea Stizenberger [in Jack, Leiner & Stizenberger, Krypt. Badens 17: 772. 1859] ex Gomont, ibid. 15: 328. 1892. —Type specimen from Germany: Heidelberg, *W. Ahles*, Jun. 1859, in Jack, Leiner & Stizenberger, Kryptogamen Badens no. 772 (W) = *Microcoleus vaginatus* (Vauch.) Gom.

*Hypheothrix Hegetschweileri Nägeli [in Kützing, Sp. Algar., p. 893. 1849] ex Gomont, loc. cit. 1892. *H. rufescens* d *Hegetschweileri* Rabenhorst [Fl. Eur. Algar. 2: 84. 1865] ex Gomont, loc. cit. 1892. *H. rufescens* fa. *Hegetschweileri* Nägeli ex Forti, Syll. Myxoph., p. 337. 1907. —Type specimen from Switzerland: Rigi, *Nägeli 439*, in herb. Kützing (L) = *Schizothrix calcicola* (Ag.) Gom.

*Hypheothrix Hilseana Rabenhorst [ibid. 2: 87. 1865] ex Gomont, Ann. Sci. Nat. VII. Bot. 15: 330. 1892. —Type specimen from Silesia, Poland: am Galgenberge bei Strehlen, *Hilse*, in Rabenhorst, Algen Europas no. 1038 (L) = *Schizothrix rubella* Gom.

*Hypheothrix incrustata Nägeli [in Kützing, ibid., p. 269. 1849] ex Gomont, loc. cit. 1892. —Type specimen from Switzerland: Zürich, *Nägeli 462*, in herb. Kützing (L) = *Microcoleus vaginatus* (Vauch.) Gom.

Hypheothrix investiens Martens [in Kurz, Proc. Asiatic Soc. Bengal 1870: 11. 1870] ex Gomont, ibid. 15: 328. 1892. —Type specimen from Bengal, India: zwischen Titalya und Silligoree, *S. Kurz*, in herb. Agardh (LD) = *Schizothrix Friesii* (Ag.) Gom.

Hypheothrix involvens Kützing [loc. cit. 1849] ex Gomont, ibid. 15: 330. 1892. *Calothrix involvens* Areschoug *pro synon.* [ex Kützing, loc. cit. 1849] ex Bornet & Flahault, Ann. Sci. Nat. VII. Bot. 3: 371. 1886. —Type specimen from Sweden: Christineberg, *Areschoug*, Aug. 1841, in herb. Kützing (L) = *Schizothrix mexicana* Gom.

Hypheothrix jassaensis Vouk, Jugosl. Akad. Prirod. Istraž. Hrvatske i Slavonije, Mat.-Prirod. Razr. 8: 7. 1916. —In the absence of the original material, the original description is here designated as the Type = *Schizothrix calcicola* (Ag.) Gom.

*Hypheothrix lateritia Kützing [Sp. Algar., p. 268. 1849] ex Gomont, Ann. Sci. Nat. VII. Bot. 15: 308. 1892. —Type specimen from Switzerland: Zürich, *Nägeli 380*, in herb. Kützing (L) = *Schizothrix calcicola* (Ag.) Gom.

Hypheothrix lateritia β turfacea Hepp [in Rabenhorst, Alg. Sachs. 67–68: 671. 1858] ex Gomont, ibid. 15: 309. 1892. [*H. coriacea* fa. *turfacea* Rabenhorst, Fl. Eur. Algar. 2: 83. 1865.] [*H. lateritia* var. *turfacea* Hepp ex Rabenhorst, loc. cit. 1865.] —Type specimen from Switzerland: Zürich, *Hepp*, Sept. 1857, in Rabenhorst, Algen Sachsens no. 671 (PH) = *Schizothrix calcicola* (Ag.) Gom.

Hypheothrix Leveilleana Kützing [ibid., p. 267. 1849] ex Gomont, ibid. 15: 328. 1892. —Type specimen from Corsica, France: Ajaccione, *Leveille,* in herb. Kützing (L) = *Schizothrix calcicola* (Ag.) Gom.

*Hypheothrix litoralis Hansgirg in Foslie, Mar. Alg. Norway 1: 166. 1890. —Type specimen from Norway: Tromsö, *M. Foslie,* Sept. 1889 (PC) = *Schizothrix calcicola* (Ag.) Gom.

*Hypheothrix longiarticulata Gardner, Mem. New York Bot. Gard. 7: 50. 1927. —Type specimen from Puerto Rico: San Juan, *N. Wille 2022a,* Mar. 1915 (NY) = *Schizothrix arenaria* (Berk.) Gom.

*Hypheothrix Naegelii Kützing [Sp. Algar., p. 268. 1849] ex Gomont, Ann. Sci. Nat. VII. Bot. 15: 328. 1892. —Type specimen from Switzerland: Zürich, *Nägeli 274,* in herb. Kützing (L) = *Schizothrix calcicola* (Ag.) Gom.

*Hypheothrix nullipora Grunow [in Rabenhorst, Fl. Eur. Algar. 2: 82. 1865] ex Gomont, loc. cit. 1892. —Type specimen from Austria: Wien, im Laxenburger Parke, *Reichardt* (W) = *Schizothrix calcicola* (Ag.) Gom.

Hypheothrix obscura Dickie [Journ. Linn. Soc. Bot. 17: 8. 1878] ex Gomont, ibid. 16: 181. 1892. —Type specimen from Ellesmere Land, Canada: Distant cape, Discovery bay, *Feilden, Moss & Hart* (BM) = *Microcoleus vaginatus* (Vauch.) Gom.

*Hypheothrix pallida Kützing [ibid., p. 893. 1849] ex Gomont, ibid. 15: 328. 1892. —Type specimen from Switzerland: Zürich, *Nägeli 426,* in herb. Kützing (L) = *Microcoleus vaginatus* (Vauch.) Gom.

*Hypheothrix panniformis Rabenhorst [Hedwigia 1853(4): 17. 1853] ex Gomont, loc. cit. 1892. —In the absence of the original material, the original description is here designated as the Type = *Schizothrix calcicola* (Ag.) Gom.

*Hypheothrix parciramosa Gardner, Mem. New York Bot. Gard. 7: 50. 1927. —Type specimen from Puerto Rico: Maricao, *N. Wille 1228* (NY) = *Schizothrix Friesii* (Ag.) Gom.

*Hypheothrix parietina Stizenberger [in Rabenhorst, Alg. Sachs. 71–72: 708. 1858] ex Bornet & Flahault, Ann. Sci. Nat. VII. Bot. 5: 100. 1887. *H. coriacea* b *parietina* Stizenberger [ex Rabenhorst, Fl. Eur. Algar. 2: 83. 1865] ex Hansgirg, Prodr. Algenfl. Böhm. 2: 92. 1892. *H. coriacea* fa. *parietina* Stizenberger ex Forti, Syll. Myxoph., p. 337. 1907. —Type specimen from Switzerland: Zürich, *Hepp,* May 1858 (L) = *Fischerella* sp.

Hypheothrix plumula Fiorini Mazzanti [Atti Accad. Pontif. N. Lincei Roma 14(4): 240. 1861] ex Forti, ibid., p. 345. 1907. —Type specimen from Italy: Terracina, *E. Fiorini* (PAD) = Bacteria.

*Hypheothrix porphyromelana Brühl & Biswas, Journ. Dept. Sci. Calcutta Univ. 4(Bot.): 8. 1922. —In the absence of the original material, the original description is here designated as the Type = *Oscillatoria lutea* Ag.

Hypheothrix pulvinata Nägeli [in Kützing, Sp. Algar., p. 268. 1849] ex

Gomont, Ann. Sci. Nat. VII. Bot. 15: 328. 1892. —Type specimen from Switzerland: Zürich, *Nägeli 191*, in herb. Kützing (L) = *Schizothrix calcicola* (Ag.) Gom.

Hypheothrix pulvinata β *confluens* Kützing [loc. cit. 1849] ex Gomont, loc. cit. 1892. —Type specimen from Switzerland: Zürich, *Nägeli 542*, in herb. Kützing (L) = *Schizothrix calcicola* (Ag.) Gom.

*Hypheothrix Regeliana Nägeli [in Kützing, ibid., p. 267. 1849] ex Gomont, loc. cit. 1892. —Type specimen from Switzerland: Zürich, *Nägeli 390*, in herb. Kützing (L) = *Schizothrix calcicola* (Ag.) Gom.

*Hypheothrix Regeliana fa. *crassior* Rabenhorst ex Forti, Syll. Myxoph., p. 331. 1907. —Type specimen from Switzerland: Zürich, *Nägeli 390*, in herb. Kützing (L) = *Schizothrix calcicola* (Ag.) Gom.

*Hypheothrix roseola Richter [Hedwigia 1879(7): 97. 1879] ex Gomont, Ann. Sci. Nat. VII. Bot. 15: 330. 1892. *Plectonema roseolum* Gomont, loc. cit. 1892. —Type specimen from Sachsen, Germany: Gewächshaus in Anger, *P. Richter*, Feb. 1879 (W) = *Schizothrix calcicola* (Ag.) Gom.

*Hypheothrix rufescens b *lardacea* Rabenhorst [Fl. Eur. Algar. 2: 84. 1865] ex Gomont, ibid. 15: 311. 1892. *H. rufescens* var. *lardacea* Rabenhorst ex Hansgirg, Physiol. & Algol. Stud., p. 95. 1887. *H. lardacea* Hansgirg ex Dalla Torre & Sarnthein, Alg. v. Tirol, Vorarlb. & Liechtenst., p. 144. 1901. —Type specimen from Piemonte, Italy: Viverone, *Cesati*, Nov. 1856, in Rabenhorst, Algen Sachsens no. 578 (PC) = *Schizothrix calcicola* (Ag.) Gom.

*Hypheothrix Sauteriana Grunow [in Rabenhorst, Fl. Eur. Algar. 2: 89. 1865] ex Hansgirg, Prodr. Algenfl. Böhm. 2: 94. 1892. —Type specimen from Austria: Salzburg, *Sauter* (W) = *Schizothrix Friesii* (Ag.) Gom.

*Hypheothrix scopulorum Kützing [Sp. Algar., p. 269. 1849] ex Gomont, ibid. 15: 328. 1892. —Type specimen from Yugoslavia: Dalmatia, in herb. Kützing (L) = *Schizothrix calcicola* (Ag.) Gom.

*Hypheothrix Sophiae Areschoug [Alg. Scand. Exs., ser. nov., 6: 288. 1866] ex Forti, Syll. Myxoph., p. 343. 1907. —Type specimen from Sweden: *S. Åkermark*, in Areschoug, Algae Scandinavicae, ser. nov., no. 288(S) = *Oscillatoria lutea* Ag.

*Hypheothrix subcontinua Kützing [ibid., p. 268. 1849] ex Gomont, loc. cit. 1892. —Type specimen from Switzerland: Zürich, *Nägeli 88*, in herb. Kützing (L) = *Schizothrix calcicola* (Ag.) Gom.

*Hypheothrix subtilis Kützing [ibid., p. 267. 1849] ex Gomont, loc. cit. 1892. [*H. lateritia* d *subtilis* Rabenhorst, Fl. Eur. Algar. 2: 85. 1865.] *H. lateritia* var. *subtilis* Rabenhorst ex Hansgirg in Kerner, Schedae ad Fl. Exs. Austro-Hung. 5: 114. 1888. —Type specimen from Switzerland: Rheinfall, *Nägeli 189*, in herb. Kützing (L) = *Schizothrix calcicola* (Ag.) Gom.

*Hypheothrix subundulata Martens [in Kurz, Proc. Asiatic Soc. Bengal 1871:

220   REVISION OF THE NOSTOCACEAE

171. 1871] ex Desikachary, Cyanoph., p. 332. 1959. —Type specimen from
Bengal, India: Rajmahal, *S. Kurz 4802*, in herb. Agardh (LD) = *Schizothrix
calcicola* (Ag.) Gom.

*Hypheothrix sudetica Nave [Verh. Naturf. Ver. Brünn 2: 39. 1864] ex
Gomont, Ann. Sci. Nat. VII. Bot. 15: 328. 1892. —Type specimen from
Moravia, Czechoslovakia: Quellen im Gesenke, *J. Nave* (W) = *Schizothrix
calcicola* (Ag.) Gom.

*Hypheothrix symplocoides Gardner, Mem. New York Bot. Gard. 7: 51. 1927.
—Type specimen from Puerto Rico: Palmer, *N. Wille 760* (NY) = *Schizothrix
Friesii* (Ag.) Gom.

Hypheothrix tenax Martens [in Kurz, ibid. 1871: 172. 1871] ex Desikachary,
ibid., p. 675. 1959. —Type specimen from India: Calcutta, *S. Kurz 3054* (BM)
= *Schizothrix mexicana* Gom.

*Hypheothrix tenax Wolle [Bull. Torrey Bot. Club 6: 282. 1879] ex Wolle,
Fresh-w. Alg. U. S., p. 319. 1887. —Type specimen from Pennsylvania: State
Quarries, *F. Wolle*, Aug. 1878 (D) = *Schizothrix calcicola* (Ag.) Gom.

Hypheothrix thermalis Rabenhorst [Fl. Eur. Algar. 2: 81. 1865] ex Gomont,
ibid. 15: 329. 1892. —Type specimen from Baden, Germany: Baden, *W. F. R.
Suringar 1071*, Sept. 1862 (L) = *Schizothrix calcicola* (Ag.) Gom.

*Hypheothrix toficola Nägeli [in Kützing, Tab. Phyc. 1: 42. 1845—49] ex
Gomont, ibid. 15: 331. 1892. —Type specimen from Switzerland: Zürich,
Nägeli 599, in herb. Kützing (L) = *Microcoleus vaginatus* (Vauch.) Gom.

Hypheothrix torulosa Grunow [in Heufler, Sitzungsber. K. K. Zool.-bot. Ges.
Wien 1858: 70. 1858] ex Bornet & Flahault, Revision, Table, p. 7. 1888.
Amphithrix janthina var. *torulosa* Bornet & Flahault, Ann. Sci. Nat. VII. Bot. 3:
344. 1886. *A. janthina* fa. *torulosa* Poliansky in Hollerbach, Kossinskaia &
Poliansky, Opred. Presnov. Vodor. SSSR 2: 608. 1953. *Homoeothrix janthina*
var. *torulosa* Bornet & Flahault ex Bourrelly, Alg. d'Eau Douce 3: 406. 1970.
—Type specimen from Austria: Berndorf, *A. Grunow*, in herb. Bornet —Thuret
(PC) = *Schizothrix calcicola* (Ag.) Gom.

*Hypheothrix variegata Nägeli [in Kützing, Sp. Algar., p. 893. 1849] ex
Gomont, Ann. Sci. Nat. VII. Bot. 15: 329. 1892. *H. lateritia* var. *variegata* Forti,
Syll. Myxoph., p. 335. 1907. —Type specimen from Switzerland: Zürich,
Nägeli 547, in herb. Kützing (L) = *Schizothrix calcicola* (Ag.) Gom.

Hypheothrix violacea Kützing [ibid., p. 267. 1849] ex Bornet & Flahault,
Revision, Table, p. 7. 1888. *Amphithrix violacea* Bornet & Flahault [Mém. Soc.
Nat. Sci. Nat. & Math. Cherbourg 25: 200. 1885] ex Gomont, ibid. 15: 330.
1892. *A. marina* Bornet & Flahault ex Frémy, Bull. Soc. Linn. Normandie, sér.
7, 9: 123. 1927. —Type specimen from France: Calvados, *De Brébisson 520*, in
herb. Kützing (L) = *Schizothrix calcicola* (Ag.) Gom.

*Hypheothrix viridula Zeller [Journ. Asiatic Soc. Bengal 42(2): 177. 1873] ex
Forti, Syll. Myxoph., p. 330. 1907. —In the absence of the original material,

the original description is here designated as the Type = *Schizothrix arenaria* (Berk.) Gom.

*Hypheothrix vulpina Kützing [loc. cit. 1849] ex Gomont, ibid. 15: 329. 1892. —Type specimen from Baden, Germany: Freiburg, *A. Braun 4*, May 1847, in herb. Kützing (L) = *Schizothrix calcicola* (Ag.) Gom.

Hypheothrix vulpina var. *tumida* Wittrock [Bih. K. Svensk. Vet.-Akad. Handl. 1(1): 69. 1872] ex Forti, ibid., p. 339. 1907. —Type specimen from Gotland, Sweden: Visby (S) = *Microcoleus vaginatus* (Vauch.) Gom.

*Hypheothrix Willei Gardner, Mem. New York Bot. Gard. 7: 52. 1927. —Type specimen from Puerto Rico: Humacao, *N. Wille 613b*, Jan. 1915 (NY) = *Porphyrosiphon Notarisii* (Menegh.) Kütz.

Hypheothrix Zenkeri b *carpatica* Rabenhorst [Fl. Eur. Algar. 2: 85. 1865] ex Gomont, Ann. Sci. Nat. VII. Bot. 15: 309. 1892. *H. Zenkeri* fa. *carpatica-sublateritia* Rabenhorst *pro synon.* [loc. cit. 1865] ex Gomont, loc. cit. 1892. —Type specimen from Slovakia, Czechoslovakia: Vorberge der Zipfer Carpathen, *C. Kalchbrenner*, in Rabenhorst, Algen Europas no. 1287 (W) = *Schizothrix calcicola* (Ag.) Gom.

*Hypheothrix Zenkeri fa. *viridis* Kützing [ex Rabenhorst, loc. cit. 1865] ex Forti, ibid., p. 338. 1907. —Type specimen from Switzerland: Zürich, *Nägeli 541*, in herb. Kützing (L) = *Schizothrix calcicola* (Ag.) Gom.

*Inactis tornata Kützing [Phyc. Gener., p. 202. 1843] ex Gomont, ibid. 15: 327. 1892. *Rivularia tornata* Wallroth *pro synon.* ex Kützing, loc. cit. 1843. —Type specimen from Sachsen, Germany: Himmelsgarten (Nordhausen), in herb. Kützing (L) = *Schizothrix calcicola* (Ag.) Gom.

*Leibleinia aequalis Kützing [Sp. Algar., p. 276. 1849] ex Gomont, ibid. 16: 152. 1892. —Type specimen from Brazil: Bahia, in herb. Kützing (L) = *Schizothrix mexicana* Gom.

*Leibleinia caerulea Montagne [Ann. Sci. Nat. III. Bot. 14: 306. 1850] ex Gomont, ibid. 16: 136. 1892. —Type specimen from French Guyana: *Leprieur 820*, in herb. Montagne (PC) = *Microcoleus lyngbyaceus* (Kütz.) Crouan.

*Leibleinia caeruleo-violacea Crouan [in Schramm & Mazé, Essai Class. Alg. Guadeloupe, p. 31. 1865] ex Gomont, ibid. 16: 137. 1892. —Type specimen from Guadeloupe, West Indies: Moule, in herb. Crouan (PC) = *Microcoleus lyngbyaceus* (Kütz.) Crouan.

Leibleinia caespitosa Crouan [in Schramm & Mazé, ibid., p. 30. 1865] ex Gomont, ibid. 16: 152. 1892. —In the absence of the original material, the original description is here designated as the Type = *Microcoleus lyngbyaceus* (Kütz.) Crouan.

*Leibleinia capillacea Kützing [Phyc. Gener., p. 221. 1843] ex Gomont, ibid. 16: 152. 1892. *Calothrix sordida* Zanardini *pro synon.* [in Kützing, Sp. Algar., p. 278. 1849] ex Bornet & Flahault, Ann. Sci. Nat. VII. Bot. 3: 370. 1886. [*Leibleinia elongata* Meneghini *pro synon.* in Kützing, loc. cit. 1849] —Type

specimen from Italy: Chioggia, *Meneghini,* in herb. Kützing (L) = *Schizothrix mexicana* Gom.

*Leibleinia cirrulus Kützing [Bot. Zeit. 5(12): 193. 1847] ex Gomont, Ann. Sci. Nat. VII. Bot. 16: 136. 1892. —Type specimen from Dalmatia, Yugoslavia: Lesina, *Botteri 114,* in herb. Kützing (L) = *Microcoleus lyngbyaceus* (Kütz.) Crouan.

[*Leibleinia cirrulus β minor* Kützing, ibid. 5(12): 194. 1847.] —Type specimen from Dalmatia, Yugoslavia: Lesina, *Botteri 107,* in herb. Kützing (L) = *Microcoleus lyngbyaceus* (Kütz.) Crouan.

*Leibleinia Corallinae Kützing [Sp. Algar., p. 276. 1849] ex Gomont, ibid. 16: 218. 1892. —Type specimen from Calvados, France: Arromanches, *Lenormand 5,* in herb. Kützing (L) = *Microcoleus lyngbyaceus* (Kütz.) Crouan.

Leibleinia gloeothrix Frauenfeld [Alg. Dalmat. Küste, p. 3. 1855] ex Forti, Syll. Myxoph., p. 294. 1907. —Type specimen from Dalmatia, Yugoslavia: *Vidovich,* in herb. Hauck (L) = *Microcoleus lyngbyaceus* (Kütz.) Crouan.

*Leibleinia gracilis Meneghini [Giorn. Bot. Ital. 1(1): 304. 1844] ex Gomont, ibid. 16: 124. 1892. —Type specimen from Dalmatia, Yugoslavia: *Vidovich,* in herb. Hauck (L) = *Schizothrix mexicana* Gom.

*Leibleinia Lenormandii Kützing [Bot. Zeit. 5: 194. 1847] ex Gomont, ibid. 16: 106. 1892. —Type specimen from France: Cherbourg, *Lenormand,* in herb. Kützing (L) = *Schizothrix mexicana* Gom.

*Leibleinia littoralis Crouan [in Schramm & Mazé, Essai Class. Alg. Guadeloupe, p. 31. 1865] ex Gomont, ibid. 16: 136. 1892. —Type specimen from Guadeloupe, West Indies: Vieux-Fort, in herb. Crouan (PC) = *Microcoleus lyngbyaceus* (Kütz.) Crouan.

[*Leibleinia maxima* Crouan in Schramm & Mazé, ibid., p. 72. 1866.] —Type specimen from Guadeloupe, West Indies: Dolé, May 1864, in herb. Crouan (PC) = *Schizothrix mexicana* Gom.

*Leibleinia Meneghiniana Kützing [Phyc. Gener., p. 222. 1843] ex Gomont, Ann. Sci. Nat. VII. Bot. 16: 125. 1892. —Type specimen from Italy: Chioggia, *Meneghini,* in herb. Kützing (L) = *Schizothrix mexicana* Gom.

*Leibleinia penicillata Kützing [Bot. Zeit. 5(12): 194. 1847] ex Gomont, ibid. 15: 305. 1892. —Type specimen from Calvados, France: Falaise, *Lenormand 78,* in herb. Kützing (L) = *Oscillatoria lutea* Ag.

*Leibleinia polychroa Meneghini [Giorn. Bot. Ital. 1(1): 304. 1844] ex Gomont, ibid. 16: 126. 1892. —Type specimen from Yugoslavia: Dalmazia, ex Meneghini (L) = *Schizothrix mexicana* Gom.

*Leibleinia torta Crouan [in Schramm & Mazé, Essai Class. Alg. Guadeloupe, p. 31. 1865] ex Gomont, ibid. 16: 143. 1892. —Type specimen from Guadeloupe, West Indies, in herb. Crouan (PC) = *Schizothrix mexicana* Gom.

*Leibleinia violacea Meneghini [loc. cit. 1844] ex Gomont, ibid. 16: 146. 1892.

Calothrix violacea Zanardini *pro synon.* ex Forti, Syll. Myxoph., p. 636. 1907. —Type specimen from Yugoslavia: Dalmatia, *Vidovich*, in herb. Hauck (L) = *Schizothrix mexicana* Gom.

Leptobasis spirulina var. *goesingensis* Palik (as "goesingense"), Index Horti Bot. Univ. Budapest 1938: 6. 1938. *L. goesingensis* Palik (as "goesingense") *pro synon.*, loc. cit. 1938. *Fortiea spirulina* var. *goesingensis* J. de Toni, Gen. & Sp. Myxoph. 1(D—L): 214. 1949. *Leptobasis spirulina* fa. *goesingensis* Poliansky (as "goesingense") in Hollerbach, Kossinskaia & Poliansky, Opred. Presn. Vodor. SSR 2: 394. 1953. *Fortiea spirulina* fa. *goesingensis* Poliansky (as "goesingense") ex Starmach, Fl. Slodkow. Polski 2: 556. 1966. —In the absence of the original material, the original description is here designated as the Type = Fungus (*Helicoon* sp.?), as suggested by D. & R. Mollenhauer in Schweiz. Zeitschr. f. Hydrol. 32(2): 532—537 (1970).

Leptochaete amara Richter in Kuntze, Rev. Gen. Pl. 3(2): 385. 1898. —In the absence of the original material, the original description is here designated as the Type = *Schizothrix calcicola* (Ag.) Gom.

Leptochaete crustacea var. *gracilis* Hansgirg, Sitzungsber. Böhm. Ges. Wiss., Math.-nat. Cl. 1892(1): 138. 1893. *L. gracilis* Geitler in Pascher, Süsswasserfl. 12: 208. 1925. —Type specimen from Tirol, Austria: Hall —S. Magdalena, *A. Hansgirg*, 1891 (W) = *Schizothrix calcicola* (Ag.) Gom.

Leptochaete Hansgirgii Schmidle, Allg. Bot. Zeitschr. 1900: 34. 1900. —In the absence of the original material, the original description is here designated as the Type = *Schizothrix calcicola* (Ag.) Gom.

Leptochaete marina Hansgirg, ibid. 1890(1): 13. 1890. —Type specimen from Istria, Yugoslavia: Pola, *Hansgirg*, Apr. 1889 (W) = *Schizothrix calcicola* (Ag.) Gom.

Leptochaete marina Hansgirg in Foslie, Mar. Alg. Norway 1: 160. 1890. —Type specimen from Norway: Bugonaes, *M. Foslie*, Aug. 1889, in Wittrock, Nordstedt & Lagerheim, Algae Exs. no. 1501 (W) = *Schizothrix calcicola* (Ag.) Gom.

Leptochaete Sect. *Xanthochaete* Forti, Syll. Myxoph., p. 599. 1907. —Type species: *Leptochaete nidulans* Hansgirg, Österr. Bot. Zeitschr. 37: 121. 1887. —Type specimen from Bohemia, Czechoslovakia: Bystric pr. Beneschau, *Hansgirg*, Aug. 1884 (W) = *Schizothrix calcicola* (Ag.) Gom.

Leptochaete parasitica Borzi [N. Giorn. Bot. Ital. 14(4): 298. 1882] ex Bornet & Flahault, Ann. Sci. Nat. VII. Bot. 3: 342. 1886. —In the absence of the original material, the original description is here designated as the Type = *Schizothrix calcicola* (Ag.) Gom.

Leptochaete rivularis Hansgirg, ibid. 38: 117. 1888. —Type specimen from Bohemia, Czechoslovakia: Pampferhütte bei Eisenstein, *Hansgirg*, Aug. 1887 (W) = *Schizothrix calcicola* (Ag.) Gom.

Leptochaete rivularis var. *rivularium* Hansgirg, Beih. z. Bot. Centralbl. 18(2):

493. 1905. *L. rivularis* var. *Rivulariarum* Hansgirg ex Forti, Syll. Myxoph., p. 600. 1907. *L. Rivulariarum* Lemmermann, Krypt.-Fl. Mark Brandenb. 3: 238. 1910. —Type specimen from Lower Austria: Puchberg ad Schneeberg, *Hansgirg*, in Musei Vindobon. Kryptogamae Exs. no. 1203 (W) = *Schizothrix calcicola* (Ag.) Gom.

Leptochaete rivularis fa. *tenuior* Huber-Pestalozzi, Arch. f. Hydrobiol. 19(4): 675. 1928. —In the absence of the original material, the original description is here designated as the Type = *Schizothrix calcicola* (Ag.) Gom.

Leptochaete tenella Gardner, Mem. New York Bot. Gard. 7: 65. 1927. —Type specimen from Puerto Rico: Maricao, *N. Wille 1128*, Feb. 1915 (NY) = *Schizothrix calcicola* (Ag.) Gom.

Leptopogon Borzi, Atti Congr. Nat. Ital. Milano, p. 372. 1907; in Forti, ibid., p. 704. 1907. —Type species: *L. Braunii* Borzi, Atti Congr. Nat. Milano, p. 375. 1907. *L. intricatus* Borzi, Studi sulle Mixof., p. 131. 1916. —Type specimen from Germany: im Berliner bot. Garten, *A. Braun*, May 1875, in Rabenhorst, Algen Europas no. 2464 (F) = *Fischerella ambigua* (Näg.) Gom.

**Leptothrix aeruginea* Kützing, Phyc. Gener., p. 198. 1843. *Hypheothrix aeruginea* Rabenhorst [Fl. Eur. Algar. 2: 78. 1865] ex Hansgirg, Prodr. Algenfl. Böhm. 2: 88. 1892. —Type specimen from Sachsen, Germany: Nordhausen, in herb. Kützing (L) = *Schizothrix calcicola* (Ag.) Gom.

Leptothrix aeruginea β *pallida* Kützing, Sp. Algar., p. 265. 1849. *Hypheothrix aeruginea* fa. *pallida* Rabenhorst [loc. cit. 1865] ex Forti, Syll. Myxoph., p. 332. 1907. —Type specimen from Baden, Germany: Freiburg, *A. Braun 8*, Jun. 1846, in herb. Kützing (L) = *Schizothrix calcicola* (Ag.) Gom.

Leptothrix amethystea Kützing, Bot. Zeit. 5(13): 219. 1847. *Amphithrix amethystea* Kützing [Sp. Algar., p. 275. 1849] ex Hansgirg, ibid. 2: 54. 1892. *Kuetzingina amethystea* Kuntze, Rev. Gen. Pl. 3(2): 411. 1898. —Type specimen from Calvados, France: Falaise, *De Brébisson 187*, in herb. Kützing (L) = *Schizothrix calcicola* (Ag.) Gom.

**Leptothrix Braunii* Kützing, Phyc. Gener., p. 198. 1843. *Hypheothrix Braunii* Kützing [Sp. Algar., p. 266. 1849] ex Gomont, Ann. Sci. Nat. VII. Bot. 16: 168. 1892. *H. Braunii* α *fasciculata* Kützing [loc. cit. 1849] ex Gomont, loc. cit. 1892. [*H. thermalis* b *fasciculata* Rabenhorst, ibid. 2: 82. 1865.] *H. thermalis* var. *fasciculata* Rabenhorst ex Hansgirg, ibid. 2: 89. 1892. *H. Braunii* var. *fasciculata* Kützing ex Elenkin, Monogr. Alg. Cyanoph., Pars Spec. 2: 1475. 1949. —Type specimen from Baden, Germany: Badensweiler, *A. Braun 188*, Aug. 1848, in herb. Kützing (L) = Fungi.

**Leptothrix calcicola* var. *opaca* Rabenhorst [Alg. Sachs. 13–14: 129. 1851] ex Gomont, ibid. 15: 307. 1892. *Hypheothrix calcicola* fa. *opaca* Rabenhorst [Fl. Eur. Algar. 2: 78. 1865] ex Gomont, loc. cit. 1892. —Type specimen from Silesia, Poland: Lauban, *R. Peck*, in Rabenhorst, Algen Sachsens no. 129 (PC) = *Schizothrix calcicola* (Ag.) Gom.

Leptothrix coriacea Kützing, Phyc. Gener., p. 198. 1843. *Hypheothrix coriacea* Kützing [Sp. Algar., p. 267. 1849] ex Gomont, ibid. 15: 309. 1892. —Type specimen from Italy: Triest, in herb. Kützing (L) = *Schizothrix calcicola* (Ag.) Gom.

Leptothrix cyanea Kützing, Bot. Zeit. 5(13): 219. 1847. *Hypheothrix cyanea* Rabenhorst [ibid. 2: 88. 1865] ex Gomont, Ann. Sci. Nat. VII. Bot. 15: 328. 1892. —Type specimen from Italy: Therm. Eugan., *Meneghini 25*, in herb. Kützing (L) = *Schizothrix calcicola* (Ag.) Gom.

*Leptothrix dalmatica Kützing, Sp. Algar., p. 265. 1849. *Hypheothrix dalmatica* Rabenhorst [ibid. 2: 89. 1865] ex Forti, Syll. Myxoph., p. 344. 1907. —Type specimen from Croatia, Yugoslavia: Cherso, in herb. Kützing (L) = *Schizothrix mexicana* Gom.

*Leptothrix dubia Nägeli in Kützing, ibid., p. 264. 1849. *Hypheothrix dubia* Nägeli [ex Rabenhorst, Alg. Sachs. 59–60: 593. 1857] ex Gomont, ibid. 15: 330. 1892. *H. Bremiana* var. *griseo-fuscescens* Kützing *pro synon.* ex Gomont, ibid. 15: 329. 1892. —Type specimen from Switzerland: Zürich, *Nägeli 433*, in herb. Kützing (L) = *Schizothrix calcicola* (Ag.) Gom.

*Leptothrix foveolarum Montagne, Ann. Sci. Nat. III. Bot. 12: 287. 1849. *Hypheothrix foveolarum* Rabenhorst [Fl. Eur. Algar. 2: 77. 1865] ex Gomont, ibid. 16: 165. 1892. —Type specimen from Seine-et-Oise, France: carrières du Marne, *Montagne*, in herb. Agardh (LD) = *Schizothrix calcicola* (Ag.) Gom.

*Leptothrix gloeophila Kützing, Bot. Zeit. 5(13): 219. 1847. *Hypheothrix gloeophila* Rabenhorst [loc. cit. 1865] ex Hansgirg, Prodr. Algenfl. Böhm. 2: 87. 1892. —Type specimen, with that of *Gloeocapsa confluens* Kütz., from Sachsen, Germany: Nordhausen, in herb. Kützing (L) = *Schizothrix calcicola* (Ag.) Gom.

Leptothrix herbacea Kützing, Sp. Algar., p. 265. 1849. *Hypheothrix herbacea* Kützing [ex Rabenhorst, ibid. 2: 79. 1865] ex Forti, Syll. Myxoph., p. 328. 1907. *Schizothrix herbacea* Cocke, Myxoph. N. Carolina, p. 70. 1967. —Type specimen from Italy: Patavii, *Meneghini*, in herb. Kützing (L) = *Schizothrix calcicola* (Ag.) Gom.

*Leptothrix kermesina Kützing, Bot. Zeit. 5(13): 220. 1847. [*Hypheothrix lateritia* b *kermesina* Rabenhorst, ibid. 2: 85. 1865.] *H. lateritia* var. *kermesina* Rabenhorst ex Forti, ibid., p. 334. 1907. —Type specimen from the Euganean springs, Italy: Battaglia, *Meneghini*, in herb. Kützing (L) = *Schizothrix calcicola* (Ag.) Gom.

Leptothrix Kohleri Nägeli in Kützing, Sp. Algar., p. 266. 1849. *Hypheothrix Kohleri* Nägeli [ex Rabenhorst, ibid. 2: 88. 1865] ex Forti, ibid., p. 343. 1907. —Type specimen from Switzerland: Zürich, *Nägeli 281*, in herb. Kützing (L) = *Schizothrix calcicola* (Ag.) Gom.

*Leptothrix Kuehniana Rabenhorst (as "Kuehneana," but corrected in the table of contents), Alg. Sachs. 29–30: 284. 1853. *Hypheothrix Kuehniana* Rabenhorst [Fl. Eur. Algar. 2: 88. 1865] ex de Toni & Trevisan, Syll.

Schizomyc., p. 925. 1889. —Type specimen in Rabenhorst, Algen Sachsens no. 284, from Silesia, Poland: Gross-Krausche bei Bunzlau, *L. Rabenhorst* (PC) = Bacteria.

Leptothrix Kuetzingiana Nägeli in Kützing, ibid., p. 265. 1849. *Hypheothrix Kuetzingiana* Rabenhorst [Fl. Eur. Algar. 2: 89. 1865] ex Forti, Syll. Myxoph., p. 343. 1907. —Type specimen from Switzerland: Rheinfall, *Nägeli 230*, in herb. Kützing (L) = *Schizothrix calcicola* (Ag.) Gom.

Leptothrix lateritia Kützing, ibid., p. 264. 1849. *Hypheothrix lateritia* Kützing [ex Rabenhorst, ibid. 2: 84. 1865] ex Hansgirg, Prodr. Algenfl. Böhm. 2: 94. 1892. *H. lateritia* c *leptotrichoides* Grunow [in Rabenhorst, ibid. 2:85. 1865] ex Forti, ibid., p. 225. 1907. *H. lateritia* var. *leptotrichoides* Rabenhorst ex Elenkin, Monogr. Alg. Cyanoph., Pars Spec. 2: 1475. 1949. —Type specimen from Sachsen, Germany: Nordhausen, in herb. Kützing (L) = *Schizothrix calcicola* (Ag.) Gom.

Leptothrix lurida Kützing, Sp. Algar., p. 264. 1849. *Hypheothrix lurida* Rabenhorst [Fl. Eur. Algar. 2: 81. 1865] ex Gomont, Ann. Sci. Nat. VII. Bot. 15: 330. 1892. —Type specimen from Germany: Stuttgart, *Martens*, in herb. Kützing (L) = *Schizothrix calcicola* (Ag.) Gom.

Leptothrix lutea Kützing, Bot. Zeit. 5(13): 220. 1847. *Hypheothrix lutea* Rabenhorst [ibid. 2: 89. 1865] ex de Toni & Trevisan, Syll. Schizomyc., p. 929. 1889. —Type specimen from Italy: Therm. Eugan., *Meneghini 32*, in herb. Kützing (L) = *Schizothrix calcicola* (Ag.) Gom.

Leptothrix lutescens var. *Streinzii* Heufler, Sitzungsber. K. K. Zool.-bot. Ges. Wien 1853: 184. 1853. *Hypheothrix lutescens* b *Streinzii* Heufler [ex Rabenhorst, ibid. 2: 76. 1865] ex Gomont, ibid. 15: 328. 1892. *H. lutescens* var. *Streinzii* Heufler ex Forti, Syll. Myxoph., p. 333. 1907. —Type specimen from Austria: Graz, *Streinz* (W) = *Schizothrix calcicola* (Ag.) Gom.

Leptothrix mamillosa Meneghini in Kützing, Sp. Algar., p. 264. 1849. *Hypheothrix mamillosa* Rabenhorst [ibid. 2: 86. 1865] ex Gomont, loc. cit. 1892. —Type specimen from Italy: Therm. Eugan., *Meneghini*, in herb. Kützing (L) = *Schizothrix calcicola* (Ag.) Gom.

Leptothrix Meneghinii Kützing, Phyc. Germ., p. 166. 1845. *Hypheothrix Meneghinii* Kützing [Sp. Algar., p. 268. 1849] ex Gomont, ibid. 15: 309. 1892. [*H. coriacea* c *Meneghinii* Rabenhorst, ibid. 2: 83. 1865.] *H. coriacea* var. *Meneghinii* Dalla Torre & Sarnthein, Alg. Tirol, Vorarlb. & Liechtenst., p. 144. 1901. *H. coriacea* fa. *Meneghinii* Kützing ex Forti, ibid., p. 337. 1907. —Type specimen from Austria: Innsbruck, *Heufler*, in herb. Kützing (L) = *Schizothrix calcicola* (Ag.) Gom.

Leptothrix miraculosa Kützing, Phyc. Gener., p. 199. 1843. *Hypheothrix miraculosa* Kützing [ex Rabenhorst, Fl. Eur. Algar. 2: 79. 1865] ex Gomont, ibid. 15: 309. 1892. —Type specimen from Sachsen, Germany: Nordhausen, 1839—48, in herb. Kützing (L) = *Schizothrix calcicola* (Ag.) Gom.

Leptothrix muralis Kützing, Phyc. Gener., p. 199. 1843. [*Hypheothrix calcicola* b *muralis* Rabenhorst, ibid. 2: 78. 1865.] *H. muralis* Richter in Hauck & Richter, Phyk. Univ. 4: 192. 1888. —Type specimen from Sachsen, Germany: Nordhausen, 1842, in herb. Kützing (L) = *Schizothrix calcicola* (Ag.) Gom.

Leptothrix olivacea Kützing, Bot. Zeit. 5(13): 220. 1847. *Hypheothrix olivacea* Rabenhorst [ibid. 2: 77. 1865] ex Gomont, Ann. Sci. Nat. VII. Bot. 15: 328. 1892. —Type specimen from Italy: Ferrara, *Meneghini*, in herb. Kützing (L) = *Schizothrix calcicola* (Ag.) Gom.

Leptothrix purpurascens Kützing, loc. cit. 1847. *Hypheothrix purpurascens* Rabenhorst [ibid. 2: 87. 1865] ex Gomont, loc. cit. 1892. —Type specimen from Calvados, France: Falaise, *De Brébisson 200*, in herb. Kützing (L) = *Schizothrix calcicola* (Ag.) Gom.

Leptothrix purpurascens β *iridaea* Kützing, loc. cit. 1847. *Hypheothrix Braunii* γ *iridaea* Kützing (as "iridea") [Sp. Algar., p. 267. 1849] ex Gomont, ibid. 16: 166. 1892. [*H. thermalis* d *iridaea* Rabenhorst (as "iridea"), ibid. 2: 82. 1865.] —Type specimen from Baden, Germany: Baden, *Lenormand 82*, in herb. Kützing (L) = *Schizothrix calcicola* (Ag.) Gom.

Leptothrix rosea Kützing, Bot. Zeit. 5(13): 220. 1847. *Hypheothrix lateritia* a *rosea* Rabenhorst [ibid. 2: 84. 1865] ex Gomont, ibid. 15: 330. 1892. *H. lateritia* var. *rosea* Kirchner ex Forti, Syll. Myxoph., p. 334. 1907. —Type specimen from Neuchâtel, Switzerland: St. Aubin, *A. Braun 190*, Sept. 1848, in herb. Kützing (L) = *Schizothrix calcicola* (Ag.) Gom.

Leptothrix rufescens Kützing, Phyc. Gener., p. 199. 1843. *Hypheothrix rufescens* Rabenhorst [ibid. 2: 83. 1865] ex Hansgirg, Prodr. Algenfl. Böhm. 2: 96. 1892. —Type specimen from Italy: Viterbo, Jun. 1835, in herb. Kützing (L) = *Schizothrix calcicola* (Ag.) Gom.

Leptothrix tenuissima Nägeli in Kützing, Sp. Algar., p. 265. 1849. *Hypheothrix tenuissima* Rabenhorst [Fl. Eur. Algar. 2: 77. 1865] ex Gomont, Ann. Sci. Nat. VII. Bot. 15: 329. 1892. —Type specimen from Switzerland: Zürich, *Nägeli 231*, in herb. Kützing (L) = *Schizothrix calcicola* (Ag.) Gom.

Leptothrix tomentosa Kützing, Phyc. Gener., p. 199. 1843. *Hypheothrix tomentosa* Rabenhorst [ibid. 2: 80. 1865] ex Gomont, loc. cit. 1892. —Type specimen from the Euganean springs, Italy: Abano, in herb. Kützing (L) = *Schizothrix calcicola* (Ag.) Gom.

Leptothrix zonata Cesati in Rabenhorst, Alg. Sachs. 57–58: 577. 1857. *Hypheothrix zonata* Rabenhorst [Fl. Eur. Algar. 2: 78. 1865] ex Gomont, ibid. 16: 167. 1892. —Type specimen from Piemonte, Italy: in concha thermarum Valderii, *Cesati*, 1856, in Rabenhorst, Algen Sachsens no. 577 (PC) = *Schizothrix calcicola* (Ag.) Gom.

Linckia dura var. *crustacea* Lyngbye, Tent. Hydroph. Dan., p. 197. 1819. —Type specimen from Denmark: Bistrupgaard, 1817, in herb. Lyngbye (C) = *Chaetophora* sp.

*Lyngbya bosniaca Hansgirg [Sitzungsber. K. Böhm. Ges. Wiss., Math.-nat. Cl. 1891(1): 348. 1891] ex Forti, Syll. Myxoph., p. 339. 1907. *Hypheothrix bosniaca* Forti, loc. cit. 1907. —Type specimen from Bosnia, Yugoslavia: bei Doboj, *A. Hansgirg*, Aug. 1890 (W) = *Schizothrix calcicola* (Ag.) Gom.

*Lyngbya calcicola var. symplociformis Hansgirg [Bot. Centralbl. 22(10): 310. 1885] ex Hansgirg, Prodr. Algenfl. Böhm. 2: 92. 1892. *Hypheothrix calcicola* var. *symplociformis* Hansgirg ex Forti, ibid., p. 328. 1907. —Type specimen from Prague, Czechoslovakia: Smichow, *A. Hansgirg* (W) = *Schizothrix calcicola* (Ag.) Gom.

*Lyngbya calcicola var. violacea Hansgirg, Prodr. Algenfl. Böhm. 2: 93. 1892. *Hypheothrix calcicola* var. *violacea* Hansgirg ex Forti, loc. cit. 1907. —Type specimen from Bohemia, Czechoslovakia: Hermanicky, *Hansgirg*, Aug. 1886 (W) = *Schizothrix calcicola* (Ag.) Gom.

*Lyngbya halophila Hansgirg [Österr. Bot. Zeitschr. 34: 355. 1884] ex Gomont, Ann. Sci. Nat. VII. Bot. 16: 151. 1892. *Hypheothrix halophila* Migula, Krypt.-Fl. Deutschl. 2(1): 78. 1907. —Type specimen from Bohemia, Czechoslovakia: Oužič pr. Kralup, *Hansgirg*, Oct. 1883 (W) = *Schizothrix calcicola* (Ag.) Gom.

*Lyngbya lateritia var. subaeruginea Hansgirg, Prodr. Algenfl. Böhm. 2: 94. 1892. *Hypheothrix lateritia* var. *subaeruginea* Hansgirg ex Forti, Syll. Myxoph., p. 335. 1907. —Type specimen from Bohemia, Czechoslovakia: Podmoran, *Hansgirg*, Jun. 1885 (W) = *Schizothrix calcicola* (Ag.) Gom.

*Lyngbya lateritia var. symplocoides Hansgirg, Prodr. Algenfl. Böhm. 2: 94. 1892. *Hypheothrix lateritia* var. *symplocoides* Hansgirg ex Forti, loc. cit. 1907. —Type specimen from Bohemia, Czechoslovakia: Pampferhütte bei Eisenstein, *Hansgirg*, Aug. 1887 (W) = *Schizothrix calcicola* (Ag.) Gom.

Lyngbya major Kützing [Phyc. Gener., p. 226. 1843] ex Gomont, ibid. 16: 132. 1892. *Calothrix major* Kützing *pro synon.* [loc. cit. 1843] ex Bornet & Flahault, Ann. Sci. Nat. VII. Bot. 3: 371. 1886. —Type specimen from Italy: Civitavecchia, in herb. Kützing (L) = *Microcoleus lyngbyaceus* (Kütz.) Crouan.

*Lyngbya major b Brignolii Rabenhorst [Fl. Eur. Algar. 2: 140. 1865] ex Gomont, Ann. Sci. Nat. VII. Bot. 16: 133. 1892. *Leibleinia major* b *Brignolii* Rabenhorst ex Forti, ibid., p. 299. 1907. —Type specimen from Italy: in sinu Speziae, *L. Dufour*, in Rabenhorst, Algen Sachsens no. 588 (W) = *Microcoleus lyngbyaceus* (Kütz.) Crouan.

Lyngbya margaritacea Kützing [Phyc. Gener., p. 226. 1843] ex Gomont, ibid. 16: 151. 1892. *Calothrix recta* Kützing *pro synon.* [loc. cit. 1843] ex Forti, ibid., p. 271. 1907. —Type specimen from Italy: Neapel, Jul. 1835, in herb. Kützing (L) = *Microcoleus lyngbyaceus* (Kütz.) Crouan.

*Lyngbya Martensiana Meneghini [Consp. Algol. Eugan., p. 330. 1837] ex Gomont, ibid. 16: 145. 1892. [*Leibleinia Mertensiana* Kützing, Bot. Zeit. 5(12):

193. 1847.] *L. Martensiana* Kützing [Sp. Algar., p. 276. 1849] ex Gomont, loc. cit. 1892. —Type specimen from Italy: in thermis Euganeis, *G. Meneghini* (FI) = *Microcoleus lyngbyaceus* (Kütz.) Crouan.

Lyngbya mauretanica β *Gaudichaudiana* Kützing [ibid., p. 284. 1849] ex Gomont, ibid. 16: 132. 1892. [*Scytonema Gaudichaudianum* Montagne *pro synon.* in Kützing, loc. cit. 1849.] —Type specimen from Admiralty islands, Oceania: *Gaudichaud*, in herb. Kützing (L) = *Microcoleus lyngbyaceus* (Kütz.) Crouan.

Lyngbya minuta Hansgirg [Sitzungsber. K. Böhm. Ges. Wiss., Math.-nat. Cl. 1890(1): 17. 1890] ex Gomont, ibid. 16: 151. 1892. *Hypheothrix minuta* Forti, Syll. Myxoph., p. 335. 1907. —Type specimen from Istria, Yugoslavia: Orsera, *A. Hansgirg*, Apr. 1889 (W) = *Schizothrix calcicola* (Ag.) Gom.

Lyngbya nigrovaginata Hansgirg [Österr. Bot. Zeitschr. 36:110. 1886] ex Gomont, Ann. Sci. Nat. VII. Bot. 16: 151. 1892. *Hypheothrix nigrovaginata* Hansgirg, Notarisia 1887: 342. 1887. —Type specimen from Bohemia, Czechoslovakia: Solopisk nächst Cernosic, *Hansgirg*, Jun. 1888 (W) = *Schizothrix calcicola* (Ag.) Gom.

Lyngbya obscura Kützing [Phyc. Gener., p. 224. 1843] ex Gomont, ibid. 16: 128. 1892. *Scytonema obscurum* Borzi [N. Giorn. Bot. Ital. 11(4): 373. 1879] ex Forti, Syll. Myxoph., p. 263. 1907. —Type specimen from Sachsen, Germany: Salze bei Nordhausen, in herb. Kützing (L) = *Microcoleus lyngbyaceus* (Kütz.) Crouan.

Lyngbya Regeliana var. *calothrichoidea* Hansgirg, Prodr. Algenfl. Böhm. 2: 96. 1892. *Hypheothrix Regeliana* var. *calothrichoidea* Hansgirg (as "calotrichoides") ex Forti, ibid., p. 331. 1907. —Type specimen from Bohemia, Czechoslovakia: zwischen Karlstein und Beraun, *Hansgirg*, Jul. 1884 (W) = *Schizothrix calcicola* (Ag.) Gom.

Lyngbya Schowiana Kützing [ibid., p. 223. 1843] ex Gomont, ibid. 16: 138. 1892. *Leibleinia Schowiana* Crouan ex Forti, ibid., p. 299. 1907. —Type specimen from Sicily: Palermo, *Schow* (F) = *Microcoleus lyngbyaceus* (Kütz.) Crouan.

Lyngbya stragulum Kützing [loc. cit. 1843] ex Gomont, ibid. 16: 152. 1892. [*Leibleinia stragulum* Kützing *pro synon.* ex Crouan in Mazé & Schramm, Essai Class. Alg. Guadeloupe, ed. 2, p. 18. 1870—77.] —Type specimen from Italy: Neapel, Jul. 1835, in herb. Kützing (L) = *Microcoleus lyngbyaceus* (Kütz.) Crouan.

Lyngbya thermalis Kützing [Phyc. Gener., p. 223. 1843] ex Gomont, loc. cit. 1892. [*Scytonema plinianum* Borzi, N. Giorn. Bot. Ital. 11(4): 373. 1879.] —Type specimen from the Euganean springs, Italy: Battaglia, *Meneghini*, in herb. Kützing (L) = *Microcoleus lyngbyaceus* (Kütz.) Crouan.

Mastigocladus flagelliformis Schmidle (as "flagelliforme"), Allg. Bot. Zeitschr. 1900(4): 53. 1900. *Scytonema phormidioides* Schmidle, Hedwigia 39: 178. 1900. *Chondrogloea flagelliformis* Schmidle (as "flagelliforme"), Bot. Jahrb. 30: 248.

1902. *Hapalosiphon flagelliformis* Forti, Syll. Myxoph., p. 567. 1907. —In the absence of the original material, the original description is here designated as the Type = Stigonemataceae.

Mastigonema Donnellii Wolle [Bull. Torrey Bot. Club 6: 283. 1879] ex Forti, ibid., p. 629. 1907. *Calothrix Donnellii* Forti, loc. cit. 1907. —Type specimen from Florida: *J. D. Smith*, Mar. 1878 (D) = *Microcoleus lyngbyaceus* (Kütz.) Crouan.

Microchaete purpurea Schmidt, Dansk Bot. Tidsskr. 22: 379. 1899. *Fremyella purpurea* J. de Toni, Noter. di Nomencl. Algol. 8:[4]. 1936. —Type specimen from the Kattegat, Denmark: Nord for Laesø, *L. K. Rosenvinge 5600*, Jan. 1895 (C) = *Schizothrix mexicana* Gom.

Microchaete tenuissima W. & G. S. West, Journ. Linn. Soc. Bot. 30: 269. 1895. *Leptobasis tenuissima* Elenkin, Izv. Imp. Bot. Sada Petra Velik. 15: 21. 1915. *Fortiea tenuissima* J. de Toni, ibid. 8: [3]. 1936. —Type specimen, the trichomes rare among *Scytonema Hofmannii* Ag. and *Schizothrix Friesii* (Ag.) Gom., from Dominica, West Indies: *Elliott 904a*, in the slide collection (BM) = *Schizothrix calcicola* (Ag.) Gom.

*Microcoleus curvatus Meneghini [in Trevisan, Prosp. Fl. Eugan., p. 56. 1842] ex Gomont, Ann. Sci. Nat. VII. Bot. 15: 361. 1892. [*Scytonema curvatum* Borzi, N. Giorn. Bot. Ital. 11(4): 373. 1879.] —Type specimen from the Euganean springs, Italy: Abano, in herb. Kützing (L) = *Microcoleus lyngbyaceus* (Kütz.) Crouan.

Myrionema Leclancheri Harvey (as "Leclancherii"), Phyc. Brit. 1: XLI A. 1846. *Rivularia Leclancheri* Chauvin (as "Leclancherii") *pro synon.* in Harvey, loc. cit. 1846. —Type specimen presumably from England: "Harvey to Mrs. Griffiths. Spec. auth." (BM) = Phaeophycea.

*[Nematococcus viridis Kützing, Linnaea 8: 381. 1833.] *Hypheothrix subtilissima* Rabenhorst [Fl. Eur. Algar. 2: 77. 1865] ex Gomont, ibid. 16: 169. 1892. —Type specimen from Germany: Merseburg, in herb. Kützing (L) = *Schizothrix calcicola* (Ag.) Gom.

Nostoc Quoyi Agardh (as "Quoji") [Syst. Algar., p. 22. 1824] ex Bornet & Flahault, Ann. Sci. Nat. VII. Bot. 4: 373. 1887. [*Undina Quoyi* Fries (as "Quoji"), Syst. Orb. Veget. 1: 348. 1825.] *Hormactis Quoyi* Bornet & Thuret [Not. Algol. 2: 173. 1880] ex Bornet & Flahault, loc. cit. 1887. *H. Farlowii* Bornet [in Farlow, Mar. Alg. N. Engl., p. 40. 1881] ex Bornet & Flahault, loc. cit. 1887. *Brachytrichia Quoyi* Bornet & Flahault [Mém. Soc. Nat. Sci. Math. & Nat. Cherbourg 25: 208. 1885] ex Bornet & Flahault, Ann. Sci. Nat. VII. Bot. 4: 373. 1886. —Type specimen from Oceania: Marianas, in herb. Agardh (LD) = Stigonemataceae.

*Oscillatoria calcicola Agardh [Disp. Alg. Suec., p. 37. 1812] ex Gomont, Ann. Sci. Nat. VII. Bot. 15: 307. 1892. *Hypheothrix calcicola* Rabenhorst [Fl. Eur. Algar. 2: 78. 1865] ex Gomont, loc. cit. 1892. —Type specimen from Sweden:

Lund, 1812, in herb. Agardh (LD) = *Schizothrix calcicola* (Ag.) Gom.

*Oscillatoria Friesii Agardh [Syn. Alg. Scand., p. 107. 1817] ex Gomont, ibid. 15: 316. 1892. *Scytonema Friesii* Montagne & Fries [in Montagne, Ann. Sci. Nat. II. Bot. 6: 327. 1836] ex Bornet & Flahault, ibid. 5: 113. 1887. —Type specimen from Sweden: "misit Fries," in. herb. Agardh (LD) = *Schizothrix Friesii* (Ag.) Gom.

*[Oscillatoria jadrensis Zanardini (as "Oscillaria"), Sagg. Class. Nat. Ficee, p. 64. 1843.] *Hypheothrix jadertina* Kützing [ex Rabenhorst, ibid. 2: 89. 1865] ex Forti, Syll. Myxoph., p. 335. 1907. —Type specimen from Dalmatia, Yugoslavia: Zara, *Meneghini*, in herb. Kützing (L) = *Porphyrosiphon miniatus* (Hauck) Dr.

*Oscillatoria laminosa Agardh [Flora 10(1:40): 633. 1827] ex Gomont, ibid. 16: 168. 1892. *Hypheothrix laminosa* Rabenhorst [ibid. 2: 79. 1865] ex Gomont, ibid. 16: 222. 1892. *H. compacta* Rabenhorst [loc. cit. 1865] ex Gomont, ibid. 16: 114. 1892. —Type specimen from Czechoslovakia: Carlsbad, in herb. Agardh (LD) = *Hapalosiphon laminosus* (Ag.) Hansg.

*Oscillatoria Mucor Agardh [Algar. Dec. 3: 36. 1814] ex Gomont, ibid. 16: 244. 1892. *Calothrix Mucor* Agardh [Syst. Algar., p. 70. 1824] ex Bornet, Bull. Soc. Bot. France 36: 154. 1889. [*Leibleinia Mucor* Zanardini, Not. Intorno alle Cellul. Mar. d. Lag. & Lit. Venezia, p. 79. 1847.] [*Calothrix Nemalionis* Zanardini *pro synon.*, loc. cit. 1847.] —Type specimen from Sweden: Båstad, in herb. Agardh (LD) = *Nodularia spumigena* Mert.

*Oscillatoria muscorum Agardh [ibid., p. 65. 1824] ex Gomont, Ann. Sci. Nat. VII. Bot. 16: 110. 1892. *Tolypothrix tenuis* var. *tingens* Kreischer *pro synon.* ex Gomont, loc. cit. 1892. —Type specimen from Sweden: Jäder, in herb. Agardh (LD) = *Oscillatoria Retzii* Ag.

Oscillatoria setigera Aptekar, Dnevn. Vses. S'ezd Bot. Leningrad 1928: 139. 1928; Zapisk. Dnepropetrovsk. Inst. Narodn. Osviti 2: 203. 1928. *Raphidiopsis setigera* Eberly, Trans. Amer. Microsc. Soc. 85(1): 135. 1966. —In the absence of the original material, the original description is here designated as the Type = *Schizothrix tenerrima* (Gom.) Dr.

*Oscillatoria tapetiformis Zenker [Linnaea 9: 125. 1835] ex Gomont, ibid. 15: 310. 1892. *Hypheothrix Zenkeri* Kützing [Sp. Algar., p. 268. 1849] ex Gomont, loc. cit. 1892. *H. Zenkeri* var. *cobaltina* Rabenhorst [Alg. Sachs. 6—7: 66. 1850] ex Gomont, ibid. 15: 309. 1892. —Type specimen from Germany: bei Jena, *Schlechtendal*, in herb. Kützing (L) = *Schizothrix calcicola* (Ag.) Gom.

*Phormidium amoenum Kützing [Phyc. Gener., p. 192. 1843] ex Gomont, Ann. Sci. Nat. VII. Bot. 16: 225. 1892. *Hypheothrix amoena* Hansgirg ex Dalla Torre & Sarnthein, Alg. Tirol, Vorarlb. & Liechtenst., p. 142. 1901. —Type specimen from Italy: Patavii, *Meneghini*, in herb. Kützing (L) = *Microcoleus vaginatus* (Vauch.) Gom.

*Phormidium inundatum Kützing [ibid., p. 193. 1843] ex Gomont, ibid. 16:

172. 1892. *Hypheothrix inundata* Hansgirg ex Dalla Torre & Sarnthein, loc. cit. 1901. —Type specimen from Sachsen, Germany: Nordhausen, in herb. Kützing (L) = *Schizothrix rubella* Gom.

Hormactis Thuret [Ann. Sci. Nat. VI. Bot. 1: 376. 1875] ex Bornet & Flahault, Ann. Sci. Nat. VII. Bot. 4: 372. 1887. —Type species: *Physactis Lloydii* Crouan [Bull. Soc. Bot. France 7: 836. 1860] and Kützing [Ostern-Progr. Realsch. Nordhausen 1862–63: 9. 1863] ex Bornet & Flahault, loc. cit. 1887. *Rivularia Balanii* Lloyd *pro synon.* in Crouan, loc. cit. 1860. *R. Lloydii* Crouan, Fl. Finistère, p. 117. 1867. *Hormactis Balanii* Thuret (as "Balani") [ibid., p. 382. 1875] ex Bornet & Flahault, loc. cit. 1887. *Brachytrichia Balanii* Bornet & Flahault (as "Balani") [Mém. Soc. Nat. Sci. Nat. & Math. Cherbourg 25: 208. 1885] ex Bornet & Flahault, loc. cit. 1887. —Type specimen from France: Belle-Ile, Sept. 1854, in Lloyd, Algues de l'Ouest de la France no. 303, in herb. Bornet—Thuret (PC) = *Brachytrichia Quoyi* (Ag.) Born. & Flah.

Plectonema adriaticum Vouk (as "adriatica"), Rad Jugosl. Akad., Mat.-prirod. Razr. 254(79): 6. 1936. —In the absence of the original material, the original description is here designated as the Type = *Schizothrix mexicana* Gom.

Plectonema africanum Borge, Hedwigia 68: 107. 1928. —Type specimen from Tanzania: *Schroeder 2* (S) = *Porphyrosiphon Notarisii* (Menegh.) Kütz.

Plectonema andinum Schwabe, N. Hedwigia 2(1–2): 257. 1960. —In the absence of the original material, the original description is here designated as the Type = *Schizothrix calcicola* (Ag.) Gom.

Plectonema araucanum Schwabe, ibid. 2(1–2): 255. 1960. —In the absence of the original material, the original description is here designated as the Type = *Microcoleus vaginatus* (Vauch.) Gom.

Plectonema Batrachospermii Starmach, Acta Soc. Bot. Polon. 26(3): 568. 1957. —In the absence of the original material, the original description is here designated as the Type = *Oscillatoria lutea* Ag.

Plectonema Battersii Gomont, Bull. Soc. Bot. France 46: 36. 1899. —Type specimen from Northumberland, England: Berwick, *E. Batters,* Oct. 1884, in herb. Gomont (PC) = *Schizothrix calcicola* (Ag.) Gom.

Plectonema Boryanum Gomont, loc. cit. 1899. —Type specimen from France: flaçon d'eau destilé, *Bory,* in herb. Gomont (PC) = *Schizothrix calcicola* (Ag.) Gom.

Plectonema Boryanum fa. *Hollerbachianum* Elenkin, Monogr. Alg. Cyanoph., Pars Spec. 2: 1787. 1949. —In the absence of the original material, the original description is here designated as the Type = *Schizothrix calcicola* (Ag.) Gom.

Plectonema calothrichoides Gomont, ibid. 46: 30. 1899. —Type specimen from Massachusetts: Nahant, *F. S. Collins,* Jun. 1889, in herb. Gomont (PC) = *Schizothrix calcicola* (Ag.) Gom.

Plectonema capitatum Jaag (as "capitata"), Ber. Schweiz. Bot. Ges. 44: 437. 1935. *P. Jaagii* Geitler in Engler & Prantl, Natürl. Pflanzenfam., ed. 2, 1b: 157.

1942. —In the absence of the original material, the original description is here designated as the Type = *Schizothrix calcicola* (Ag.) Gom.

Plectonema capitatum Lemmermann, Engler Bot. Jahrb. 38: 353. 1907. —In the absence of the original material, the original description is here designated as the Type = *Microcoleus lyngbyaceus* (Kütz.) Crouan.

Plectonema Cloverianum Drouet, Field Mus. Bot. Ser. 20(6): 134. 1942. —Type specimen from Utah: above Dark Canyon rapids, Colorado river, *E. U. Clover & L. Jotter 36*, Jul. 1938 (MICH) = *Schizothrix calcicola* (Ag.) Gom.

Plectonema congregationis Schwabe, Rev. Algol., N. S. 6(2): 87. 1962. —Type specimen from Switzerland: culture from Julierpass, *P. Bourrelly* (collection of G. H. Schwabe) = *Schizothrix calcicola* (Ag.) Gom.

Plectonema crispatum Playfair, Proc. Linn. Soc. New South Wales 40: 350. 1916. —Type specimen from New South Wales, Australia: Lismore, *Playfair 210* (NSW) = *Schizothrix calcicola* (Ag.) Gom.

Plectonema Dangeardii Frémy, Arch. de Bot. Caen 3(Mém. 2): 175. 1929. —In the absence of the original material, the original description is here designated as the Type = *Schizothrix calcicola* (Ag.) Gom.

Plectonema desmidiacearum Noda, Sci. Rep. Niigata Univ., ser. D (Biol.), 8: 61. 1971. —Type specimen from Niigata pref., Japan: pond, Yoshida, *M. Noda*, Oct. 1969 (collection of M. Noda) = *Schizothrix calcicola* (Ag.) Gom.

Plectonema diplosiphon Voronikhin, Not. Syst. Inst. Crypt. Horti Bot. Petropol. 2(8): 114. 1923. —In the absence of the original material, the original description is here designated as the Type = *Schizothrix calcicola* (Ag.) Gom.

Plectonema endolithicum Ercegović, Rad Jugosl. Akad., Mat.-prirod. Razr. 244(75): 159. 1932. —In the absence of the original material, the original description is here designated as the Type = *Schizothrix calcicola* (Ag.) Gom.

Plectonema flexuosum Gardner, Mem. New York Bot. Gard. 7: 47. 1927. —Type specimen from Puerto Rico: Coamo Springs, *N. Wille 272b*, Jan. 1915 (NY) = *Schizothrix rubella* Gom.

Plectonema Fortii Frémy, Arch. de Bot. Caen 3(Mém. 2): 171. 1929. —In the absence of the original material, the original description is here designated as the Type = *Porphyrosiphon Notarisii* (Menegh.) Kütz.

Plectonema gloeophilum Borzi, N. Notarisia 3: 40. 1892. —In the absence of the original material, the original description is here designated as the Type = *Schizothrix calcicola* (Ag.) Gom.

Plectonema Golenkinianum Gomont, Bull. Soc. Bot. France 46: 35. 1899. —Type specimen from Italy: culture, Naples, *Golenkin*, 1894 (PC) = *Schizothrix calcicola* (Ag.) Gom.

Plectonema Golenkinianum fa. *Anissimovianum* Elenkin, Monogr. Alg. Cyanoph., Pars Spec. 2: 1800. 1949. —In the absence of the original material, the original description is here designated as the Type = *Schizothrix calcicola* (Ag.) Gom.

Plectonema gracillimum fa. *aquaticum* Elenkin, ibid. 2: 1793. 1949. —In the absence of the original material, the original description is here designated as the Type = *Schizothrix calcicola* (Ag.) Gom.

Plectonema Hansgirgii Schmidle, Hedwigia 39: 186. 1900. —In the absence of the original material, the original description is here designated as the Type = *Schizothrix rubella* Gom.

Plectonema indicum Dixit (as "indica"), Proc. Indian Acad. Sci. 3(1): 99. 1936. —In the absence of the original material, the original description is here designated as the Type = *Schizothrix Friesii* (Ag.) Gom.

Plectonema litorale Anand (as "litoralis"), Journ. of Bot. 75(Suppl. 2): 44. 1937. —In the absence of the original material, the original description is here designated as the Type = *Schizothrix calcicola* (Ag.) Gom.

Plectonema Monodii Compère, Bull. Inst. Franç. Afr. Noire 22(A, 1): 32. 1970. —Type specimen from Massif de l'Ennedi, Tchad: bords de l'Enneri Ouro Gale, *T. Monod 13879*, Jan. 1967 (BR) = *Porphyrosiphon Notarisii* (Menegh.) Kütz.

Plectonema murale Gardner, Mem. New York Bot. Gard. 7: 47. 1927. —Type specimen from Puerto Rico: Fajardo, *N. Wille 732c*, Jan. 1915 (NY) = *Schizothrix calcicola* (Ag.) Gom.

Plectonema norvegicum Gomont, Bull. Soc. Bot. France 46: 34. 1899. —Type specimen from arctic Norway: Ingö, *M. Foslie*, Jul. 1891 (PC) = *Schizothrix calcicola* (Ag.) Gom.

Plectonema Nostocorum Bornet [in Bornet & Thuret, Not. Algol. 2: 137. 1880] ex Gomont, Ann. Sci. Nat. VII. Bot. 16: 102. 1892. —Type specimen from France: Paris, été 1880, in herb. Bornet—Thuret (PC) = *Schizothrix calcicola* (Ag.) Gom.

Plectonema Nostocorum fa. *discolor* Cedercreutz, Mem. Soc. Fauna & Fl. Fenn. 5: 155. 1929. —In the absence of the original material, the original description is here designated as the Type = *Schizothrix calcicola* (Ag.) Gom.

Plectonema Nostocorum fa. *majus* Pandey (as "major"), N. Hedwigia 10(1—2): 199. 1965. —In the absence of the original material, the original description is here designated as the Type = *Schizothrix calcicola* (Ag.) Gom.

Plectonema notatum Schmidle in Simmer, Allg. Bot. Zeitschr. 7(5): 84. 1901. *P. terebrans* var. *notatum* Schmidle (as "notata") *pro synon.* in Simmer, loc. cit. 1901. —Type specimen from Kärnten, Austria: Kl. Knoten, Kreuzeckgebiet, *H. Simmer*, Jun. 1898 (W) = *Schizothrix calcicola* (Ag.) Gom.

Plectonema notatum var. *africanum* Fritsch & Rich, Trans. R. Soc. S. Afr. 18: 90. 1929. —Type specimen from Griqualand West, South Africa: tank near Kimberley, *M. Wilman 822*, Jan. 1923 (BM) = *Schizothrix calcicola* (Ag.) Gom.

Plectonema notatum fa. *longearticulatum* Geitler & Ruttner, Arch. f. Hydrobiol. Suppl.-Bd. 14(6): 444. 1935. —In the absence of the original material, the original description is here designated as the Type = *Schizothrix calcicola* (Ag.) Gom.

Plectonema notatum var. *Saegeri* Drouet, Bot. Gaz. 95(4): 699. 1934.

—Type specimen from Missouri: culture, Columbia, *A. C. Saeger & F. Drouet 1050*, Dec. 1932 (D) = *Schizothrix calcicola* (Ag.) Gom.

Plectonema notatum fa. *Woronichinianum* Elenkin, Monogr. Alg. Cyanoph., Pars Spec. 2: 1790. 1949. —In the absence of the original material, the original description is here designated as the Type = *Schizothrix calcicola* (Ag.) Gom.

Plectonema polymorphum Schwabe, N. Hedwigia 2(1–2): 251. 1960. —Type specimen from Cautin, Chile: Termas Lonquén, culture, 1958 (collection of G. H. Schwabe) = *Schizothrix calcicola* (Ag.) Gom.

Plectonema polymorphum var. *viride* Schwabe (as "viridis"), Österr. Bot. Zeitschr. 107(3–4): 291. 1960. —Type specimen from Atacama, Chile: culture from Caldera, *1289,3010* (collection of G. H. Schwabe) = *Schizothrix calcicola* (Ag.) Gom.

Plectonema purpureum Gomont, Ann. Sci. Nat. VII. Bot. 16: 101. 1892. —Type specimen from Hérault, France: Courpoiran pr. Montpellier, *J. Huber*, 1892 (PC) = *Schizothrix calcicola* (Ag.) Gom.

Plectonema purpureum fa. *edaphicum* Melnikova, Bot. Mat. Otd. Sporov. Rast. Bot. Inst. Komarova Akad. Nauk SSSR 9: 70. 1953. —In the absence of the original material, the original description is here designated as the Type = *Oscillatoria lutea* Ag.

Plectonema purpureum fa. *pauciramosum* Anissimova in Elenkin, Monogr. Alg. Cyanoph., Pars. Spec. 2: 1798. 1949. —In the absence of the original material, the original description is here designated as the Type = *Oscillatoria lutea* Ag.

Plectonema puteale fa. *edaphicum* Elenkin, ibid. 2: 1782. 1949. *P. edaphicum* Vaulina, Bot. Mat. Otd. Sporov. Rast. Bot. Inst. Komarova Akad. Nauk SSSR 12(1): 19. 1959. —In the absence of the original material, the original description is here designated as the Type = *Schizothrix calcicola* (Ag.) Gom.

Plectonema puteale fa. *muscicola* Elenkin, loc. cit. 1949. —In the absence of the original material the original description is here designated as the Type = *Porphyrosiphon Notarisii* (Menegh.) Kütz.

Plectonema rhenanum Schmidle, Hedwigia 36: 19. 1897. *Tolypothrix rhenana* Schmidle ex Royers, Jahres-Ber. Naturw. Ver. Elberfeld 10: 83. 1903. —In the absence of the original material, the original description is here designated as the Type = *Microcoleus vaginatus* (Vauch.) Gom.

Plectonema rhenanum fa. *scytonemiforme* Schwabe (as "scytonemaeforme"), N. Hedwigia 1(1–2): 256. 1960. —In the absence of the original material, the original description is here designated as the Type = *Schizothrix mexicana* Gom.

Plectonema rugosum Jao, Bot. Bull. Acad. Sinica 1: 71. 1947. —In the absence of the original material, the original description is here designated as the Type = *Schizothrix calcicola* (Ag.) Gom.

Plectonema Schmidlei Limanowska, Arch. f. Hydrobiol. & Planktonk. 7: 364. 1911. —In the absence of the original material, the original description is here designated as the Type = *Schizothrix calcicola* (Ag.) Gom.

Plectonema spelaeoides Čado, Zborn. na Rabotite Prirod.-Mat. Fak. Univ.

Skopje Khidrobiol. Zavod, Okhrid, 7(5): 3, 6. 1959. —In the absence of the original material, the original description is here designated as the Type = *Schizothrix calcicola* (Ag.) Gom.

Plectonema spirale Gardner, Mem. New York Bot. Gard. 7: 46. 1927. —Type specimen from Puerto Rico: Maricao, *N. Wille 1276b*, Feb. 1915 (NY) = *Schizothrix calcicola* (Ag.) Gom.

Plectonema spirale Schwabe & El Ayouty, N. Hedwigia 10(3–4): 528. 1965. —In the absence of the original material, the original description is here designated as the Type = *Schizothrix calcicola* (Ag.) Gom.

Plectonema spongiosum Schwabe, N. Hedwigia 4(3–4): 523. 1962. —In the absence of the original material, the original description is here designated as the Type = *Schizothrix calcicola* (Ag.) Gom.

Plectonema subtile Gayral & Seizilles de Mazancourt, Bull. Soc. Bot. France 105(7–8): 344. 1958. —In the absence of the original material, the original description is here designated as the Type = *Schizothrix calcicola* (Ag.) Gom.

Plectonema tauricum Voronikhin, Journ. Bot. de l'URSS 17(3): 313. 1932. —In the absence of the original material, the original description is here designated as the Type = *Microcoleus vaginatus* (Vauch.) Gom.

Plectonema tenue Thuret [in Bornet & Thuret, Not. Algol. 2: 137. 1880] ex Gomont, Ann. Sci. Nat. VII. Bot. 16: 101. 1892. —Type specimen from France: Antibes, *G. Thuret*, Dec. 1874 (PC) = *Porphyrosiphon Notarisii* (Menegh.) Kütz.

Plectonema tenuissimum Gardner, Mem. New York Bot. Gard. 7: 47. 1927. —Type specimen from Puerto Rico: Maricao, *N. Wille 1049b*, Feb. 1915 (NY) = *Schizothrix calcicola* (Ag.) Gom.

Plectonema c *Terebra* Forti, Syll. Myxoph., p. 497. 1907. —Type species: *Plectonema terebrans* Bornet & Flahault, Bull. Soc. Bot. France 36(Congr. Bot.): clxiii. 1889. —Type specimen from Nièvre, France: Cosne, Aug. 1889, in herb. Bornet–Thuret (PC) = *Schizothrix calcicola* (Ag.) Gom.

Plectonema terebrans fa. *Hansgirgianum* Forti, ibid., p. 498. 1907. —Type specimen from Dalmatia, Yugoslavia: Ragusa, *A. Hansgirg*, 1891 (W) = *Schizothrix calcicola* (Ag.) Gom.

Plectonema Tomasinianum var. *gracile* Hansgirg, Sitzungsber. Böhm. Ges. Wiss., Math.-nat. Cl. 1891: 338. 1891. *P. Tomasinianum* fa. *gracile* Poliansky in Hollerbach, Kossinskaia & Poliansky, Opred. Presn. Vodor. SSSR 2: 595. 1953. —Type specimen from Dalmatia, Yugoslavia: Gruda–Castellnuovo, *A. Hansgirg*, 1891 (W) = *Porphyrosiphon Notarisii* (Menegh.) Kütz.

Plectonema Tomasinianum fa. *hyalinum* Chu (as "hyalina"), Sinensia 15(1–6): 156. 1944. —In the absence of the original material, the original description is here designated as the Type = *Porphyrosiphon Notarisii* (Menegh.) Kütz.

Plectonema Tomasinianum var. *indicum* Sampaio, Garcia de Orta (Lisbôa) 10(2): 334. 1962. —Type specimen from Portuguese India: planalto de Mormugão, *C. Teixeira*, Jul. 1960 (PO) = *Porphyrosiphon Notarisii* (Menegh.) Kütz.

Plectonema Tomasinianum var. *vandalurense* Desikachary, Cyanoph., p. 618. 1959. —In the absence of the original material, the original description is here designated as the Type = *Porphyrosiphon Notarisii* (Menegh.) Kütz.

*Plectonema Wollei Farlow [Bull. Bussey Inst. 1877: 77. 1877] ex Gomont, Ann. Sci. Nat. VII. Bot. 16: 98. 1892. —Type specimen from Pennsylvania: Bethlehem, *F. Wolle* (FH) = *Oscillatoria princeps* Vauch.

Plectonema Wollei fa. *gracile* Frémy (as "gracilis"), Arch. de Bot. Caen 3(Mém. 2): 168. 1929. —In the absence of the original material, the original description is here designated as the Type = *Oscillatoria princeps* Vauch.

Plectonema Wollei fa. *robustissimum* Frémy (as "robustissima"), loc. cit. 1929. —In the absence of the original material, the original description is here designated as the Type = *Oscillatoria princeps* Vauch.

Plectonema Wollei fa. *robustissimum* subfa. *violaceum* Frémy (as "violacea"), ibid. 3(Mém. 2): 171. 1930. —In the absence of the original material, the original description is here designated as the Type = *Oscillatoria princeps* Vauch.

Plectonema Wollei fa. *robustum* G. S. West (as "robusta"), Journ. Linn. Soc. Bot. 38: 173. 1907. —Type specimen from Tanzania: Komba bay, Lake Tanganyika, *Cunnington 134*, Oct. 1904, in the slide collection (BM) = *Oscillatoria princeps* Vauch.

Plectonema Wollei fa. *robustum* subfa. *violaceum* Frémy (as "violacea"), ibid. 3(Mém. 2): 170. 1929. —In the absence of the original material, the original description is here designated as the Type = *Oscillatoria princeps* Vauch.

Plectonema yellowstonense Prát, Studie o Biolith., p. 99. 1929. —Type specimen, here designated, from near the original locality in Yellowstone national park, Wyoming: White Elephant Caves, Mammoth hot springs, *W. A. Setchell 2008*, Aug. 1898 (D) = *Schizothrix calcicola* (Ag.) Gom.

Racodium turfaceum Persoon, Mycol. Eur. 1: 68. 1822. *Scytonema turfaceum* Berkeley [in Smith & Sowerby, Suppl. Engl. Bot. 3: 2826, fig. 1. 1843] ex Bornet & Flahault, Ann. Sci. Nat. VII. Bot. 5: 74. 1887. *Hassallia turfosa* Hassall [Hist. Brit. Freshw. Alg. 1: 232. 1845] ex Bornet & Flahault, loc. cit. 1887. —Type specimen from Switzerland: "Racodium turfaceum," in herb. Persoon (L) = *Stigonema ocellatum* (Dillw.) Thur.

Rivularia confervoides Roth, Catal. Bot. 1: 213. 1797. —Type specimen presumably from Germany: "R. confervoides ex Mertens," in herb. Hooker (K in BM) = *Chaetophora incrassata* (Huds.) Haz.

Rivularia Roth, ibid. 1: 212. 1797; Bemerk. ü. d. Stud. d. Crypt. Wassergew., p. 55. 1797. *Polycoma* Palisot de Beauvois ex Desvaux, Journ. de Bot. 1: 124. 1808. —Type species: *Rivularia cornu-damae* Roth, Catal. Bot. 1: 212. 1797. —Type specimen from Germany: "R. cornu-damae Roth, dedit Roth ipse," in herb. Agardh (LD) = *Chaetophora incrassata* (Huds.) Haz.

Rivularia cylindrica Wahlenberg ex Hooker, Journ. of a Tour in Iceland, p. 71. 1811. —Type specimen from Iceland: 1809, in herb. Hooker (K in BM) = *Tetraspora cylindrica* (Wahlenb.) Ag.

Rivularia elegans Roth, N. Beytr. z. Bot. 1: 269. 1802. —Type specimen from Germany: "R. elegans dedit Roth ipse," in herb. Agardh (LD) = *Chaetophora elegans* (Roth) Ag.

Rivularia elongata Roth, Catal. Bot. 3: 332. 1806. —Type specimen presumably from England: "R. elongata, *Turner*," in herb. Agardh (LD) = *Chaetophora incrassata* (Huds.) Haz.

Rivularia endiviifolia Roth (as "endiviaefolia"), Arch. f. d. Bot. 1(3): 51. 1798. —Type specimen from Germany: "R. endiviaefolia, dedit Roth ipse," in herb. Agardh (LD) = *Chaetophora incrassata* (Huds.) Haz.

Rivularia fucicola Roth, Catal. Bot. 3: 334. 1806. —In the absence of the original material, the original description is here designated as the Type = Rhodophycea.

Rivularia Halleri Lamarck & De Candolle, Fl. Franç., ed. 3, 2: 5. 1805. —Type specimen from France: Doubs, *Chaillet 196*, in herb. De Candolle (G) = *Palmella myosurus* (Ducluz.) Lyngb.

Rivularia multifida Weber & Mohr, Naturh. Reise Durch e. Th. Schwedens, p. 193. 1804. *R. Zosterae* Mohr in Weber, Beitr. z. Naturk. 2: 367. 1810. *R. frondosa* Roth *pro synon.* ex Weber, loc. cit. 1810. —In the absence of the original material, the original description is here designated as the Type = Rhodophycea.

Rivularia pisiformis Roth, N. Beytr. z. Bot. 1: 272. 1802. —In the absence of the original material, the original description is here designated as the Type = *Chaetophora* sp.

Isactis Thuret, Ann. Sci. Nat. VI. Bot. 1: 376. 1875. *Rivularia* Sect. *Isactis* Thuret in Bornet & Thuret, Not. Algol. 2: 165. 1880. —Type species: *Rivularia plana* Harvey in Hooker, Brit. Fl. 2(Engl. Fl. 5): 394. 1833. [*Scytochloria plana* Harvey *pro synon.* in Hooker, loc. cit. 1833.] *Dasyactis plana* Kützing [Tab. Phyc. 2: 23. 1852] ex Bornet & Flahault, Ann. Sci. Nat. VII. Bot. 4: 344. 1887. *Mastigonema planum* Rabenhorst (as "plana") [Fl. Eur. Algar. 2: 226. 1865] ex Bornet & Flahault, loc. cit. 1887. *Isactis plana* Thuret, ibid. 1: 382. 1875. *Calothrix plana* Poliansky in Kossinskaia, Opred. Morsk. Sinez. Vodor., p. 148. 1948. —Type specimen from Ireland: Malbay, in herb. Harvey (TCD) = *Petrocelis* sp.

Rivularia rosea Suhr, Flora 17(1:14): 209. 1834. —Type specimen from Germany: Gellinger Bucht, Ostsee, *Suhr*, in herb. Kützing (L) = Rhodophycea.

Rivularia rugosa Roth, ibid. 1: 280. 1802. —Type specimen from Germany: "Linckia rugosa Trentepohl," in herb. Agardh (LD) = *Anacystis montana* fa. *minor* (Wille) Dr. & Daily.

Rivularia stellata Suhr, ibid. 17(1:14): 210. 1834. —Type specimen: Mare baltico, *Suhr* (C) = Phaeophycea.

Rivularia tuberculosa Roth, N. Beytr. f. Bot. 1: 285. 1802. —In the absence of the original material, the original description is here designated as the Type = *Chaetophora* sp.

Rivularia tuberiformis Smith & Sowerby, Engl. Bot., pl. 1956. 1808. —In the absence of the original material, the original description is here designated as the Type = Phaeophycea.

Rivularia vermiculata Smith & Sowerby, ibid., pl. 1818. 1807. —Type specimen, a later collection from the original locality by the original collector, from Ireland: Larne, *J. L. Drummond*, Jul. 1833 (BM) = Rhodophycea.

Rivularia verrucosa Roth, ibid. 1: 281. 1802. —Type specimen from Germany: prope Oldenburg, *Trentepohl*, in herb. Agardh (LD) = *Nostoc* sp.

*Schizosiphon affinis Meneghini [in Kützing, Sp. Algar., p. 327. 1849] ex Bornet & Flahault, Ann. Sci. Nat. VII. Bot. 3: 366. 1886. —Type specimen from Italy: Patavii, *Meneghini*, in herb. Kützing (L) = *Schizothrix calcicola* (Ag.) Gom.

Schizosiphon apiculatus Kützing [loc. cit. 1849] ex Forti, Syll. Myxoph., p. 633. 1907. —Type specimen from Italy: Neapel, in mari, in herb. Kützing (L) = *Microcoleus lyngbyaceus* (Kütz.) Crouan.

Schizosiphon lasiopus Kützing [Phyc. Gener., p. 234. 1843] ex Bornet & Flahault, ibid. 3: 359. 1886. —Type specimen from Italy: Civitavecchia, in herb. Kützing (L) = *Microcoleus lyngbyaceus* (Kütz.) Crouan.

[*Schizosiphon littoralis* Crouan, Bull. Soc. Bot. France 7: 836. 1860.] —Type specimen, here designated, from Finistère, France: Pantarvelin, Sept. 1848, in herb. Crouan (Lab. Mar. Biol., Concarneau) = *Schizothrix arenaria* (Berk.) Gom.

Schizosiphon major Kützing [Sp. Algar., p. 328. 1849] ex Forti, ibid., p. 269. 1907. —Type specimen from the Netherlands: Zeeland, *van den Bosch*, in herb. Kützing (L) = *Microcoleus lyngbyaceus* (Kütz.) Crouan.

Schizosiphon nigrescens Hilse [Jahres-Ber. Schles. Ges. f. Vaterl. Cultur 1864: 94. 1865] ex Bornet & Flahault, ibid. 3: 373. 1886. —Type specimen from Silesia, Poland: Schottwitz bei Breslau, *Hilse*, Oct. 1864, in Rabenhorst, Algen Europas no. 1835 (W) = *Schizothrix Friesii* (Ag.) Gom.

[*Schizosiphon sabulicola* fa. *ericetorum* Rabenhorst, Fl. Eur. Algar. 2: 236. 1865.] —Type specimen, here designated, from Silesia, Poland: auf feuchter Erde, Gross-Lauden bei Strehlen, *Hilse*, 1860, in Rabenhorst, Algen Europas no. 1040 (FH) = *Fischerella ambigua* (Näg.) Gom.

Schizothrix affinis Lemmermann, Abh. Nat. Ver. Bremen 18(1): 153. 1904. *Hypheothrix affinis* Forti, Syll. Myxoph., p. 341. 1907. —In the absence of the original material, the original description is here designated as the Type = *Schizothrix calcicola* (Ag.) Gom.

Schizothrix calida De Wildeman, Ann. Jard. Bot. Buitenzorg, Suppl. 1: 36. 1897. *Hypheothrix calida* Forti, ibid., p. 340. 1907. —In the absence of the original material, the original description is here designated as the Type = *Schizothrix calcicola* (Ag.) Gom.

Schizothrix delicatissima W. & G. S. West, Journ. of Bot. 35: 269. 1897. *Hypheothrix delicatissima* Forti, ibid., p. 341. 1907. —Type specimen from Angola: Pungo Andongo, *F. Welwitsch 151,* Jan. 1857 (BM) = *Schizothrix calcicola* (Ag.) Gom.

Schizothrix ericetorum Lemmermann, Krypt.-Fl. Mark Brandenb. 3(1): 153. 1907. *Scytonema coerulescens* Dickie *pro synon.* ex Durairatnam & Amaratunga, Bull. Fish. Res. Sta. Ceylon 20(2): 154. 1969. —Type specimen from Brandenburg, Germany: Triglitz, *O. Jaap 672,* Apr. 1897 (BREM) = *Schizothrix Friesii* (Ag.) Gom.

Schizothrix fuscescens Kützing [Phyc. Gener., p. 230. 1843] ex Gomont, Ann. Sci. Nat. VII. Bot. 15: 324. 1892. *Scytonema muscicola* Dickie *pro synon.* ex Durairatnam & Amaratunga, loc. cit. 1969. —Type specimen from Jutland, Denmark: Markerup-Moor. *Frölich,* Feb. 1831, in herb. Kützing (L) = *Schizothrix Friesii* (Ag.) Gom.

Schizothrix Heufleri Grunow [in Rabenhorst, Fl. Eur. Algar. 2: 270. 1865] ex Gomont, ibid. 15: 325. 1892. *Hypheothrix Heufleri* Dalla Torre & Sarnthein, Alg. Tirol, Vorarlb. & Liechtenst., p. 142. 1901. —Type specimen from Tirol, Austria: Kienbachsthal bei Kufstein, *Heufler,* in herb. Grunow (W) = *Schizothrix calcicola* (Ag.) Gom.

Schizothrix lateritia fa. *lyngbyacea* Schmidle, Hedwigia 39: 186. 1900. *Hypheothrix lateritia* fa. *lyngbyacea* Schmidle ex Forti, Syll. Myxoph., p. 334. 1907. —In the absence of the original material, the original description is here designated as the Type = *Schizothrix tenerrima* (Gom.) Dr.

Schizothrix Lenormandiana Gomont, ibid. 15: 312. 1892. *Hypheothrix Lenormandiana* Forti, ibid., p. 341. 1907. —Type specimen, presumably from Calvados, France: *Lenormand,* in herb. Gomont (PC) = *Schizothrix calcicola* (Ag.) Gom.

Schizothrix rubra Crouan [Fl. Finistère, p. 118. 1867] ex Gomont, ibid. 15: 329. 1892. *Calothrix rubra* Bornet & Flahault [Mém. Soc. Nat. Sci. Nat. & Math. Cherbourg 25: 200. 1885] ex Bornet & Flahault, Ann. Sci. Nat. VII. Bot. 3: 347. 1886. *Homoeothrix rubra* Kirchner in Engler & Prantl, Natürl. Pflanzenfam. 1(1a): 87. 1900. —Type specimen, here designated, from the original locality in Finistère, France: Brest, *Le Dantec,* Apr. 1886, in herb. Bornet—Thuret (PC) = *Schizothrix calcicola* (Ag.) Gom.

Scytonema adnatum Montagne [Ann. Sci. Nat. III. Bot. 14: 305. 1850] ex Bornet & Flahault, ibid. 5: 110. 1887. —Type specimen from French Guyana: Cayenne, *Leprieur 1109,* in herb. Montagne (PC) = *Porphyrosiphon Notarisii* (Menegh.) Kütz.

Scytonema ambiguum Nägeli [in Kützing, Sp. Algar., p. 894. 1849] ex Bornet & Flahault, ibid. 5: 100. 1887. *S. mirabile* var. *ambiguum* Playfair, Proc. Linn. Soc. New South Wales 37(3): 533. 1913. —Type specimen from Switzerland: Zürich, *Nägeli 409*, in herb. Kützing (L) = *Fischerella ambigua* (Näg.) Gom.

Scytonema arenarium Berkeley [Ann. of Nat. Hist. 3: 327. 1839] ex Bornet & Flahault, ibid. 5: 112. *Schizosiphon arenarius* Kützing [Tab. Phyc. 2: 18. 1850—52] ex Bornet & Flahault, ibid. 3: 372. 1886. *Hypheothrix arenaria* Forti, Syll. Myxoph., p. 343. 1907. —Type specimen from Tasmania, Australia: Van Dieman Land, on seasand (K in BM) = *Schizothrix arenaria* (Berk.) Gom.

Scytonema atrovirens β ocellatum Agardh (as "ocellata") [Disp. Alg. Suec., p. 39. 1812] ex Bornet & Flahault, Ann. Sci. Nat. VII. Bot. 5: 70. 1887. *S. panniforme* Agardh [Syn. Alg. Scand., p. 116. 1817] ex Bornet & Flahault, ibid. 5: 71. 1887. —Type specimen from Sweden: Lessjöfors, in herb. Agardh (LD) = *Stigonema* sp.

Scytonema badium Wolle [Bull. Torrey Bot. Club 6(35): 184. 1877] ex Bornet & Flahault, ibid. 5: 111. 1887. —Type specimen from New York: Herkimer county, *C.F. Austin*, 1868 (D) = *Fischerella ambigua* (Näg.) Gom.

Scytonema collinum Kützing [Sp. Algar., p. 305. 1849] ex Bornet & Flahault, loc. cit. 1887.—Type specimen from Barbados, West Indies: top of Forster hill, *R. Schomburgk*, in herb. Kützing (L) = *Microcoleus lyngbyaceus* (Kütz.) Crouan.

Scytonema coloratum De Wildeman, Ann. Jard. Bot. Buitenzorg Suppl. 1: 42. 1897. —Type specimen from Indonesia: Java, *J. Massart 737*, in herb. Bornet—Thuret (PC) = *Porphyrosiphon Notarisii* (Menegh.) Kütz.

Scytonema dubium Wood [Smithson. Contrib. Knowl. 241: 63. 1872] ex Bornet & Flahault, loc. cit. 1887. —In the absence of the original material, the original description is here designated as the Type = Stigonematacea.

Scytonema fecundum Zopf (as "fecunda") [Zur Morphol. d. Spaltpfl., p. 53. 1882] ex Bornet & Flahault, loc. cit. 1887. —In the absence of the original material, the original description is here designated as the Type = *Nostoc* sp.

Scytonema fuscum Zeller [Journ. Asiatic Soc. Bengal 42(2): 182. 1873; in Rabenhorst, Alg. Eur. 225—226: 2342. 1873] ex Bornet & Flahault, Ann. Sci. Nat. VII. Bot. 5: 113. 1887. —Type specimen from Burma: Pegu, Yomah, *S. Kurz 3153* (L) = *Porphyrosiphon Notarisii* (Menegh.) Kütz.

Scytonema granulatum Martens [in Kurz, Proc. Asiatic Soc. Bengal 1871: 172. 1871] ex Bornet & Flahault, ibid. 5: 111. 1887. —Type specimen from India: Calcutta, salt lakes, *S. Kurz 3044* (BM) = *Microcoleus lyngbyaceus* (Kütz.) Crouan.

Scytonema Gunnerae Reinke [Nachr. K. Ges. Wiss. & Univ. z. Göttingen 1871(25): 626. 1871] ex Bornet & Flahault, loc. cit. 1887. —In the absence of the original material, the original description is here designated as the Type = *Nostoc* sp.

Scytonema hormoides Kützing [Phyc. Gener., p. 215. 1843] ex Bornet & Flahault, ibid. 5: 68. 1887. —Type specimen from Calvados, France: Falaise, *Binder*, in herb. Kützing (L) = *Stigonema* sp.

*Scytonema Kaernbachii Hennings, Bot. Jahrb. 15(Beibl. 33): 8. 1892. —Type specimen from New Guinea: Mole-Insel, *L. Kärnbach*, 1888, in herb. Gomont (PC) = *Porphyrosiphon Notarisii* (Menegh.) Kütz.

*Scytonema minutum Agardh [Syn. Alg. Scand., p. 117. 1817] ex Bornet & Flahault, ibid. 5: 72. 1887. —Type specimen from Västmanland, Sweden: Jäder, in herb. Agardh (LD) = *Stigonema* sp.

Scytonema myochrous β inundatum Agardh [Syn. Alg. Scand., p. 114. 1817] ex Bornet, Bull. Soc. Bot. France 36: 152. 1889. —Type specimen from Sweden: af Sjön Opdämpen vid Rümmen, in herb. Agardh (LD) = *Stigonema* sp.

Scytonema natans De Brébisson [in Kützing, Sp. Algar., p. 306. 1849] ex Bornet & Flahault, Ann. Sci. Nat. VII. Bot. 5: 113. 1887. —Type specimen from Calvados, France: Falaise, *A. De Brébisson 194*, in herb. Kützing (L) = *Porphyrosiphon Notarisii* (Menegh.) Kütz.

*Scytonema Notarisii Meneghini [in Kützing, ibid., p. 307. 1849] ex Bornet & Flahault, ibid. 5: 114. 1887. *S. sanguineum* Cesati *pro synon.* [ex Rabenhorst, Fl. Eur. Algar. 2: 251. 1865] ex Bornet & Flahault, loc. cit. 1887. —Type specimen from Italy: am Lago Maggiore, Tutra, *Meneghini*, in herb. Kützing (L) = *Porphyrosiphon Notarisii* (Menegh.) Kütz.

*Scytonema palmarum Martens [in Kurz, Proc. Asiatic Soc. Bengal 1870: 11. 1870] ex Brühl & Biswas, Journ. Dept. Sci. Calcutta Univ. 5(Bot.): 11. 1923. —Type specimen from India: Calcutta, *S. Kurz 1789* (L) = *Porphyrosiphon Notarisii* (Menegh.) Kütz.

Scytonema Parlatorii Fiorini Mazzanti [Atti Accad. Pontif. N. Lincei Roma 16(5): 631. 1863] ex Bornet & Flahault, ibid. 5: 112. 1887. —Type specimen from near Rome, Italy: nelle acque minerali di Terracina (FI) = *Microcoleus lyngbyaceus* (Kütz.) Crouan.

Scytonema peguanum Martens [Journ. Asiatic Soc. Bengal 40(2): 463. 1871] ex Bornet & Flahault, ibid. 5: 114. 1887. —Type specimen from Burma: Yomah, Pegu, *S. Kurz 1855* (L) = *Porphyrosiphon Notarisii* (Menegh.) Kütz.

*Scytonema Perrottetii Montagne [Syll. Gen. Sp. Crypt., p. 466. 1856] ex Bornet & Flahault, loc. cit. 1887. —Type specimen from Senegambia: ad Erpodium, in herb. Montagne (PC) = *Porphyrosiphon Notarisii* (Menegh.) Kütz.

Scytonema repens Agardh [Syst. Algar., p. 38. 1824] ex Bornet & Flahault, loc. cit. 1887. —In the absence of the original material, the original description is here designated as the Type = *Ulothrix* sp., according to Bornet in Bull. Soc. Bot. France 36: 150 (1889).

Scytonema stygium Heufler [Verh. Zool.-bot. Ver. Wien 2(1852, Abh.: Drei neue Algen): 6. 1853] ex Bornet & Flahault, loc. cit. 1887. —Type specimen from eastern Transylvania, Romania: ad montem Büdös, *R. L. von Heufler*, May 1850 (FH) = Lichen. W. G. Farlow determined this as *Cystocoleus ebeneus* Thw.

Scytonema tenue Gardner, Mem. New York Bot. Gard. 7: 78. 1927. *S. Gardneri* J. de Toni, Noter. di Nomencl. Algol. 1: 7. 1934. —Type specimen from Puerto Rico: Rio Piedras, *N. Wille 106*, Dec. 1914 (NY) = *Fischerella ambigua* (Näg.) Gom.

Scytonema tenuissimum Nägeli [in Kützing, Sp. Algar., p. 893. 1849] ex Forti, Syll. Myxoph., p. 342. 1907. *Hypheothrix tenuissima* Rabenhorst [Fl. Eur. Algar. 2: 292. 1865] ex Gomont, Ann. Sci. Nat. VII. Bot. 15: 329. 1892. —Type specimen from Switzerland: Zürich, *Nägeli 410*, with the Type of *Scytonema crassum* Näg. in herb. Kützing (L) = *Schizothrix calcicola* (Ag.) Gom.

*Scytonema thelephoroides Montagne [Ann. Sci. Nat. II. Bot. 12: 45. 1839] ex Bornet & Flahault, Ann. Sci. Nat. VII. Bot. 5: 114. 1887. —Type specimen from Brazil: *A. Saint-Hilaire*, in herb. Montagne (PC) = *Schizothrix Friesii* (Ag.) Gom.

Scytonema variegatum Liebmann [Naturh. Tidskr. 2(5): 488. 1838–39] ex Bornet & Flahault, ibid. 5: 70. 1887. —Type specimen from Denmark: Lyngby Mose (C) = *Stigonema* sp.

Scytonema varium β Schomburgkii Kützing [Sp. Algar., p. 307. 1849] ex Bornet & Flahault (as "Schomburkii"), ibid. 5: 114. 1887. —Type specimen from the West Indies: Barbadoes, in herb. Kützing (L) = *Microcoleus lyngbyaceus* (Kütz.) Crouan.

Scytonema velutinum Wallroth [Fl. Crypt. Germ. 2: 56. 1833] ex Bornet & Flahault, loc. cit. 1887. —Type specimen from Thüringen, Germany: Heringen, an Sandstein, *Wallroth*, in herb. Kützing (L) = *Fischerella ambigua* (Näg.) Gom.

Scytosiphon intestinalis γ cornucopiae Lyngbye, Tent. Hydroph. Dan., p. 67. 1819. *Scytonema intestinale β cornucopiae* Lyngbye ex Bornet & Flahault, Ann. Sci. Nat. VII. Bot. 5: 113. 1887. —Type specimen from the Faeroes: ad Qvalböe, in herb. Lyngbye (C) = *Enteromorpha intestinalis* (L) Link, det. F. Børgesen.

Seguenzaea minor Geitler in Rabenhorst, Krypt.-Fl., ed. 2, 14: 701. 1932. —In the absence of the original material, the original description is here designated as the Type = *Fischerella ambigua* (Näg.) Gom.

Seguenzaea Borzi, Malpighia 1: 78. 1887. —Type species: *S. sicula* Borzi, Studi sulle Mixof., p. 150. 1917. —In the absence of the original material, the original description is here designated as the Type = *Fischerella ambigua* (Näg.) Gom.

Sokolovia Elenkin, Not. Syst. Inst. Crypt. Hort. Bot. Petropol. 4(7): 89. 1926. —Type species: *S. Neumaniae* Elenkin, loc. cit. 1926. —In the absence of the original material, the original description is here designated as the Type =

244 REVISION OF THE NOSTOCACEAE

filamentous Bacteria, as suspected by Geitler & Ruttner in Arch. f. Hydrobiol. Suppl.-Bd. 14(6): 430 (1935).

Sokolovia Neumaniae var. *gracilis* Elenkin, ibid. 4(7): 91. 1926. *S. Neumaniae* fa. *gracilis* Elenkin, Monogr. Alg. Cyanoph., Pars Spec. 2: 1836. 1949. —In the absence of the original material, the original description is here designated as the Type = filamentous Bacteria.

Spelaeopogon Populii González Guerrero, Mem. R. Soc. Españ. Hist. Nat. 15: 437. 1929. —In the absence of the original material, the original description is here designated as the Type = *Torula* sp., as suspected by Geitler in Rabenhorst, Krypt.-Fl., ed. 2, 14: 697 (1932).

[*Symphyosiphon coriaceus* Wood (as "coriacea"), Proc. Amer. Philos. Soc. 11: 131. 1869.] —In the absence of the original material, the original description is here designated as the Type = *Microcoleus lyngbyaceus* (Kütz.) Crouan.

*Symphyosiphon Kützing [Phyc. Gener., p. 218. 1843] ex Gomont, Ann. Sci. Nat. VII. Bot. 15: 292. 1892. —Type species: *S. dentatus* Kützing [loc. cit. 1843] ex Gomont, ibid. 15: 316. 1892. *Scytonema dentatum* Rabenhorst [Krypt.-Fl. Deutschl. 2(2): 86. 1847] ex Bornet & Flahault, Ann. Sci. Nat. VII. Bot. 5: 111. 1887. *S. Hofmannii* b *dentatum* Rabenhorst [Fl. Eur. Algar. 2: 260. 1865] ex Bornet & Flahault, loc. cit. 1887. *S. Hofmannii* var. *dentatum* Rabenhorst ex Forti, Syll. Myxoph., p. 515. 1907. *S. Hofmannii* fa. *dentatum* Kossinskaia in Elenkin, Monogr. Alg. Cyanoph., Pars Spec. 1: 913. 1938. —Type specimen from Sachsen, Germany: Steigerthal, Nordhausen, in herb. Kützing (L) = *Schizothrix Friesii* (Ag.) Gom.

Symphyosiphon Hegetschweileri Nägeli [in Kützing, Sp. Algar., p. 894. 1849] ex Forti, ibid., p. 539. 1907. —Type specimen from Switzerland: Rigi, *Hegetschweiler*,in herb. Kützing (L) = Lichen.

Symphyosiphon intertextus Kützing [ibid., p. 323. 1849] ex Gomont, ibid. 15: 321. 1892. *Calothrix intertexta* De Brébisson *pro synon.* [in Kützing, loc. cit. 1849] ex Forti, ibid., p. 511. 1907. *Scytonema intertextum* Rabenhorst [Fl. Eur. Algar. 2: 263. 1865] ex Bornet & Flahault, loc. cit. 1887. *Calothrix Kuetzingii* Beck in Becker, Hernstein in Niederösterr. 2: 269. 1886. *Scytonema guyanense* var. *intertextum* Sampaio, Publ. Inst. Bot. Fac. Ciênc. Univ. Porto, ser. 2, no. 48: 252. 1962. —Type specimen from Calvados, France: Falaise, *Lenormand 191*, in herb. Kützing (L) = *Schizothrix Friesii* (Ag.) Gom.

Symphyosiphon leucocephalus Kützing [ibid., p. 324. 1849] ex Gomont, Ann. Sci. Nat. VII. Bot. 16: 128. 1892. *Scytonema leucocephalum* Rabenhorst [ibid. 2: 260. 1865] ex Bornet & Flahault, Ann. Sci. Nat. VII. Bot. 5: 111. 1887. —Type specimen from Tambov region, Russia: Kosloff (now Michurinsk), Lenormand comm., in herb. Kützing (L) = *Microcoleus lyngbyaceus* (Kütz.) Crouan.

*Symphyosiphon minor Hilse [Jahres-Ber. Schles. Ges. Vaterl. Cultur 1864: 94. 1865; in Rabenhorst, Alg. Eur. 177—178: 1776. 1865] ex Forti, Syll. Myxoph., p. 539. 1907. *Calothrix tenuissima* A. Braun [in Rabenhorst, Fl. Eur. Algar. 2:

271. 1865] ex Bornet & Flahault, ibid. 5: 128. 1887. *Dillwynella minor* Kuntze, Rev. Gen. Pl. 2: 892. 1891. —Type specimen from Silesia, Poland: Kavallen bei Breslau, *Hilse*, Nov. 1864, in Rabenhorst, Algen Europas no. 1776 (PH) = *Hydrocoryne spongiosa* Schwabe.

*Symploca elegans Kützing (as "Synploca") [Phyc. Gener., p. 201. 1843] ex Gomont, ibid. 16: 116. 1892. [*Calothrix elegans* Meneghini *pro synon.* in Kützing, loc. cit. 1843.] —Type specimen from Italy: Therm. Eugan., *Meneghini*, in herb. Kützing (L) = *Schizothrix calcicola* (Ag.) Gom.

Symploca Harveyi Le Jolis [Liste Alg. Mar. Cherbourg, p. 29. 1863] ex Gomont, ibid. 16: 107. 1892. *Schizosiphon Harveyi* Crouan [Fl. Finistère, p. 116. 1867] ex Bornet & Flahault, ibid. 3: 373. 1886. *Calothrix Harveyi* Lloyd [Alg. de l'Ouest de la France no. 386. 1881] ex Gomont, loc. cit. 1892. —Type specimen in Le Jolis, Algues Marines de Cherbourg no. 139, from France, in herb. Bornet–Thuret (PC) = *Schizothrix mexicana* Gom.

Tapinothrix Sauvageau, Bull. Soc. Bot. France 39(1892): cxxiii. 1893. —Type species: *T. Bornetii* Sauvageau (as "Borneti"), loc. cit. 1893. *Homoeothrix Bornetii* Mabille, Rev. Algol., N. S. 1(1): 11. 1954. —Type specimen from Algeria: source d'Ain-Oumach près Biskra, *Sauvageau* (PC) = *Schizothrix calcicola* (Ag.) Gom.

Tapinothrix mucicola Borge, Ark. f. Bot. 18(10): 3. 1923. —Type specimen from Sweden: Omberg, *O. Borge* (S) = *Schizothrix calcicola* (Ag.) Gom.

Thiochaete Welsh, N. Hedwigia 3(1): 39. 1961. —Type species: *T. Chutteri* Welsh, loc. cit. 1961. —Type specimen from Transvaal, South Africa: Vaal river at Standerton, *F. M. Chutter*, Jul. 1960 (BM) = Bacteria.

Tolypothrix amphibica Zopf [Ber. Deutsch. Bot. Ges. 1: 319. 1883] ex Bornet & Flahault, Ann. Sci. Nat. VII. Bot. 5: 125. 1887. —In the absence of the original material, the original description is here designated as the Type = a mixture of *Oscillatoria Retzii* Ag. and *Nostoc* sp.

Tolypothrix binata Zeller (as "Polypothrix") [Journ. Asiatic Soc. Bengal 42(2:3): 184. 1873] ex Bornet & Flahault, loc. cit. 1887. —Type specimen from Burma: Pegu, Kya Eng, *Kurz 3203*, in herb. Kützing (L) = *Hapalosiphon* sp.

Tolypothrix Brebissonii Kützing (as "Tolypothria") [Bot. Zeit. 5(11): 179. 1847] ex Bornet & Flahault, ibid. 5: 121. 1887. *Hapalosiphon Brebissonii* Kützing [ex Rabenhorst, Fl. Eur. Algar. 2: 284. 1865] ex Bornet & Flahault, ibid. 5: 61. 1887. —Type specimen from Calvados, France: Falaise, *De Brébisson 180*, in herb. Kützing (L) = *Hapalosiphon* sp.

Tolypothrix extensa Crouan [in Mazé & Schramm, Essai Class. Alg. Guadeloupe, p. 37. 1870–77] ex Bornet & Flahault, ibid. 5: 126. 1887. —Type specimen from Guadeloupe, West Indies: Vieux-Fort, in herb. Crouan (PC) = *Schizothrix calcicola* (Ag.) Gom.

Tolypothrix fuscescens De Brébisson [in Kützing, Sp. Algar., p. 313. 1849] ex Bornet & Flahault, ibid. 5:61. 1887. *Hapalosiphon fuscescens* Kützing [ex

Rabenhorst, ibid. 2: 283. 1865] ex Bornet & Flahault, loc. cit. 1887. —Type
specimen from Calvados, France: Falaise, *De Brébisson 165*, in herb. Kützing
(L) = *Hapalosiphon* sp.

   Tolypothrix implexa Martens [Proc. Asiatic Soc. Bengal 1870: 183. 1870] ex
Bornet & Flahault, ibid. 5: 125. 1887. —Type specimen from Java, Indonesia:
im Flusse Tjiliwong, in herb. Agardh (LD) = *Oscillatoria Retzii* Ag.

   Tolypothrix irregularis Berkeley [in Harvey in Hooker, Fl. New Zealand 2:
265. 1855] ex Forti, Syll. Myxoph., p. 556. 1907. —Type specimen from New
Zealand: *Colenso 2641* (BM) = *Microcoleus lyngbyaceus* (Kütz.) Crouan.

   Tolypothrix lyngbyacea Grunow [in Rabenhorst, Alg. Eur. 227—228: 2269.
1872] ex Bornet & Flahault, loc. cit. 1887. —Type specimen from Sardinia:
Lagunen von Cagliari, *A. Grunow*, Nov. 1871, in Rabenhorst, Algen Europas
no. 2269 (W) = *Microcoleus lyngbyaceus* (Kütz.) Crouan.

   Tolypothrix majuscula Itzigsohn [in Rabenhorst, Alg. Sachs. 11—12: 119.
1851] ex Bornet & Flahault, Revision, Index etc., p. 126. 1888. [*T. pulchra*
Itzigsohn in Rabenhorst, loc. cit. 1851.] —Type specimen from Brandenburg,
Germany: an den Neudammer Teichen, *Itzigsohn & Rothe*, Jul. 1851, in
Rabenhorst, Algen Sachsens no. 119 (BM) = *Oscillatoria Retzii* Ag.

   Tolypothrix mucosa Meneghini [in Kützing, Sp. Algar., p. 314. 1849] ex
Bornet & Flahault, Ann. Sci. Nat. VII. Bot. 5: 125. 1887. —Type specimen
from Italy: Napoli, *Meneghini*, in herb. Kützing (L) = *Microcoleus lyngbyaceus*
(Kütz.) Crouan.

   Tolypothrix pumila Kützing [Phyc. Gener., p. 227. 1843] ex Bornet &
Flahault, ibid. 5: 61. 1887. [*Calothrix muscicola* c *pumila* Rabenhorst, Deutschl.
Krypt.-Fl. 2(2): 84. 1847.] *Schizosiphon sphagnicola* A. Braun *pro synon*. ex Bornet
& Flahault, ibid. 3: 373. 1886. —Type specimen from Thüringen, Germany:
Schleusingen, in herb. Kützing (L) = *Hapalosiphon pumilus* (Kütz.) Kirchn.

   [*Tremella palustris* Weber, Spicil. Fl. Goettingens., p. 279. 1778.] *Rivularia
lubrica* Lamarck & De Candolle, Synops. Pl. in Fl. Gall. Descr., p. 2. 1806.
—Type specimen from England: "Tremella palustris, vulgari marina similis,
sed minor & tenerior," in herb. Dillenius (OXF) = Chlorophyceae.

   Tremella adnata Linnaeus [Fl. Suec., ed. 2, p. 430. 1755] ex Agardh, Syst.
Algar., p. 28. 1824. *Rivularia adnata* Linnaeus ex Fries, Fl. Scan., p. 324. 1835.
—In the absence of the original material, the original description is here
designated as the Type = Phaeophyceae.

   [*Trichodesmium flos-aquae* Ehrenberg, Ann. d. Phys. & Chem. 18(4): 506.
1830.] [*Limnochlide Harveyana* Kützing, Sp. Algar., p. 286. 1849.] *L. flos-aquae*
var. *Harveyana* Kützing ex Bornet & Flahault, Ann. Sci. Nat. VII. Bot. 7: 241.
1888. —Type specimen, here designated, from the probably original locality in
Sachsen, Germany: Lindenauer Teich bei Leipzig, *Kunze*, 1834, in herb.
Kützing (L) = *Microcoleus lyngbyaceus* (Kütz.) Crouan.

   Ulva gelatinosa Vaucher, Hist. d. Conf. d'Eau Douce, p. 244. 1803. *Rivularia tubulosa* Lamarck & De Candolle, Fl. Franç., ed. 3, 2: 5. 1805. —In the absence of the original material, the original description is here designated as the Type = Chlorophycea.

Ulva incrassata Hudson, Fl. Anglica, ed. altera, 2: 572. 1778. *Rivularia incrassata* Sowerby in Smith & Sowerby, Engl. Bot., Gen. Indexes, Alphab. Index. 1814. —Type specimen from England: "Tremella palustris gelatinosa, Damae cornuum facie," in herb. Dillenius (OXF) = *Chaetophora incrassata* (Huds.) Haz.

Ulva pruniformis Linnaeus, Sp. Pl. 2: 1164. 1753. *Rivularia pruniformis* Sowerby in Smith & Sowerby, loc. cit. 1814. —In the absence of the original material, the original description is here designated as the Type = *Nostoc* sp.

Ulva verticillata Withering, Syst. Arr. Brit. Pl., ed. 4, 4: 125. 1801. *Rivularia verticillata* Smith & Sowerby, Engl. Bot., pl. 2466. 1812. —Type specimen from England: "Ulva verticillata. *Major Velley*" (K in BM) = Rhodophycea.

Literature Cited

Blinks, L. R. 1951. Physiology and biochemistry of algae. In G. M. Smith (ed.), Manual of Phycology, p. 263—291. Waltham.

Bornet, E. 1873. Recherches sur les gonidies des lichens. Ann. Sci. Nat. V. Bot. 17: 45—110.

Bornet, E., and C. Flahault. 1886—88. Revision des Nostocacées hétérocystées contenues dans les principaux herbiers de France. Ann. Sci. Nat. VII. Bot. 3: 323—381 (1886), 4: 343—373 (1887), 5: 51—129 (1887), 7: 177—262 (1888).

Borzi, A. 1914. Studi sulle Mixoficee, I. N. Giorn. Bot. Ital., N. S. 21(4): 307—360.

Bowen, C. C., and T. E. Jensen. 1965. Blue-green algae: fine structure of gas vacuoles. Science, N. S. 147(3664): 1460—1462.

Canabaeus, L. 1929. Über die Heterocysten und Gasvakuolen der Blaualgen und ihre Beziehung zueinander. Pflanzenforsch. 13: 1—48.

Darley, J. 1968. Contribution à l'étude systématique et biologique des Rivulariacées marines. Le Botaniste, Fasc. I—VI, sér. LI, p. 141—220.

Dodd, J. D. 1954. A note on the increase in flake size of Aphanizomenon flos-aquae (L.) Ralfs. 1954. Proc. Iowa Acad. Sci. 60: 117—118.

Drouet, F. 1951. Cyanophyta. In G. M. Smith (ed.), Manual of Phycology, p. 159—166. Waltham.

Drouet, F. 1962. Gomont's ecophenes of the blue-green alga, *Microcoleus vaginatus*. Proc. Acad. Nat. Sci. Philadelphia 114(6): 191—205.

Drouet, F. 1968. Revision of the classification of the Oscillatoriaceae. Monogr. Acad. Nat. Sci. Philadelphia 15: 1—370.

Drouet, F., and W. A. Daily. 1956. Revision of the coccoid Myxophyceae. Butler Univ. Bot. Stud. 12: 1—218.

Fan, K. C. 1956. Revision of *Calothrix* Ag. Delineation of species. Rev. Algol., N. S. 2: 154—178.

Fay, P., and N. J. Lang. 1971. The heterocysts of blue-green algae. I. Ultrastructural integrity after isolation. Proc. R. Soc. London, ser. B, 178(1051): 185—192.

Fogg, G. E. 1944. Growth and heterocyst production in *Anabaena cylindrica* Lemm. New Phytol. 43(2): 164—175.

Fogg, G. E. 1949. Growth and heterocyst production in *Anabaena cylindrica* Lemm., II. In relation to carbon and nitrogen metabolism. Ann. of Bot., N. S. 13: 241—259.

249

Geitler, L. 1932. Cyanophyceae. In L. Rabenhorst, Kryptogamen-Flora von Deutschland, Österreich und der Schweiz, ed. 2, 14: 1—1196.

Geitler, L. 1934. Beiträge zur Kenntnis der Flechtensymbiose, IV, V. Arch. f. Protistenk. 82(1): 51—85.

Jaag, O. 1943. Scytonema myochrous (Dillw.) Ag., Formenkreis und Variabilität einer Blaualge. Boissiera 7: 437—454.

Kirchner, O. 1900. Schizophyceae. In A. Engler and K. Prantl, Die natürlichen Pflanzenfamilien 1(1a): 45—92.

Klebahn, H. 1895. Gasvacuolen, ein Bestandtheil der Zellen der wasserblüthebildenden Phycochromaceen. Flora oder Allg. bot. Zeit. 80: 241—282.

Lang, N. J. 1965. Electron microscopic study of heterocyst development in Anabaena Azollae Strasburger. Journ. of Phycol. 1: 127—134.

Lang, N. J. 1971. Ultrastructure of the heterocyst polar region. Journ. of Phycol. 7(Suppl.): 9.

Lang, N. J., and P. Fay. 1971. The heterocysts of blue-green algae. II. Details of ultrastructure. Proc. R. Soc. London, ser. B, 178(1051): 193—203.

Lang, N. J., and K. A. Fisher. 1969. Variation in the fixation image of "structured granules" in Anabaena. Arch. f. Mikrobiol. 67: 173—181.

Lanjouw, J., and F. A. Stafleu. 1964. Index herbariorum, part I. The herbaria of the world. Regn. Veget. 31: 1—251.

Lemmermann, E. 1900. Beiträge zur Kenntniss der Planktonalgen. Ber. deutsch. bot. Ges. 18: 135—143.

Lewin, R. A. (ed.). 1962. Physiology and Biochemistry of Algae. New York and London.

Maxwell, T. F. 1971. Hormogones in the developmental morphology of Gloeotrichia echinulata. Journ. of Phycol. 7(Suppl.): 14.

Pearson, J. E., and J. M. Kingsbury. 1966. Culturally induced variation in four morphologically diverse blue-green algae. Amer. Journ. Bot. 53(2): 192—200.

Prud'homme van Reine, W. F., and C. van den Hoek. 1966. Isolation of living algae growing in the shells of molluscs and barnacles with EDTA (ethylenediaminetetraacetic acid). Blumea 14(2): 331—332.

Rose, E. T. 1934. Notes on the life history of Aphanizomenon flos-aquae. Univ. Iowa Stud. Nat. Hist. 16(2): 129—140.

Royers, H. 1906. Zum Polymorphismus der Cyanophyceen. Jahres-Ber. Naturw. Ver. Elberfeld 11: 3—40.

Schwabe, G. H. 1964. Lagerbildungen hormogonaler Blaualgen in thermalen und anderen extremen Biotopen. Verh. Int. Ver. Limnol. 15: 772—781.

Stein, J. 1963. Morphological variation of Tolypothrix in culture. Brit. Phycol. Bull. 2(4): 206—209.

Stewart, W. D. P., A. Haystead, and H. W. Pearson. 1969. Nitrogenase activity in heterocysts of blue-green algae. Nature 224(5216): 226—228.

Thuret, G. 1875. Essai de classification des Nostochinées. Ann. Sci. Nat. VI. Bot. 1: 372—382.

Umezaki, I. 1958. Revision of Brachytrichia Zanard. and Kyrtuthrix Erceg. Mem. Coll. Agric. Kyoto Univ., Fish. Ser., Spec. No., p. 55—67.

Walsby, A. E. 1969. Studies on the physiology of gas-vacuoles. Brit. Phycol. Journ. 4(2): 216.

Wolk, C. P. 1965. Heterocyst germination under defined conditions. Nature 205: 201—202.

Illustrations

Scytonema Hofmannii Ag., Figs. 1—14.

Fig. 1. Trichomes in my collection no. 11088 (D) from Hesperides, Florida, showing twin-branching of the sheath. Fig. 2. Above, intact tips of a trichome and a sheath; below, the terminal cell and the tip of the sheath destroyed; these are from a culture of my collection no. 14992 (PH) from Niagara Falls, Ontario, Sept. 18—Oct. 20, 1969. Fig. 3. Tip of a trichome of the Type specimen of *Arthrosiphon densus* A. Br. (L). Fig. 4. A fungus hypha invading a sheath in a culture of my collection no. 15008 (PH) from Niagara Falls, Ontario, Sept. 22—Oct. 14, 1969. Fig. 5. Parts of trichomes of the Type specimen of *Tolypothrix limbata* Thur. (PC). Fig. 6. Fragments of trichomes issuing from the broken tip of a sheath, in a culture of my collection no. 15008 (PH) from Niagara Falls, Ontario, Sept. 22—Oct. 9, 1969. Fig. 7. Parts of two trichomes of the Type specimen of *T. tenuis* Kütz. (L). Fig. 8. Tip of a trichome of the Type specimen of *Scytonema byssoideum* Ag. (LD). Fig. 9. Tip of a trichome of the Type specimen of *Symphyosiphon involvens* A. Br. (L). Fig. 10. Tip of a trichome in a culture of my collection no. 14992 (PH) from Niagara Falls, Ontario, Sept. 18—24, 1969, showing deposition of successive layers of sheath material. Fig. 11. Tip of a trichome of the Type specimen of *Scytonema polycystum* Born. & Flah. (PC). Fig. 12. Vegetative cells of an anchored trichome of my collection no. 14910 (PH) from Stroudsburg, Pennsylvania, showing elongation and the initial step in the formation of U-shaped branching of the sheath. Fig. 13. A trichome of the Type specimen of *Diplocolon Heppii* Näg. (PH). Fig. 14. Tip of a trichome of the Type specimen of *Oscillatoria alata* Carm (LD).

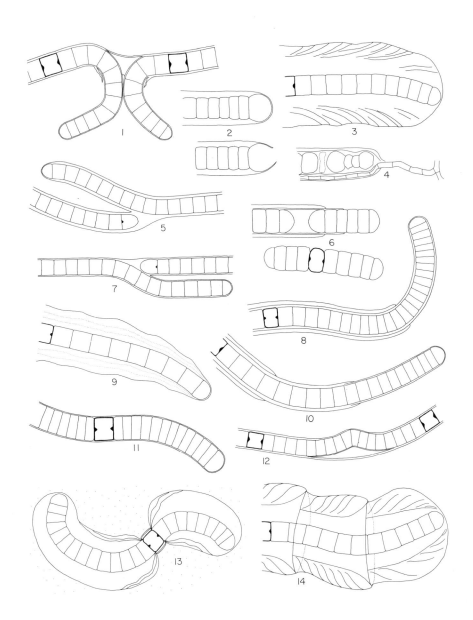

Scytonema Hofmannii Ag., Figs. 15–27.

Fig. 15. Tip of a trichome of the Type specimen of *Scytonema crassum* Näg. (L). Fig. 16. Trichomes of the Type specimen of *Calothrix olivacea* Hook. f. & Harv. (TCD). Fig. 17. Tip of a trichome of the Type specimen of *Scytonema crustaceum* Ag. (LD). Fig. 18. Tip of a trichome of the Type specimen of *Calothrix pilosa* Harv. (BM). Fig. 19. Tip of a trichome of my collection no. 15007 (PH) from Niagara Falls, Ontario, showing shrinking of cells within a thick sheath. Fig. 20. Tip of a trichome of the Type specimen of *Scytonema figuratum* Ag. (LD). Fig. 21. Tip of a trichome of material determined as *S. myochrous* (Dillw.) Ag. by E. Bornet, from Gorge-du-Loup (Alpes-Maritimes), France, collected by G. Thuret, March 1870 (PC). Fig. 22. A trichome of a culture of my collection no. 14910 (PH) from Stroudsburg, Pennsylvania, showing great range in diameters of parts. Fig. 23. Looping growth of a trichome of the Type specimen of *S. crustaceum* Ag. (LD). Fig. 24. Tip of a trichome of the Type specimen of *Symphyosiphon velutinus* Kütz. (L). Fig. 25. Parts of trichomes of the Type specimen of *Aulosira implexa* Born. & Flah. (PC), with and without spores. Fig. 26. Looped trichome of the Type specimen of *Scytonema incrustans* Kütz. (L). Fig. 27. Tips of trichomes of the Type specimen of *Thorea Wrangelii* Ag. (LD).

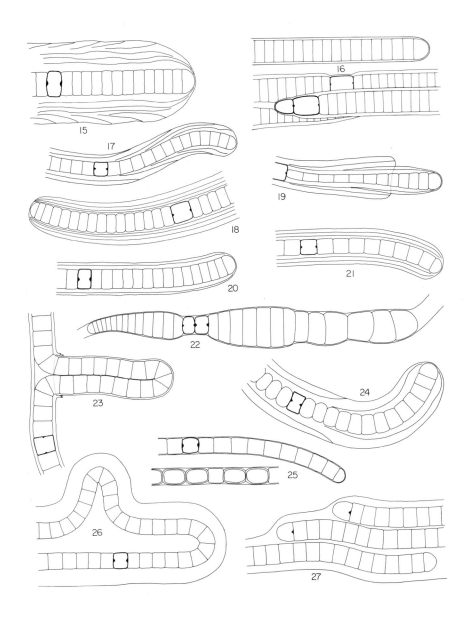

Calothrix parietina (Näg.) Thur., Figs. 28–39.

Fig. 28. Two trichomes of the Type specimen of *Batrachospermum haematites* Lam. & DC. (PC). Fig. 29. A trichome of Mougeot & Nestler, Stirpes Cryptogamae Vogeso-rhenanae no. 796 (D). Fig. 30. A trichome of the Type specimen of *Oscillatoria subulata* Corda (PC). Fig. 31. A trichome of the Type specimen of *Gloeotrichia Rabenhorstii* Born. & Thur. (PC). Fig. 32. A trichome of the Type specimen of *G. punctulata* Thur. (PC). Fig. 33. A trichome of the Type specimen of *Rivularia salina* Kütz. (L). Fig. 34. Part of a trichome of the Type specimen of *Aphanizomenon holsaticum* Richt. (NY), showing a heterocyst, a young spore, and colorless terminal cells. Fig. 35. A trichome of the Type specimen of *Calothrix Braunii* Born. & Flah. (PC). Fig. 36. Three trichomes of the Type specimen of *Anabaena unispora* Gardn. (NY). Fig. 37. Two trichomes of the Type specimen of *Aulosira laxa* Kirchn. (W). Fig. 38. Left: trichomes showing vegetative cells, resting spores, germinating spores, and growth of new trichomes; above: an intercalary heterocyst; right: spores of various shapes; all of these are from a culture of my collection no. 3036 (D) from Hermosillo, Mexico. Fig. 39. Spores germinating in a culture of material collected by Dr. H. S. Forest in Conesus lake, New York (PH).

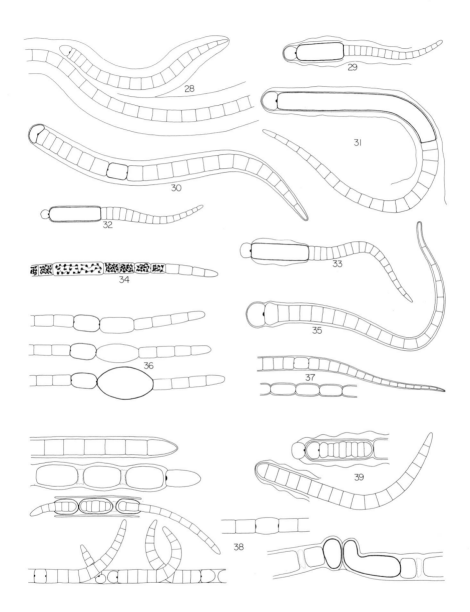

Calothrix parietina (Näg.) Thur., Figs. 40–52.

Fig. 40. A trichome of the Type specimen of *Schizosiphon parietinus* Näg. (L). Fig. 41. Two trichomes of the Type specimen of *S. Bauerianus* Grun. (W). Fig. 42. A trichome of the Type specimen of *S. gypsophilus* Kütz. (L). Fig. 43. Two trichomes of the Type specimen of *Scytonema compactum* Ag. (LD). Fig. 44. Two trichomes of the Type specimen of *Euactis rufescens* Näg. (L). Fig. 45. Two trichomes of the Type specimen of *Rivularia Biasolettiana* Menegh. (FI). Fig. 46. Two trichomes of the Type specimen of *Limnactis minutula* Kütz. (L). Fig. 47. Two trichomes of the Type specimen of *Calothrix adscendens* Born. & Flah. (PC). Fig. 48. Three trichomes of the Type slide of *Ammatoidea Normanii* W. & G. S. West (BM). Fig. 49. A trichome of the Type specimen of *Lyngbya juliana* Menegh. (L). Fig. 50. Two trichomes of the Type specimen of *Dichothrix Nordstedtii* Born. & Flah. (PC). Fig. 51. Four trichomes of the Type specimen of *Mastigonema Orsinianum* Kütz. (L). Fig. 52. Two trichomes of the Type specimen of *Lyngbya juliana* Menegh. (L).

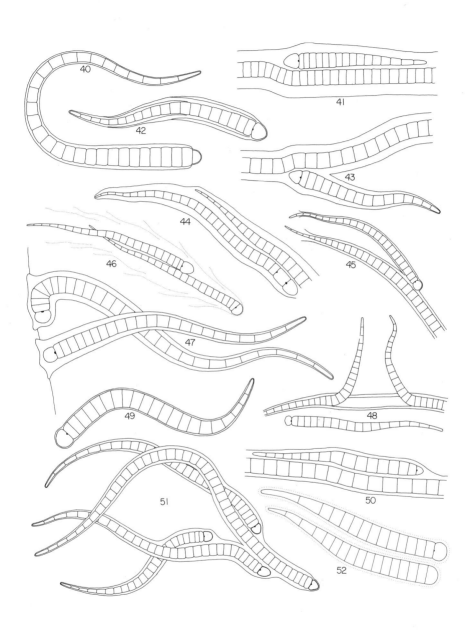

Calothrix parietina (Näg.) Thur., Figs. 53—60.

Fig. 53. Spores germinating in a culture of material (PH) collected by Dr. H. S. Forest in Conesus lake, New York. Fig. 54. Short trichomes, one issuing from a sheath, in a culture of my collection no. 14977 (PH) from Pottstown, Pennsylvania. Fig. 55. Germinating spores, and a portion of a trichome with an intercalary heterocyst, in a culture of my collection no. 14602 (D) from Madera Canyon, Arizona, July 10—Sept. 25, 1969. Fig. 56. Germinating spores in a culture of Dr. H. C. Bold's collection in Sept. 1966 from Austin, Texas (D), cultivated July 17—24, 1969. Fig. 57. A trichome in a stock culture (D) from the University of Wisconsin showing spiral growth within a sheath. Fig. 58. Parts of two trichomes of my collection of a waterbloom, no. 5500 (D), on Lake Monona in Madison, Wisconsin. Fig. 59. Two trichomes of the Type specimen of *Euactis Beccariana* De Not. (FH). Fig. 60. A trichome of the Type specimen of *Calothrix balearica* Born. & Flah. (PC).

Raphidiopsis curvata Fritsch & Rich, Figs. 61—66.

Fig. 61. Parts of three trichomes in a culture from Lake Warren, Florida (D), by Dr. D. E. Henley, grown July—Dec. 1971. Fig. 62. Two trichomes of the preceding culture, July 1971 (D). Fig. 63. A trichome of the Type specimen of *Aphanizomenon americanum* Reinh. (D). Fig. 64. Parts of four trichomes of material collected by Dr. V. W. Proctor near Ashland, Missouri (D). Fig. 65. Two trichomes of the Type slide of *Raphidiopsis curvata* Fritsch & Rich (BM). Fig. 66. A trichome from the original locality of *Anabaenopsis philippinensis* W. R. Taylor, San Pablo City, Philippines, collected by Dr. G. T. Velasquez no. 678 (D).

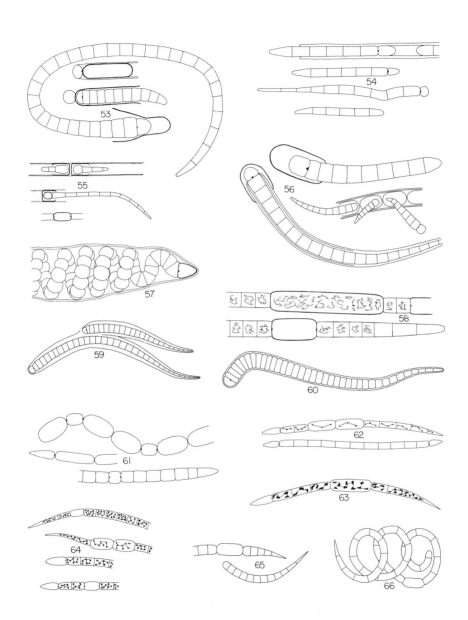

Calothrix crustacea Schousb. & Thur., Figs. 67–83.

Fig. 67. A trichome of the Type specimen of *C. fusco-violacea* Crouan (Lab. Marit., Concarneau). Fig. 68. Four trichomes of the Type specimen of *Diplotrichia polyotis* J. Ag. (LD). Fig. 69. Three trichomes of the Type specimen of *Heteractis mesenterica* Kütz. (L). Fig. 70. A trichome of the Type specimen of *Leibleinia aeruginea* Kütz. (L). Fig. 71. Two trichomes of the Type specimen of *Calothrix prolifera* Born. & Flah. (PC). Fig. 72. A trichome of the Type specimen of *Rivularia parasitica* Chauv. (CN). Fig. 73. Two trichomes of the Type specimen of *Calothrix vivipara* Harv. (TCD). Fig. 74. A trichome of the Type specimen of *Conferva Mucor* Roth (OXF). Fig. 75. Two trichomes of the Type specimen of *Rivularia atra* Roth (PC). Fig. 76. Two trichomes of the Type specimen of *Zonotrichia hemispherica* J. Ag. (LD). Fig. 77. Two trichomes of the Type specimen of *Rivularia nitida* Ag. (LD). Fig. 78. Two trichomes of the Type specimen of *Calothrix crustacea* Schousb. & Thur. (PC). Fig. 79. Two trichomes of the Type specimen of *Dichothrix penicillata* Zanard. (PC). Fig. 80. Trichomes of the Type specimen of *Microcoleus corymbosus* Harv. (TCD). Fig. 81. Two trichomes of the Type specimen of *Rivularia australis* Harv. (BM). Fig. 82. Three trichomes of the Type specimen of *Brachytrichia maculans* Gom. (PC). Fig. 83. Two trichomes of the Type specimen of *Ulva bullata* Poir. (PC).

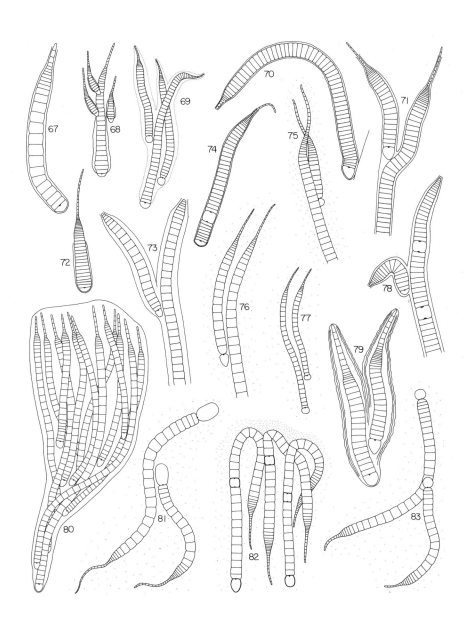

Additions and Corrections

Page 21, after line 6 from top, insert the new paragraph:

Tolypothrix Sect. *Diplocoleopsis* Hollerbach, Bot. Mat. Inst. Sporov. Rast. Bot. Sada Petrograd 2(12): 183 (misprinted as 175). 1923. —Type species: *T. Elenkinii* Hollerb.

Page 47, line 20 from bottom, after "(PC)." insert: A duplicate Type specimen is in the Hy herbarium (COLO).

Page 62, after line 19 from top, insert the new paragraph:

Microchaete transvaalensis Cholnoky, Bot. Not. 1954(3): 292. 1954.

Page 132, after line 20 from bottom, insert the new paragraph:

Aulosira kotoensis Okada, Bull. Biogeogr. Soc. Japan 3(1): 56. 1932.

Page 174, line 22 from top, insert after "1887.": [*Rivularia pulchra* Cramer ex Rabenhorst, Fl. Eur. Algar. 2: 209. 1865.] *Potarcus pulcher* Kuntze (as "Portacus"), Rev. Gen. Pl. 2: 912. 1891.

Page 211, after line 8 from top, insert the new paragraph:

**Conferva majuscula* Dillwyn, Brit. Conf., p. 40. 1809. *Leibleinia pacifica* Kützing, *L. crispa* var. *violacea* Desmazières, *L. ferruginea* Desmazières, *L. prasina* Montagne, *L. rigidissima* Zanardini, and *L. aeruginosa* β *ferruginea* Rabenhorst ex Forti *pro synon.*, Syll. Myx., p. 199. 1907. —Type specimen from England, in herb. Agardh (LD) = *Microcoleus lyngbyaceus* (Kütz.) Crouan.

Page 214, before line 16 from bottom, insert the new paragraph:

**Hydrocoleum Brebissonii* Kützing [Bot. Zeit. 5(13): 220. 1847] ex Gomont, Ann. Sci. Nat. VII. Bot. 15: 343. 1892. [*Leibleinia atroviolacea* Brébisson in Kützing pro synon., loc. cit. 1847.] —Type specimen from France: Falaise, *De Brébisson 162* (L) = *Microcoleus lyngbyaceus* (Kütz.) Crouan.

Page 214, line 10 from bottom, after "1827." insert: *Calothrix lutescens* Meneghini [in Kützing *pro synon.*, Phyc. Gener., p. 198. 1843] ex Hansgirg, Prodr. Algenfl. Böhm. 2: 85. 1892.

Page 215, before line 13 from bottom, insert new paragraph:

Hypheothrix caerulea Carter [Ann. & Mag. Nat. Hist., ser. 5, 2: 164. 1878] ex Forti, Syll. Myx., p. 344. 1907. —In the absence of the original specimen, the original description is here designated as the Type = *Schizothrix calcicola* (Ag.) Gom.

Page 222, line 20 from top, insert after "1892.": [*L. Lenormandiana* Kützing, loc. cit. 1847.]

Page 229, after line 16 from top, insert the new paragraph:

**Lyngbya nigrovaginata* var. *microcoleiformis* Hansgir [Sitzungsber. k. Böhm. Ges. Wiss., Math.-nat. Cl. 1890(2): 98. 1891] ex Hansgirg, Prodr. Algenfl. Böhm. 2: 92. 1892. *Hypheothrix nigrovaginata* var. *microcoleiformis* Hansgirg ex Geitler *pro synon.* in Rabenh. Krypt.-Fl., ed. 2, 14: 1108. 1932. —Type specimen from Bohemia: Sct. Prokop nächst Prag, *Hansgirg,* May 1890 (W) = *Schizothrix calcicola* (Ag.) Gom.

Page 229, before line 7 from bottom, insert the new paragraph:

Lyngbya subcyanea Hansgirg, Prodr. Algenfl. Böhm. 2: 88. 1892. *Hypheothrix subcyanea* Forti, Syll. Myx., p. 332. 1907. —Type specimen from Bohemia: Selc nächst Prag, *Hansgirg,* Oct. 1885 (W) = *Schizothrix calcicola* (Ag.) Gom.

Page 232, after line 3 from top, insert the new paragraph:

**Phormidium rupestre* ƴ *tingens* Kützing [Sp. Algar., p. 255. 1849] ex Gomont, ibid. 16: 192. 1892. *Calothrix putida* Suhr [ex Kützing *pro synon.*, loc. cit. 1849] ex Gomont, ibid. 16: 177. 1892. —Type specimen from Switzerland: Zürich, *Nägeli 193* (L) = *Microcoleus vaginatus* (Vauch.) Gom.

Page 239, after line 13 from top, insert the new paragraph:

**Schizodictyon purpurascens* Kützing, Phyc. Gener., p. 230. 1843. [*Calothrix purpurascens* Kunze ex Kützing *pro synon.*, loc. cit. 1843.] —Type specimen from Surinam, in herb. Kützing (L) = *Porphyrosiphon Notarisii* (Menegh.) Kütz.

Index

Camptylonema indicum Schmidle, 49
Camptylonemopsis Desik., 97
  Danilovii (Hollerb.) Desik., 61
  Iyengarii Desik., 135
  lahorensis (Ghose) Desik., 123
  minor Desik., 135
  pulneyensis Desik., 135
Campylonema Schmidle, 20
  Danilovii Hollerb., 61
  Godwardii Pras. & Srivast., 62
  indicum Schmidle, 49
    var. allhabadii Gupta, 62
  lahorense Ghose, 123
  umidum Cado, 136
Ceramium pulvinatum Mert., 165
Chaetophora aeruginosa Ag., 166
  atra (Roth) Ag., 164
  chlorites Schousb., 165
  crustacea Schousb., 164
  haematites (Lam. & DC.) Bory, 100
  lumbricalis Schousb., 167
Chalaractis Kütz., 96
  mutila Kütz., 104
  villosa Kütz., 104
Chamaenema Kütz., 209
  carneum Kütz., 209
Chondrogloea flagelliformis Schmidle, 229
Chroococcus Zopfii Hansg., 57
Chroolepus janthinus Mont., 210
Chrysostigma Kirchn., 20
  cincinnatum (Kütz.) Kirchn., 30
Clavatella viridissima Bory, 165
Coenocoleus Berk. & Thw., 210
  cirrhosus Berk. & Thw., 112
  Smithii Berk. & Thw., 210
Coleodesmium Borzi, 20
  floccosum Borzi, 56
  Lievreae (Frémy) Geitl., 61
  Scottianum Welsh, 55
  swazilandicum Welsh, 55
  Wrangelii (Ag.) Borzi, 23
    var. Gajanum J. de Toni, 60
    fa. majus (Frémy) J. de Toni, 59
    var. minus W. West, 57
Coleospermum Kirchn., 20
  diplosiphon (Gom.) Elenk., 119
  Goeppertianum Kirchn., 56
    var. minus Hansg., 46
  tenerum (Thur.) Elenk., 45
Collema velutinum Ach., 22
Conferva aegagropila L., 210
  atro-virens Dillw., 210
  canescens Ag., 22
  chthonoplastes Mert., 210
  comoides Dillw., 210
  compacta (Ag.) Sommerf., 101
  confervicola Dillw., 163
  cyanea Sm. & Sow., 24
  distorta Müll., 22
  echinata Sm. & Sow., 100
  echinulata Sm. & Sow., 100
  fasciculata Schousb., 166
  foetida Vill., 210
  fontinalis L., 210

haematites Ram., 100
Hofmannii Ag., 210
intexta Poll., 24
limosa Dillw., 211
majuscula Dillw., 264
mirabilis Dillw., 22
Mucor Roth, 163
muscosa Begg., 26
myochrous Dillw., 22
  ß ocellata Dillw., 211
myurus Brouss., 206
nivea Dillw., 211
ocellata Dillw., 211
radicans Dillw., 211
scopulorum Web. & Mohr, 164
seriata Wahlenb., 211
vellea Roth, 211
Conishymene Schousb., 95
  tingitana Schousb., 164
Cornicularia hispidula Ach., 211
  velutina Ach., 22
Croatella Erceg., 21
  lithophila Erceg., 59
Cylindrospermum doryophorum Br. & Bisw., 198
  hepaticum Opiz, 127
  Kaufmannii (Schmidle) Hub., 198
  polyspermum Kütz., 212
Dasyactis Kütz., 96
  Biasolettiana Kütz., 169
  brunnea Näg., 112
  crustacea (Kütz.) Crouan, 107
  fissurata Crouan, 174
  hospita (Kütz.) Crouan, 204
  Kunzeana Kütz., 106
  minutula Kütz., 169
  mollis Wood, 116
  Naegeliana Kütz., 110
  plana (Harv.) Kütz., 238
  pulchra Näg., 114
  rivularis Näg., 110
  Saccorhizae Crouan, 174
  salina Kütz., 105
  torfacea Näg., 111
Desmarestella confervicola (Dillw.) Gaill., 164
Desmonema Berk. & Thw., 20
  cirrhosum (Carm.) Berk. & Thw., 26
  Dillwynii (Harv.) Berk. & Thw., 28
  floccosum (Borzi) Born. & Flah., 56
  Lievreae Frémy, 61
  Wrangelii (Ag.) Born. & Flah., 23
    var. Geitleri J. de Toni, 60
    fa. majus Frémy, 59
    var. majus Geitl., 60
    var. minus W. West, 57
Dichothrix Zanard., 96
  austrogeorgica Carls., 129
  Baueriana (Grun.) Born. & Flah., 115
    var. africana Choln., 136
    var. crassa Godw., 125
    var. hibernica W. & G. S. West, 128
    var. minor Hansg., 120